INTERNATIONAL TECHNOLOGICAL UNIVERSITY
This Book is Donated by:
PROF. WAI-KAI CHEN

Date:

Prospects in Modern
ACOUSTICS–EDUCATION AND DEVELOPMENT

The proceedings of the ICA Conference on
Prospects in Modern
ACOUSTICS–EDUCATION AND DEVELOPMENT

Edited by
Antoni ŚLIWIŃSKI
University of Gdańsk (Poland)

Gustaw K. E. BUDZYŃSKI
Technical University of Gdańsk (Poland)

World Scientific
Singapore • New Jersey • Hong Kong

Published by

World Scientific Publishing Co. Pte. Ltd.
P.O. Box 128, Farrer Road, Singapore 9128

U.S.A. office: World Scientific Publishing Co., Inc.
687 Hartwell Street, Teaneck NJ 07666, USA

PROSPECTS IN MODERN ACOUSTICS — EDUCATION AND DEVELOPMENT

Copyright © 1987 by World Scientific Publishing Co Pte Ltd.

All rights reserved. This book, or parts thereof, may not be reproduced in any form or by any means, electronic or mechanical, including photocopying, recording or any information storage and retrieval system now known or to be invented, without written permission from the Publisher.

ISBN 9971-50-379-4

Printed in Singapore by Utopia Press.

CONTENTS

Preface 3

1. **GENERAL INFORMATION** 7

   1.1. List of Scientific Advisory Committee Members 9
   1.2. List of Organizing Committee Members 11
   1.3. Conference Place 11
   1.4. Conference Programme (see Appendix 1 - p. 441) 12
   1.5. Letter to the ICA (see Appendix 2 - p. 449) 12
   1.6. Information on Participants (see Index - p. 453) 12

2. **INAUGURAL ADDRESSES** 13

3. **INVITED LECTURES** 21

   3.1. R. D. AYERS
        Fourier Methods In Physics: A Course In Waves At The
        Senior Level 23
   3.2. V. N. BINDAL, ASHOK KUMAR
        Status Of Teaching Acoustics In India 35
   3.3. L. BJØRNØ
        Advanced Education In Industrial Acoustics In
        Denmark 57
   3.4. G. BUDZYŃSKI, M. SANKIEWICZ, G. PAPANIKOLAOU
        How To Teach Sound-Engineering? 69
   3.5. Z. ENGEL
        Education In Vibroacoustics 79
   3.6. L. FILIPCZYŃSKI, A. KOWALSKI
        Ultrasound And Medicine 85
   3.7. P. FRANÇOIS
        Towards Forming Of A Data Bank About Didactic
        Materials And Systems, As A Means To Increase The
        Efficiency Of Education In Acoustics 93
   3.8. M. HECKL
        The Use Of Films In Teaching Acoustics 101

3.9. A. J. M. HOUTSMA, T. D. ROSSING
  A New Set Of Recorded Auditory Demonstrations 117
3.10. Z. JAGODZIŃSKI
  Specific Problems In Education And Development In Underwater Acoustics 129
3.11. H. W. JONES
  Apparatus For A Practical Course In Acoustics For Undergraduates.................................. 137
3.12. R. LEHMANN
  L'Enseignement Superieur De L'Acoustique En France.. 151
3.13. W. LØCHSTØER
  Acoustics And Physics 169
3.14. I. MALECKI
  Place Of Acoustics In The Technical Universities........... 173
3.15. W. MAJEWSKI, J. ZALEWSKI
  Academic Programs In Acoustics At The Technical University Of Wrocław............................. 185
3.16. R. MILLNER, H. J. HEIN
  Acoustical Foundations Of Training Medical Students And Postgraduates 193
3.17. A. RAKOWSKI
  Interest In Music And Progress In Acoustics 203
3.18. D. SETTE
  Physical Acoustics At The Fermi Summer School In Varenna..................................... 213
3.19. G. SCHOMMARTZ
  TE-LAB — A Successful Philosophy Of Unified Lab's At The Department Of Technical Electronics 225
3.20. W. J. STRONG
  Computers In Modern Acoustics Education And Research 227
3.21. A. S. ŚLIWIŃSKI
  Acoustical Education And Industrial Application 247
3.22. T. TARNOCZY
  The Situation Of The Acoustical Education In Hungary 265
3.23. R. A. WALKLING
  Acoustics For Everyone: A General Course In Acoustics....................................... 273

3.24.	WEI RONGJUE	
	The Impact Of Acoustics To Physics Education and Research	279
3.25.	B. WYRZYKOWSKA, R. WYRZYKOWSKI	
	The Phase Velocity Of Sound In The Acoustical Field Of A General Spherical Source	299
3.26.	A. ZAREMBOWITCH	
	The Use Of Analogies In The Teaching Of Ultrasonics	311
4.	**CONTRIBUTED PAPERS**	**321**
4.1.	K. G. BREITSCHWERDT	
	Phonon Propagation In Disordered Media	323
4.2.	V. CHALUPOVÁ, K. SOBOTKOVÁ	
	The Teaching Of Acoustics At The Faculty Of Electrical Engineering	345
4.3.	H. HARAJDA	
	Selected Problems Of Teaching Acoustics To Students Of The Lutery Faculty In Secondary And Higher Schools	349
4.4.	A. KULOWSKI	
	The Ray Method As A Teaching Tool	355
4.5.	H. LASOTA	
	Diffraction As Time-Space Phenomenon: Educational Aspects	363
4.6.	D. RUSER	
	Hydroacoustical Laboratory At The Technical University In Gdansk	371
4.7.	A. WASILEWSKI	
	Students Classes In Conditions Of Minimum Laboratory Equipment	373
5.	**POSTER PRESENTATIONS**	**381**
5.1.	W. BANDERA, Z. TRUMPAKAJ, T. ZALESKI	
	Personal Computer Application To The Demonstration Of Mechanical Impedance Analysis Of Viscoelastic Specimens	383

5.2.	G. CYPUKOW, J. DYŻEWSKI	
	Educational Recording Studio	391
5.3.	E. HOJAN, U. JORASZ	
	Research At The Institute Of Acoustics Adam Mickiewicz University In Poznań	395
5.4.	A. KACZMAREK	
	Microcomputer Application To Organ Sound Analysis	399
5.5.	E. KOTLICKA	
	Proposals Of Curriculum For Health And Environment Protection Against Noise For Elementary School	403
5.6.	J. MOTYLEWSKI	
	Technical Courses In The Domain Of An Environment And Work-Place Protection Against Noise And Vibration	409
5.7.	Z. PERUCKI, B. KOSTEK	
	Computer Modelling Of The Bowed String Oscillations	413
5.8.	E. SOCZKIEWICZ	
	Application Of The Local Kramers-Kronig Dispersion Relationship To The Calculation Of The Ultrasonic Waves Attenuation In Highly Viscous Liquids	417
5.9.	A. WITKOWSKI, M. CWIECZKOWSKI	
	Acoustical Modelling Laboratory	421
5.10.	S. ZDRAL, K. ŚRODECKI, Z. GARNUSZ	
	Personal Computer Application To The Calculation Of Kuttruff's 'Temporal Diffusion' Of A Room In The Preparation Of Students' Master Theses	425
6.	ROUND-TABLE DISCUSSIONS AND CONCLUSIONS	431
7.	APPENDIX 1	441
8.	APPENDIX 2	449
9.	INDEX	453

Prospects in Modern
ACOUSTICS−EDUCATION AND DEVELOPMENT

PREFACE

International Congresses on Acoustics being organized every three years under the patronage of the International Commission on Acoustics have become for over thirty years a powerful means, accelerating the progress of science and technology in the broad domain of acoustics and of its applications in almost every branch of human activities over the world. The acceleration of progress is mainly achieved owing to direct contacts and discussions among the participants and to an exchange of information and experience, enhanced by meetings of many scientists-acousticians from various countries.

Such useful meetings, congresses and symposia, devoted to the whole area of acoustics, are inevitably divided into many, rather narrow, professional ranges of topics, which is highly unfavourable for any broader discussion on more general problems, concerning acoustics as an entity. Thus, besides regular congresses and symposia, special conferences, devoted to some selected, yet general problems are highly desirable.

As the ICA had already organized world-wide specialized conferences, the idea of such a conference dedicated to the educational problems in acoustics, proposed by the former ICA President, Polish acoustician Professor I. Malecki, has been accepted. Thus, Polish acousticians have been entrusted with the task of organizing the conference under a provisional name "How to teach acoustics", which later on has been modified into an actual title *"Prospects in Modern Acoustics — Education and Development"*.

The actual ICA member from Poland, Professor A. Śliwiński has been put, by the Committee on Acoustics of the Polish Academy of Sciences, in charge of all the arrangements necessary to hold the Conference in Gdańsk. The Maritime Centre of Poland has been chosen as a Conference site not only for its historical importance and architectural beauty but most of all in consideration of the recent development in acoustical education at both the University of Gdańsk and the Technical University of Gdańsk, as well as at other Gdańsk educational institutions at university level.

The task of the organizers was not easy, due mainly to financial limitations and other restrictions. So, among others, they were compelled to shift the site of accomodation and debates from the city of Gdańsk to a resort-place Jastrzebia Góra ("Hawkshill"), seventy kilometers northwest of Gdańsk. To compensate that drawback, an enormous effect has been made to provide transport facilities and organization means in order to enable all the participants to visit both the acoustical laboratories in the City of Gdańsk and its most valuable monuments of culture. Thus, a full day excursion has been put into the Conference program.

The organizers' duty consisted, obviously, in creating the best possible conditions for the Conference debates, as well as a friendly atmosphere, enhancing fruitful discussions. The results of the Conference depended obviously on the contributions of all the participants, first of all, of those most experienced. The subsequent report of the Conference enables the reader to appreciate those results both in general, and in particular aspects.

The Organizing Committee is deeply grateful to all the scientists-acousticians who agreed to participate in the Advisory Scientific Committee of the Conference, which helped to create a true international forum for discussion. Furthermore all the participants who arrived, even from the other hemisphere, and in spite of various difficulties, deserve many sincere thanks from all their colleagues here, and from the international community of acousticians. Thanks to their contributions, the Conference successfully achieved the aimed tasks and provided, let us hope, a full satisfaction to all its participants.

The report, in a book form, contains, first of all the papers delivered during the debates, partly in a full text, partly in an abbreviated version, as they were prepared by the authors. The texts are reproduced by permission of the authors in original versions.

There is a number of texts, describing in a summarized form the contents of several poster-papers. Poster presentations were admitted by the Organizers to enhance the exchange of scientific information mainly between younger Polish acousticians and those more experienced, outstanding acousticians and those more experienced, outstanding acousticians from abroad. As that kind of information exchange and discussion was functioning well, posters have been included into this report.

Discussions during debates were animated, thus several hours of recordings have been made and became available to organizers. However, the editing of that material would become a difficult and lengthy task. Moreover, such a report would required an unusual capacity, trespassing normal limits acceptable for a one volume book. Thus only abstracts of the discussions, held during debates, are given here. At any rate, audio- and video-recordings made during the debates are at the Editors' disposal, and they will eventually serve, for a separate audiovisual edition.

Additional information connected with the Conference accompanies the above mentioned texts. It is limited to items helpful to readers concerned with the topic of the Conference.

The Editors hope that this volume will appear in print in the nearest future, thanks to the efficiency of the Publishing House, and will become a tool in the hands of all academic teachers and other people concerned with the educational problems of acoustics. The circle of interested people is undoubtedly larger. For example, organizers of the professional formation of acoustical engineerings, noise engineers, recording engineers, ultrasound diagnostics engineers, underwater sound engineers, and many other specialists, who must be thoroughly educated in acoustics, should all know how to teach acoustics in the face of actual development of its foundations and, most of all, its omnipresent applications.

It is a pleasant duty of the Editors to express their gratitude to all sponors, who helped in the realization of this Conference. The International Union of Pure and Applied Physics together with the International Commission on Acoustics, the Committee on Acoustics of the Polish Academy of Sciences, the Polish Acoustical Society, the Polish Noise Abatement League, as well as, the two Gdansk Universities: the University of Gdańsk, and the Technical University of Gdańsk - they all well deserve the thankfulness of all acousticians. Finally, cordial thanks must be expressed to all members of the organizing Committee, who had been working hard for several months before the Conference, during its duration, and long after it. Without their enthusiastic willingness this enterprise could have never been actualized.

Gdańsk, the 8th June 1987   The Editors

1. GENERAL INFORMATION

The aim of the Conference was presentation, discussion and a wide exchange of experience in teaching acoustics in many different world centres which have elaborated programmes and methods in various specializations and in different relations to the educational disciplines in various schools, universities, musical schools etc. Also, it was expected to discuss what proper and up to date acoustics ought to be taken into account in the basic, intermadiate and advanced courses in physics, technology and other studies.

1.1. List of Scientific Advisory Committee Members

INTERNATIONAL COMMISSION ON ACOUSTICS

COMMISSION C7 OF IUPAP

CHAIRMAN
H. Myncke (Belgium)

SECRETARY
H. Kuttruff (FRG)

MEMBERS
A. Alippi (Italy)
D. T. Blackstock (USA)
L. Brekhovskikh (USSR)
N. H. Fletcher (Australia)
S. Kameswaran (India)
W. Løchstøer (Norway)
P. Lord (Great Britain)
Z. Maekawa (Japan)
A. Śliwiński (Poland)

ASSOCIATE MEMBERS
F. Kolmer (Czechoslovakia)
M. J. Lighthill (Great Britain)
D. Y. Maa (China)
E. A. G. Shaw (Canada)
J. J. Zwislocki (USA)

LIAISON WITH THE EXECUTIVE COUNCIL
D. Wilkinson (Great Britain)

PRESIDENT OF THE POLISH NATIONAL
IUPAP LIAISON COMMITTEE
Prof. J. Werle
Institute of Theoretical Physics,
University of Warsaw

INTERNATIONAL ADVISORY COMMITTEE

1. Prof.J.Akira Ikushima(Japan),University of Tokyo
2. Dr V.N.Bindal(India),National Physical Laboratory,New Delhi
3. Prof.L.Bjørnø(Denmark),Technical Univ. of Denmark,Lyngby
4. Prof.M.A.Breazeale(USA),The University of Tennessee
5. Prof.L.M.Brekhovskikh(USSR), Institute of Oceanology,Moscow
6. Doc.Dr G.Budzyński(Poland),Technical University of Gdańsk
7. Prof.Z.Engel(Poland),Tech. Univ. of Mining and Metallurgy
8. Prof.L.Filipczyński(Poland),Polish Acad. of Sciences,Warsaw
9. Prof.P.François(France),GALF,Paris
10. Prof.J.Gruber(West Berlin),Technische Universität,Berlin
11. Prof.M.Heckl(West Berlin),Technische Universität,Berlin
12. Prof.F.Ingreslev(Denmark),Technical Univ. of Denmark,Lyngby
13. Prof.H.W.Jones(Canada),Technical University of Nova Scotia
14. Prof.Z.Jagodziński(Poland),Technical University of Gdańsk
15. Prof.F.Kolmer(CSSR),Research Inst. of Sound & Picture,Praha
16. Prof.W.Kraak(GDR),Tachnische Universität,Dresden
17. Prof.W.A.Krasilnikov(USSR),Moskowskij Gosudarstviennyj Univ
18. Prof.H.Kuttruff(FRG),Institut für Technische Akustik,Aachen
19. Prof.L.M.Lyamshev(USSR),N.N.Andreyev Acoust. Inst. of USSR Academy of Sciences,Moscow
20. Prof.R.Lehmann(France),Universite du Main,Le Mans
21. Prof.W.Lǫchstøer(Norway),Uniwesitet Oslo
22. Prof.I.Malecki(Poland),Polish Academy of Sciences,Warsaw
23. Prof.R.Millner(GDR),Martin Luther Universität,Halle
24. Doc.Dr W.Majewski(Poland),Technical University of Wrocław
25. Prof.W.G.Mayer(USA),Georgetown University,Washington
26. Prof.H.Moller(Denmark),University of Aalborg
27. Prof.H.Myncke(Belgium),Katholieke Universiteit Leuven
28. Prof.A.Opilski(Poland),Silesian Technical University
29. Doc.Dr E.Ozimek(Poland),A.Mickiewicz University,Poznań
30. Prof.G.Papanikolaou(Greece),University ofThessaloniki
31. Prof.J.Ranachowski(Poland) Inst.of Fundam.Technol.Res.Warsaw
32. Prof.A.Rakowski(Poland),Musical Academy,Warsaw
33. Doc.Dr J.Renowski(Poland),Technical University of Wrocław
34. Prof.Rongjue Wei(China), University of Nanjing
35. Prof.J.Rudnick(USA),University of California,Los Angeles
36. Prof.H.Ryffert(Poland),A.Mickiewicz University,Poznań
37. Prof.J.Sadowski(Poland),Institute of Building Technology
38. Prof.M.R.Schroeder(FRG),Universität Göttingen
39. Prof.D.Sette(Italy),Universta di Roma "La Sapienza"
40. Prof.G.Schommartz(GDR),W.Pieck Universität,Rostock
41. Prof.R.Stephens(GB),Chelsea College,London
42. Doc.Dr W.Straszewicz(Poland),Technical University of Warsaw
43. Prof.W.J.Strong(USA),Brigham Young University,Provo
44. Prof.A.Śliwiński(Poland),University of Gdańsk
45. Prof.T.Tarnoczy(Hungary),Hun. Academy of Sciences,Budapest
46. Prof.Ch.Taylor(GB),University College,Cardiff
47. Prof.E.Tzekakis(Greece),University of Thessaloniki
48. Prof.R.Wyrzykowski(Poland),Pedagogical High School,Rzeszów
49. Prof.A.Zarembowitch(France),Université Paris VI

1.2. List of Organizing Committee Members

The Conference is organized by:
University of Gdańsk, Institute of Experimental Physics and the Environmental Acoustics Laboratory in cooperation with the Institute of Telecommunication of the Technical Unversity of Gdańsk, the Gdańsk Division of Polish Acoustical Society and the Institute of Fundamental Technological Research of the Polish Academy of Sciences.

THE LOCAL ORGANIZING COMMITTEE

President	Professor A. Śliwiński
V-ce President	Doc. Dr G. Budzyński
V-ce President	Doc. Dr M. Sankiewicz
Secretary	Dr B. Linde
Members	Dr W. Bandera
	Dr M. Borysewicz
	Doc. Dr J. Gudel
	Dr M. Kosmol
	Dr A. Kulowski
	Dr H. Lasota
	Dr L. Lipiński
	Dr J. Sułocki
	Dr I. Wojciechowska

ADDRESS OF THE ORGANIZING COMMITTEE

Institute of Experimental Physics
University of Gdańsk
Wita Stwosza 57
80-952 Gdańsk
POLAND
Telex: 0512706 pl
Phone: 41-52-41 ext. 248 or 213

1.3. Conference place

Conference took place in Poland at Jastrzębia Góra, 65km north-west from Gdańsk, in the recreation centre Celuloza on May 19-21.

1.4. Conference programme (see Appendix 1 - p. 441)

1.5. Letter to the ICA (see Appendix 2 - p. 449)

1.6. Information on participants (see Index - p. 453)

2

INAUGURAL ADDRESSES

INAUGURAL ADDRESSES

(The Chairman of the Organizing Committee of the Conference Professor Antoni Śliwiński welcomes the participants:)

Ladies and Gentlemen,

I have a great honour and pleasure to welcome you as participants of the Conference on Prospects in Modern Acoustics - Education and Development at Jastrzębia Góra. I am glad to welcome our distinguished guests:
Professor Alfred Czermiński, v-ce Rector of the University of Gdańsk,
Professor Wiesław Pudlik, v-ce Rector of the Technical University of Gdańsk,
Professor Józef Werle, the President of the Polish National IUPAP Liaison Committee - as the official representative of the main sponsor of the Conference.
I welcome Professor Ignacy Malecki - the Honourary President of the Committee on Acoustics of the Polish Academy of Sciences, and
Professor Leszek Filipczyński - the actual President of this Committee,
Professor Zenon Jagodziński - the President of the Polish Acoustical Society, and
Professor Witold Straszewicz - the President of the Polish Noise Abatement League,
as representatives of the other sponsoring bodies.

Among our guests from abroad I welcome Professor Daniel Sette from Rome - the former v-ce President of the Executive Council of IUPAP, Professor William Løchstøer from Oslo and Professor Felix Kolmer from Prague - the members of the ICA.
I welcome all participants from abroad and from this country.
I regret I am not able to welcome here visitors whom we expected and who could not come for various reasons:
Professor Henry Myncke - the actual President of the ICA,
Professor Heinrich Kuttruff - the Secretary of the ICA.
They have sent telex and letters of the following texts:
Professor Myncke: "The Chairman of the International Commission on Acoustics, Prof. H. Myncke, regrets being unable for personal reasons to attend the Conference on Education and Development in Acoustics. He congratulates the organizers and more especially his colleague Śliwiński for the initiative they have taken and wishes them all a successful meeting. With his best regards to all participants".
Professor Kuttruff: "I wish you a successful conference and many fruitful discussions".

We received Greetings for the Conference from the General Secretary of IUPAP prof. J. Nielson from Gøteborg. Here is a fragment of his letter: "I take this opportunity to thank you and your collaborators for the services you have rendered the Union and the Physics Community".

Also, we received greetings for the Conference from some other distinguished colleagues, among them I should mention Professor Stephens from London, Professor Ingerslev from Kopenhagen, Professor Krasilnikov and Professor Lyamshev from Moscow, Prof. Rudnick from Los Angeles, Prof. Lehmann from Le Mans, Prof. Kraak from Dresden, Prof. Rong-jue Wei from Nanjing, Prof. Pimonow from Paris, Prof. J. Akira Ikushima from Tokyo, Prof. Breazeale from Tennessee. Their papers are included in our programme but unluckily they have not arrived.

In relation to the topic of our conference one can realize that it is the first world meeting of that kind devoted to education and development in acoustics. It should be noticed, however, that many problems connnected with education and development were discussed occasionally in separate sessions during ICA congresses and at various conferences. I am glad to mention that the first initiative for such a conference on "How to teach acoustics" was made, a few years ago by Professor Malecki. In the last years during my membership term in the ICA and later on, the IUPAP have accepted the idea. Advised by the IUPAP we changed the title of the conference into the actual form.

I take this opportunity to express my thanks to the members of the Advisory Committee for their cooperation and valuable remarks related with the organization of our meeting. I also express my thanks to all my Colleagues in the Organizing Committee from the Technical University of Gdańsk and the University of Gdańsk for their tedious work and efforts to prepare the conference.

I like to wish the conference to be successful and to you all to have a good time in our country, particularly at Jastrzębia Góra.

(Allocution of The President of the Polish National IUPAP Liaison Committee — Professor Józef Werle:)

On behalf of the Polish — IUPAP Liaison Committee I would like to welcome all the participants of this conference. It is my honour and pleasure to be invited to say a few words of a general introduction.

This conference is devoted to various aspects of education and recent development of acoustics. When in the seventies I was a member of the IUPAP Commission on Physics Education we came to the conclusion that teaching physics is a very difficult and subtle task indeed. The results achieved by a teacher of physics depend on numerous factors. These results depend not only on the scope and depth of the teacher's knowledge of physics and his pedagogical abilities and zest. Many teachers are severely limited by unsatisfactory laboratory aids, inadequate but obligatory programs and textbooks. Frequently they are torn apart by quickly changing and too often contradictory ideas, trends and demands which originates not only from the rapid progress of physics but also from psychology, pedagogy, sociology, technology, — even from politics.

A capable, interested and eager student can attain a profound knowledge of physics even if his teachers are far from being excellent. However, even the best teacher cannot help when the pupils or students are not capable or not interested in physics. Thus, the main aim of a teacher of physics should consist in arousing interest in physics. Physics is too difficult to be learned by passive and disinterested students. I wonder whether these problems become simpler if you restrict your attention to teaching acoustics.

As you know I am not doing any research in acoustics. Nevertheless being a physicist I am not a complete outsider. Acoustics is definitely a branch or part of physics, though a very peculiar and funny part that is also interdisciplinary. Acoustics is in fact one of the oldest of physical sciences. It started at least 25 centuries ago when Pythagoras or his followers investigated the sounds emitted by vibrating strings of different lenghts and tensions. The Pythagoreans discovered also the first mathematical relations hidden in the sense of musical harmony. For many subsequent centuries acoustics was so closely related to music that it existed rather as some art or craft than a science. It was practised — sometimes with great success — not only by composers, singers and musicians but also by architects and constructors of various musical instruments.

The rapid developments of Newtonian mechanics in XVIII century brought definite understanding of sound as mechanical waves of various frequencies propagating in material media. The first basic problems of quantitative, scientific acoustics consisted in finding the velocity of sound, the laws of reflection, diffraction, dispersion or — more generally — the laws of propagation of sound waves. The basic mechanical laws were found relatively quickly and in XIX-th century scientific acoustics was regarded as a not very interesting and rather closed side-track of physics.

The next important incentive came from biology or - more precisely - from the physiology of human and animal senses. In this respect the situation of acoustics is very peculiar and unique. In fact acoustics deals with a sense for which the human body has two very important and subtle organs: one that is a sensitive receiver and analyzer (ear) and the second that is a rich and accurate emitter of sounds (mouth and throat). Compare this situation with the second important and also very subtle human sense i.e. vision. Human eye is an excellent detector and analyzer of light. However, the human body has no light emitting organ. It can only reflect light coming from external sources. Without external sources of light we are invisible and blind as well.

Thus, it is no wonder that, because of this peculiarity of our bodies, the human voice developed into the most important means of communication, i.e. of immediate exchange of information between human beings. The most subtle and sophisticated thoughts, ideas, concepts, wishes, feelings etc. can be immediately exchanged with the help of human voices, with the help of one of the many natural or national languages. The same applies, of course, to the sound-based arts such as music, song, declamation, theatre etc.

Appreciating fully the cultural importance of sound we should not forget its several disadvantages and drawbacks:
1. Sound does not last long. It spreads with rapidly falling amplitude or gets absorbed and vanishes forever.
2. Because of the natural attenuation the range of good applicability of our voice is rather short, of the order of 100 meters.
3. The speed of sound in air is also relatively small - - about 300 m/s.

The first of the mentioned drawbacks has been partially removed long ago due to a cute interplay with vision. With the help of script the human speech and music could be transformed into time-resistant sequences of visible signs put on stone, clay, leather, papyrus or paper. Men could write letters, send written orders, compose music, write poems and books, develop mathematics, accounting and bureaucracy. Nevertheless, the method of registering sounds with the help of visible signs (letters, notes, symbols) has still many serious drawbacks. The method itself is very slow, requires much manual labor, and the result is always partially distorted and ambiguous due to the unavoidable simplifications at the registration.

The next breakthrough in acoustics started about 100 years ago after the discovery of the basic electromagnetic phenomena and construction of Maxwell's electrodynamics. Several methods of mutual conversion of acoustic waves into electric currents or electromagnetic waves have been discovered and are still being enriched and improved. It started with the construction of microphone, and telephone; then came the gramophone, radio, magnetophone, television, compact disc, numerical registration, laser technology etc. One should remember many mainly electromagnetic devices that analyze, measure, synthesize, register and reproduce any sound. The interplay of the acoustic (i.e. mechanical) and the electromagnetic phenomena lead to the removal of the three mentioned drawbacks of sound as a means of exchange of information. Moreover, due to this help the realm of acoustics expanded far beyond the region of frequencies and amplitudes and accuracies perceived by the human ear.

I shall not list all the important applications of acoustics in medicine and biology, architecture and industrial production as well as its importance for all the human culture. I wanted only to emphasize the most fundamental relations between acoustics and other branches of physics which made possible the astounding evolution of acoustics in this century. It is not possible to teach, to do successful research, or apply modern acoustics without a profound experimental and theoretical knowledge of almost all branches of physics: Newtonian mechanics of continuous media, statistical mechanics and thermodynamics, and in particular electrodynamics (including optics). Recently even quantum phenomena are taken into account at some acoustical problems. Although nobody is expecting from acoustics any discovery of new fundamental laws of nature, many different basic phenomena and laws of nature are being used in acoustics. In this sense acoustics is indeed a very wide, interdisciplinary science.

At the end I would like to convey you best wishes of interesting lectures, stimulating discussions and inspiring exchange of opinions and experiences.

(Further inaugural addresses were delivered by:)

Prof. A. Czerminski - on behalf of the Rector of the University of Gdansk.

Prof. W. Pudlik - on behalf of the Rector of the Technical University of Gdansk.

Prof. L. Filipczynski - as President of the Committee on Acoustics of the Polish Academy of Sciences.

Prof. Z. Jagodzinski - as President of the Polish Acoustical Society.

3

INVITED LECTURES

3

INVITED LECTURES

FOURIER METHODS IN PHYSICS:
A COURSE IN WAVES AT THE SENIOR LEVEL

R. Dean Ayers
Department of Physics-Astronomy
California State University, Long Beach
Long Beach, California 90840 USA

ABSTRACT

An elective course on Fourier transforms and the physics of vibrations and waves has been taught for seven years. The goal of this course is to provide students with a thorough grounding in the wave description of nature in a classical setting. The first half of the semester lays the mathematical foundation, and the second half covers some physical applications, which are mostly acoustical. The original approach to the transform (through the series) tended to focus too much attention on mathematical details, so that the organizing structure of the subject was not apparent. Better results have been obtained through an approach that emphasizes convolution and pictorial manipulations. An optional laboratory illustrates the ideas of the lecture course with concrete examples. Several students have gone on to do master's theses in acoustics.

FOURIER METHODS IN PHYSICS:
A COURSE IN WAVES AT THE SENIOR LEVEL

R. Dean Ayers
Department of Physics-Astronomy
California State University, Long Beach
Long Beach, California 90840 USA

I. INSTITUTIONAL BACKGROUND

California State University, Long Beach is located in the greater Los Angeles area. With an enrollment of 33,000 students (22,000 Full Time Equivalent), it is the second largest in a system of nineteen campuses. Unlike the more well-known University of California, this system focusses on the teaching of undergraduates. The highest degree that it can award is the master's, which usually represents two years of study after the baccalaureate. Very few of our students reside on campus, most of them are employed at least part-time, and their average age is 23 years.

The Department of Physics-Astronomy at CSULB is currently allocated 29 faculty positions (21 held by tenured individuals). The university has a large school of engineering, so our lower division courses that use calculus make up 30% of the department's total enrollment, which is 670 FTE. Physics majors number about 100 at the undergraduate level and 50 at the graduate level. Roughly 20% of those who graduate with a bachelor's degree continue on for their master's with us, and about the same number go directly to other universities to pursue doctoral degrees. Many of our graduate students have received their bachelor's degrees elsewhere and migrated to Southern California for employment. Most are employed full-time, so we offer their required courses at night. About 15% of those awarded master's degrees continue on for the Ph.D.

Required courses at the upper division level (last two years of the baccalaureate) include mechanics, thermodynamics, electricity and magnetism (full year), quantum mechanics, applied mathematics (full year), and two courses with laboratories from a list that includes optics, electronics, modern physics, the present course, and ionizing radiation. Three or four elective courses must also be selected from an extensive list that includes this course.

II. ORIGINS OF THE IDEA FOR THIS COURSE

I first became aware of the power and efficiency of the Fourier transform for dealing with wave phenomena in solid state physics, the area in which I received my Ph.D. Like other students of that subject, I had struggled with Miller indices, the reciprocal lattice, Brillouin zone boundaries, Bloch functions, and so on. Just as I was beginning to teach that material in the late 1960's, I came across a very useful monograph by H. Lipson and C. A. Taylor [1]. Even though it dealt exclusively with the problem of crystal structure determination, I was struck by the way that its approach could be used to provide an organizing framework for the much broader topic of waves in periodic structures.

It was a bit difficult to supplement a standard solid state course with a substantial introduction to Fourier transforms. One factor which helped to make this possible was the excellent book by Ronald Bracewell[2], which I used as a reference for that course. Still, I felt that it would have been much better if the students could have had a thorough grounding in the transform and its physical applications <u>before</u> they took the solid state course.

Early in the 1970's, a colleague and I developed a general education course on the physics of music for liberal arts students. Through teaching that course, I came to realize that there are many interesting questions yet to be answered in this area. In 1978-'79, I spent a sabbatical leave at the Pennsylvania State University, where some of Eugen Skudrzyk's students were developing an experimental technique that they called acoustical holography. I found that the spatial transform in two dimensions provided a very convenient tool for understanding the directional radiation and reconstructed images of the planar sources that were being examined. Sitting in on an introductory course in acoustics then convinced me that the best introduction to physical applications of Fourier concepts would be through that subject.

This course has been offered seven times, beginning in 1980, and since 1982 it has had an optional laboratory associated with it. Both the lecture class and the lab meet three hours each week for one semester. They can be taken for undergraduate or graduate credit. One side effect of teaching this course that I have particularly enjoyed is a steady supply of graduate students who are interested in and well prepared for doing research in acoustics.

III. EVOLUTION OF THE COURSE

The subject matter and the order of topics in this course have been modified to varying degrees almost every year. I started off with a very ambitious syllabus and tried to follow the order of topics in Bracewell's textbook quite closely, at least until we got into the physical applications.

Table 1: Syllabus for 1980.

B = the textbook by Bracewell[2]
Notes = class notes prepared by the author

Week	Readings	Topics
1	B1; Notes	Complex exponentials; Fourier series
2	B2,3	Fourier transforms; convolution
3	B4,5	Useful functions; δ and its relatives
4	B6,7,8	Theorems; doing transforms; the 2 domains
5	--	Review; first exam
6	B9; Notes	Damped oscillations; resonance
7	Notes	Waves in one dimension
8	B10,12	Series again; transforms in 2 & 3 dim.
9	Notes	Waves in two dimensions
10	--	Review; second exam
11	Notes	Waves in three dimensions
12	B13; Notes	Antennas; directivity functions
13	Notes	Holography; imaging in general
14	Notes	Waves in periodic structures: solid state
15	--	Review for final exam (comprehensive)

The supplementary notes on Fourier series for the first week were supposed to be a review of material that I thought would be familiar to the students and provide an intuitive background for the introduction of the transform. I had anticipated that the complex exponential form of the series would be new to them, so I took some time to demonstrate its equivalence to an expansion in terms of sinusoids as well as its considerable economy of notation. I was disappointed to find that previous exposure to the Fourier series had failed to provide the students with any intuitive understanding of its physical significance. They had not picked up on the notion that there were simple, systematic relationships among the coefficients for many periodic functions. Using theorems to obtain the coefficients for one function from another that was simply related to it, or even using symmetry considerations to simplify a direct calculation, seemed quite foreign to them. The homework problems should have taken very few steps to solve, but the students' solutions were amazingly tedious and roundabout (and often wrong). After trying to modify this approach a few times, I concluded reluctantly that it was the wrong way to begin the course. The effect that it had on the students was simply to reinforce a widespread view of this subject as a mysterious "black box," involving complicated, abstract calculations.

The most recent order of topics shown below seems to work much better in helping the students to see the physical significance and the beautiful structure of this subject. By assuming nothing beyond introductory calculus as specific background (plus some intellectual toughness acquired in upper division physics courses), it encourages the students to expect the material to really make sense to them. They now show more confidence in approaching new topics, and they are more willing to take a larger view of the subject, because they are able to see the efficiency that results from doing so.

Table 2: Syllabus for 1986.

Week	Readings	Topics
1	Notes; B4	Useful functions and operations
2	Notes; B2,5	Symmetry considerations; δ and III
3	Notes; B(9),3	LSI systems; convolution
4	Notes; B(10),2	Periodic functions; Fourier transforms
5	B6,(10)	Basic theorems; Fourier series
6	B5	δ' and other generalized functions

7	--	Review, first exam
8	B7, 8	Evaluation techniques; the two domains
9	Notes	Damped oscillations and resonance
10	Notes	Waves in one dimension
11	B12	2 dimensions: convolution, transforms
12	--	Review, second exam
13	B12	Sections, projections; Hankel transform
14	B12	Tomography; three dimensions
15	--	Review for final exam (comprehensive)

Notice that in this new approach, we work entirely in the direct domain of time or position until the middle of the fourth week. This gives the students time to get used to an aspect of this subject which is essential if it is to become a practical tool for them: the use of pictorial reasoning as a guide and a check for algebraic manipulations.[3] We begin by defining some standard functions, both piecewise continuous and infinitely smooth, that we shall use throughout the course: Heaviside step, rectangle, triangle, signum, ramp, ramped step, Gaussian, sinc, and complex exponential. We then play some games with these, manipulating them to get a desired behavior through some combination of basic operations: addition, multiplication, shifting, flipping, and rescaling. (The last three are all with respect to the independent variable.) I call this activity "plumbing with functions," because it is very much like the process by which a plumber figures out how to run a pipe from one point to another in a complicated situation. The students are encouraged to look for a variety of ways to represent the same functional dependence. This activity is made more concrete for them by referring to distributions of mass density, charge density, electrical potential, or optical transmissivity that they have seen before.

The Dirac δ is introduced as the derivative of the Heaviside step, using the idea of limited resolution in measurements of the independent variable. An examination of our primitive notions of "point" masses or charges provides some physical intuition about this symbol or generalized function. The only other symbol that I introduce at this point is the "shah" function (also called "comb" or "picket fence"):

$$III(x) = \sum_{-\infty}^{\infty} \delta(x - n). \qquad (1)$$

This allows us to discuss discrete versus continuous descriptions of nature and the process of shifting back and forth between them. We can multiply a continuous function by $III(x/\Delta x)$ to <u>discretize</u> it, yielding a regularly spaced sequence of weighted deltas for purposes of integration (e.g. in a computer). If we take the limit as $\Delta x \rightarrow 0$, keeping in mind the limits of resolution for physical measurements, we have returned to what appears to us to be a continuous description. References to the kinetic theory of gases and the macroscopic/microscopic views of dielectric polarization are particularly helpful here.

I introduce next a concept that Bracewell postpones until the ninth chapter of his book -- that of a linear, shift-invariant (LSI) system. We begin by characterizing such a system by its impulse response (output in the direct domain resulting from application of δ as input). Motivation for the convolution integral is then provided by the following sequence of steps: 1. Discretize a continuous input to yield a regularly spaced sequence of deltas. 2. Invoke the linearity and shift-invariance of the system to obtain an approximation for the output as a summation of appropriately weighted and shifted impulse responses. 3. Take the discretization interval Δx to the limit to obtain a continuous input and the convolution integral for the corresponding output. Once the convolution integral has been defined, it provides a very convenient way to represent a periodic function:

$$f(x) = f_b(x) * (1/X) \; III(x/X), \qquad (2)$$

where * stands for convolution, X = the period of f(x), and $f_b(x)$ is a suitable __basis__ function.

The manner in which the Fourier transform is introduced may seem a bit abstract to the students at this stage of the semester, but it emphasizes the role that the transform will play in the later applications. After we have had some practice at putting simple input functions into LSI systems with simple impulse responses, we consider what would happen if we used a complex exponential function as the input. We find that we get the same comlex exponential as output, except that it has picked up a complex multiplier. Thus any LSI system has eigenfunctions which are complex exponentials. The complex eigenvalue for a given eigenfunction is just the Fourier analysis integral evaluated at the frequency (or wave number) for that complex exponential. Thus our first view of the transform is as the continuous collection of (complex) eigenvalues of an LSI system, expressed as a function of the parameter that distinguishes the different eigenfunctions. (In other words, it is the transfer function of the LSI system.)

We do need to find transforms of other functions beside the impulse reponses of LSI systems, for instance the functional dependences of their inputs and outputs. To do this, we invoke the abstract concept of an LSI system that has a single complex exponential as its impulse reponse. We then imagine that we can vary continuously the frequency of that impulse response, making the LSI system a tunable filter, or else we set up a large number of them with different frequencies side by side to make a multichannel (perhaps FFT) spectrum analyzer.

Some of the later topics in the original syllabus have been omitted in order to accommodate the changes already discussed. The material that remains is also covered more slowly than before in order to encourage mastery by the students. In the spring of 1988, I will offer for the first time a sequel to this course which covers the omitted material as well as some new applications that have emerged in my work with thesis students.

IV. THE OPTIONAL LABORATORY

One very nice aspect of this subject is that it is easy to illustrate with demonstrations. This makes the course interesting and encourages the development of physical intuition. Even more valuable for the students are opportunities to get their own hands on the apparatus, so that they can manipulate the controls to find answers to their own questions. About half the students in the lecture class elect to take the optional laboratory. Since we have only one each of the more expensive pieces of equipment, and there is a definite order in which to do the experiments, the students arrange to work in pairs at different times. The amount of lecture time spent on direct space considerations at the beginning of the semester has made things a bit awkward for the laboratory schedule. I do not like to start the experiments as late as the third week, because by then the students have become used to schedules that do not include this activity. In future semesters, I will probably start them off earlier with an introduction to the controls of the digital function generator and the FFT spectrum analyzer, even though that gets ahead of the lecture content.

Table 3: Laboratory schedule for 1986.

Weeks	Activities
1,2	Organizing partners and meeting times
3,4	Convolution by an RC circuit; approaching δ
5,6	Analog Fourier analysis and synthesis of periodic waveforms
7,8	Digital (FFT) Fourier analysis
9,10	Illustrations of theorems on transforms
11,12	Impulse responses and transfer functions
13	No lab (two-day holiday break this week)
14,15	Two-dimensional transforms; spatial filtering

Table 4: Major pieces of equipment.

Wavetek model 275, 12 MHz programmable arbitrary/function generator (digital)
Hewlett-Packard model 3582A, spectrum analyzer (FFT)
PASCO model 9302, waveform analyzer (analog)
PASCO model 9307, Fourier synthesizer (9 exact harmonics)
Optical bench, He-Ne laser, and split-screen optics for simultaneous display of a diffraction slide (enlarged) and its diffraction pattern

---

[1] H. Lipson and C. A. Taylor, Fourier Transforms and X-Ray Diffraction (G. Bell and Sons, Ltd., London, 1958).
[2] R. N. Bracewell, The Fourier Transform and Its Applications (McGraw-Hill Book Company, New York, 1978), 2nd ed.
[3] Photocopies of the transparencies for this paper may be obtained from the author.

Status of Teaching Acoustics in India

V.N. BINDAL and ASHOK KUMAR
National Physical Laboratory
New Delhi-110 012
INDIA

ABSTRACT

The subject acoustics is taught in India, at various levels, to the students of basic and applied sciences. At present, it is being taught in engineering, medicine and architecture disciplines also. The present paper reports various aspects of acoustics teaching activities in India based on a survey conducted recently.

In the present report, the details of syllabus used at various levels, literature and experiments etc. available at teaching institutions have been analysed and discussed.

It has been observed that acoustics as separate branch of Physics is introduced at pre-university level and is also taught to the undergraduate students.

At the post-graduate level the topics covered are generally devoted to ultrasonics only. The number of universities where acoustics is included in detail is very small. Topics such as 'underwater acoustics' and 'non-linear interactions' have also been included recently in the curriculum at a number of places.

Looking at professional courses, it is noticed that engineering, medical and architecture students at many institutions study acoustics/ultrasonics in greater detail. Although Indian music is age old and very rich, there exists no serious effort of teaching basic acoustics in music courses.

As far as experiments are concerned the growth has been rather slow. The experiments such as sonometer, resonance tube, interferometer and different apparatus for the velocity measurements, are still the main experiments. No regular experiments for study of various other acoustic phenomena, such as attenuation, are being offered to students.

Regarding the availability of text books on acoustics to the students, it is seen that besides many foreign books, research and technical reports, standards issues by ISI are also being used.

It has also been seen that in addition to the teaching institutions, a number of societies, industries and hospitals are also offering special short term courses on selected topics of acoustics and ultrasonics.

The report also gives the status of work at Ph.D. level and comments on duration of courses, syllabus and modernization of experimental work.

1. INTRODUCTION

Application of acoustics in India dates back to several hundred years. Testing the genuinity of coins and soundness of glass bangles by simulated vibrations can be forwarded as examples of it uses in practice from ancient days. Musical pillars, made of granite, producing musical notes when struck by a light hammer and also wind instruments made of stone were built in the prayer halls of Hindu temples[1,2]. Corridors propagating the whisper from one end to other end, at scores of yards away, are common sights in ancient buildings. However, the knowledge of such architectural acoustics and acoustic non-destructive testing diffused amongst people like a family trade. Though the University of Nalanda and

Takshila existed in India, more than thousand years back, no information about the teaching of acoustics in those universities is available.

The acoustics teaching on a regular basis started in India with the turn of this century, perhaps. The subject of acoustics crept into the curriculum under the cover of mechanics or vibrations and appears to continue at many of the teaching institutions even in modern days.

With the advent of sophistication in measurement systems of sound, the acoustics began to acquire its separate identity, atleast in building designs. Isolated groups in India started feeling the importance of acoustics in other fields also and teaching of acoustics began as a separate topic at some institutions.

Presently, the acoustics teaching in India is spread over a number of institutions having varied interests. A survey was recently conducted by the authors to collect the data on status of teaching of acoustics in India. A response of about 50% was received as more than 70 universities and institutions responded out a total of about 150 contacted. The data so collected, has been presented in this paper giving information on syllabus, text books and experiments used, hours of teaching acoustics at various levels, etc. A few suggestions have also been put forward with a view to improve the teaching of acoustics in India.

2. TEACHING LEVELS IN GENERAL IN INDIA

Before taking up the topic of teaching acoustics, it

would be worth while to have a look at the levels of teaching, in general, in India. These levels can be broadly classified as given below.

2.1 Pre-University (School)

Twelve years of education is necessary at the pre-university level, before entering the university. Education upto this level is controlled by 'State Education Boards' as far as the syllabus and examinations are concerned. Normal age at which a student passes 12th class is about 17 years.

2.2 Graduate

After passing the 12 classes, the student is enrolled to an university or an institution deemed as university. At this level, the specialization in a branch also starts deciding the career of student as a scientist or an engineer or a medical doctor etc. Depending upon the specialization, the number of years necessary for graduation (Bachelor's degree) may vary from three to five years.

2.3 Post Graduate

After graduation the student can be enrolled for Master's degree in any of the main subjects persued at graduate level. The duration of Master's degree courses is two years.

At some universities, there exists a one year course, called M.Phil, which can be taken up after Master's degree.

2.4 Doctorate

Degree of Ph.D. in science can be awarded after a

minimum of two years research work after post graduation. In engineering and medicine, the doctorate degree can be obtained after a minimum of three years after graduation. For post-graduates in engineering, the minimum period for Ph.D. is two years.

3. ACOUSTICS AT VARIOUS TEACHING LEVELS

The number of teaching institutions offering acoustics is given in Table 1 as percentage of total institutions, according to the survey conducted. The duration of the total course and the hours devoted to acoustics teaching per week are also given along with the number of students at each level studying acoustics. A histogram is presented in Figure 1 showing percentage of institutions offering modern acoustics teachings, total hours devoted to acoustics teachings and approximate number of students at each level.

3.1 Pre-University Level

At most of the places, the subject of acoustics is taught in higher secondary schools (Pre-university) under the topic of waves and oscillations. Weightage given is 10% to 15% in a full question paper of Physics.

3.2 University Level

3.2.1 Basic and applied sciences: At University level also, like in schools, acoustics is generally taught as a part of some other paper like mechanics or vibrations, both at graduate as well as at post graduate level. However, there are atleast three universities in India, where acoustics is offered as a separate paper to post-graduate (Physics) students.

In many universities, there is a provision of an optional paper besides the other compulsory papers for the post graduate students. Here, acoustics is being offered as an optional paper.

Recently, some universities in India have started post graduation in applied sciences, for example, in material science, marine science, computer science, etc. Acoustics has figured as one of the important topics in material science and marine science disciplines.

3.2.2 Engineering : At graduate level in engineering, very little acoustics is taught at most of the places. However, at one university the students of Civil Engineering have one full paper. In architecture engineering, acoustics is a full paper at most of the places.

3.2.3 Music: India has a very rich heritage in music. A number of musical instruments have their origin in India. Some of them are given in Table 2.

Special auditoriums were built in ancient days to perform, enjoy and appreciate the fineness of music. Various well established schoools were engaged in teaching music. The number of these music schools grew with time. But it appears that acoustics as a separate subject itself is not being taught to the musicians in modern times in India.

3.2.4 Medical: The education of medicine in India is given in mainly four different systems, viz., Allopathic, Homeopathic, Ayurvedic and Unani. The latter two systems of treatment are age old with ayurvedic being of Indian origin. It appears, however, that acoustics is not generally taught in any of the systems to the graduate level students[3]. However, at post graduate level (M.D. in radiography), there exists a few topics in ultrasonics in the syllabus. Also, the students of speech and hearing course, have a course to cover on electro-acoustics.

3.3 Doctorate Level

Besides these regular courses, acoustics is taken up by the students of Ph.D. as a part of their research programme at several universities. Number of institutions where R&D in acoustics is being carried out is given in Table 3. It may be seen that about 20% universities and 25% of CSIR (Council of Scientific & Industrial Research) laboratories are engaged in R&D in the area of acoustics.

3.4 Learned Societies

There is a number of learned societies which have direct interest in the area of acoustics. Some of these are also taking up the teaching of acoustics through short term training courses and workshops. Table 4 gives information about these societies. It can be seen that three of these societies are involved in teaching of acoustics. Also, three societies are running their own journals, in which the research papers in the area of acoustics are published.

4. TOPICS IN ACOUSTICS AT VARIOUS TEACHING LEVELS

India is a very large country having the total population of about 800 million and area 3.3 million sq.km. It is divided into 23 states and union territories. Each of these states and union territories have their own local boards. These boards are responsible for education upto 12th class.

The universities and state education boards in India have the freedom to modify or change slightly their own syllabus for various teaching levels. The choice is generally based on the demand of the topics by students, availability of equipment and faculty, and requirement by the constitution of the university (or board) concerned. It is not possible to give detailed syllabus of acoustics at each level due to lack of space but a brief account is given below here.

The preuniversity and graduate science students, throughout the country, follow almost the same syllabus in acoustics. The topics covered give basic knowledge of wave motion, superposition of waves, stationary waves, beats, Doppler effect, string vibrations, stationary waves, damped vibrations, reverberations, velocity measurement, etc. Applications of acoustics and ultrasonics in modern times are also introduced at this level.

Recently, one or two universities have started offering optional papers in acoustics at graduate level also. For example, Poona University (situated near the Institute of Film & Television Training) offers two optional papers in acoustics - one in second year and the other in third year. These courses cover studio Acoustics and Sound Recording in a greater detail.

At post graduate (M.Sc.) level, the syllabus differs from place to place to a greater extent. Only a few universities are covering a wide range of acoustics and offer full paper in this subject. Some of the main topics covered in acoustics along with the number of institutions are given in Table 5. It may be seen that while some topics are covered by most of the places, these are a few topics which are followed by one or two institutions.

5. ACOUSTICS EXPERIMENTS AT VARIOUS LEVELS

A number of experiments have been devised in acoustic (Annexure 1). However, at Pre-university and graduate levels, only limited number of experiments are available, as shown under A, Annexure 1. Out of these, the universities select a few for graduates and boards select a few for pre university in a particular state. It is not uncommon to observe that an experiment suggested in one state of the Country for pre-university is offered to graduate students in other states.

It is surprising to note that at postgraduate level,; most of the universities are offering mainly two experiments only (see under B, Annexure 1), according to the survey conducted. It is only at those universities where acoustics is taught as an optional or full paper, the number of experiments offered to students is quite large as given under C, Annexure 1. It may be seen that most of these experiments are from applied acoustics, such as performance evaluation of loud-speakers and microphones, measurement of noise level, etc.

There appears hardly any experiment concerning basic acoustics phenomena or which could throw more light in understanding the fundamental concepts of acoustics.

6. TEXT BOOKS AND LITERATURE USED

The books in India are mostly published by private publishers. For education in schools, Govt. of India has established an institution N.C.E.R.T. (National Council of

Educational Research & Training). One of the responsibility of NCERT is to produce and publish books at very low prices and in many languages. These NCERT books are written by a group of highly experienced teachers from all over the country.

Besides this, a large number of books are available today in India on the subject of acoustics. These books can be classified in following three broad categories.

(i) Books on general Physics in which acoustics forms a small part.
(ii) Books which deal only with acoustics.
(iii) Books on specialised topics of acoustics, such as Sound Recording, Noise, Speech Synthesis, Underwater Acoustics, etc.

6.1 School level

The books of category (i) are used at school or pre-university level. Although many authors have written such books, the contents and coverage for acoustics does not differ much from one book to another. The students at this level seem to be fully satisfied with available text books.

6.2 University level

a) Graduate

Since acoustics is only a small part of the syllabus at most of the places at graduate level, both category (i) as well as category (ii) books are used. But as the syllabus of acoustics may vary from university to university, the books selected at each university are also different.

b) Postgraduate

At postgraduate level, there acoustics is not a full paper, the books of category (ii) are found to cover the syllabus. However, there acoustics appear as a full paper, optional or compulsory, category (iii) books are often used. The books used are generally by foreign authors and are highly priced putting students to some difficulty in owning personal copies.

At times, some universities even recommend the use of research or technical reports, standards from ISI (Indian Standard Institution) and other such literature to supplement the reading material on specialised topics.

7. SOME FINDINGS

From the data available from survey and other sources, a number of points given below, about the teaching of acoustics in India, come to the surface.

7.1 General

i) There is in general, a lack of interest in teaching of acoustics. It is seen that relatively less weightage is given in acoustics by the teachers due to which the students find this subject at bit dull.

ii) There is not a much job potential for acousticians.

iii) In industries and hospitals, where there is great need to man the available acoustic/ultrasonic equipment specialisedtrainingand and teaching of acoustics is lying quite neglected.

7.2 Syllabus

i) It is seen that teaching is generally according to the capability and liking of teachers, as in most of the places, the same teachers are paper setters also. This situation is further aggravated by the fact that the universities are independently setting up syllabi and they can afford to increase or decrease their own teaching hours on any topic they like.

ii) As explained above the weightage to acoustics given by various universities varies. At many places, the course offered is only taught by giving some experiments in acoustics, and the basic concepts are not made clear. It is shocking to note that it is possible to be a science graduate without studying acoustics at all.

7.3 Experiments

i) The time devoted to the practical work in acoustics in many courses is very low which could be attributed to the lack of proper experiments also at a particular institution.

ii) At those universities, where a full paper on acoustics is offered, there is a good number of experiments. Many of these experiments are of applied nature.

iii) Fabrication type exercises are missing almost everywhere at all levels.

7.4 Text Books

The status of text books at pre-university and graduate level can be considered to be satisfactory.

8. SOME RECOMMENDATIONS

As seen by the finding reported earlier, efforts are needed to attract more attention for teaching acoustics in India. A few recommendations as an attempt to improve the acoustics teaching have been given below.

8.1 Syllabus

i) Acoustics may be offered as separate optional paper wherever possible at postgraduate level.
ii) Certain minimum topics of acoustics may be included in the main stream, besides the optional paper.
iii) The syllabus for optional paper may be uniform throughout the country so that better effort can be put on experiments/apparatus, books, etc.
iv) Schemes of training to industrial workers, medical doctors and nurses, architects, musicians should be formulated.
v) There is a need to introduce acoustics properly in many disciplines. For example, in music a clear understanding of sound level, tuning of instruments, frequency, resonance, echo and reverberations, octave, etc., should help the students understand and perform better.

vi) Syllabus should be formed aiming at increasing the job potential.

8.2 Experiments

i) Depending upon the syllabus, the experiments should be same throughout the country for a particular course.

ii) Experimental setups should be easily available and non-expensive with fabrication type exercises incorporated in each experiments.

iii) Exhibitions of experiments which involve some principle of acoustics and also setups showing some applications of acoustics may be organised.

9. ACKNOWLEDGEMENT

The authors are thankful to the various teachers and other authorities who have extended their support by the way of sending information required in the survey proforma.

10. REFERENCES

1. Kameswaran,S., Rajendra Kumar, P.V. and Ranagasayee, R., Temples Acoustics in India, X ICA Sydney,$\underline{1}$ 131 (1980).

2. Modak, H.V. and Parmeswaran, S., Some Acoustical Measurements of Musical Pillars, ibid, p. 132.

3. Gupta, P.N., Ultrasonic Equipment in Industry and Medicine, An Analysers Report, Electronics-Information and Planning, $\underline{7}$ 83-103 (1979).

Table-1 LEVELS AT WHICH ACOUSTICS IS TAUGHT IN INDIA

Sl. No.	Level	Total Course (Year)	Duration for Acoustics teaching(year)	Teaching hours for acoustics per week	Approximate number of pupils studying acoustics	Percentage of institutions teaching acoustics	Acoustics as a separate separate paper
	SCIENCE						
1.	Pre-University	12	1	–	100,000	95	No
2.	Graduate	3	1	2	10,000	60	No
3.	Post-Graduate	2	1	2	200	30	At 3place
4.	M.Phil	1	1	4	10	10	Yes
	ENGINEERING						
5.	Graduate	4	1/2	2	10,000	90	At 1place
6.	Postgraduate	2	1/2	2	50	30	No
	ARCHITECTURE						
7.	Graduate	4	1/2	4	500	100	Yes

Table-2. SOME OF THE MUSICAL INSTRUMENTS OF INDIAN ORIGIN.

Sl.No.	Instrument	Type	Excitation
1.	Veena	String instrument	Plucking
2.	Sitar	-do-	-do-
3.	Tanpura	-do-	-do-
4.	Shahnai	Wind instrument	Blowing
5.	Murli	-do-	-do-
6.	Mridangam	Percussion instrument	Beating
7.	Damru	-do-	-do-
8.	Phakhavaj	-do-	-do-
9.	Jaltarang	Water column	Light striking

Table 3   RESEARCH AND DEVELOPMENT IN
              ACOUSTICS IN INDIA

Sl. no.	Organisation	Total	R&D In Acoustics	
1.	Universities & Institutes	140	27	(20%)
2.	Laboratories			
	a) CSIR	40	11	(25%)
	b) Others	110	8	(1%)
3.	In-house R&D establishments in industries	1000	10	(1%)

Table 4   SOCIETIES HAVING INTEREST IN ACOUSTICS

Sl. No.	Society	Membership	Head office	Journal	% Interest in acoustics	Training courses
1	Ultrasonic Society of India	300	Delhi	Yes*	100%	Yes
2	Acoustical Society of India	250	Hyderabad	Yes**	100%	No
3	NDT Society of India	1200	Bombay	Yes***	20%	Yes
4	Indian Society for Medical Ultrasound	N/A	Bombay		20%	No
5	Indian Institute of Non-destructive Inspection Engineers	N/A	Madras		20%	Yes

* Journal of Pure and Applied Ultrasonics.

** Journal of Acoustical Society of India.

***Quality Evaluation.

Table 5. TOPICS COVERED IN ACOUSTICS

Sl.No.	Topics	Level	Number of Institutions
1.	Basics of acoustics	Pre-univ. & B.Sc.	All those offering acoustics
2.	Physics of sound propagation	M.Sc. (Physics)	2 only
3.	Architectural acoustics	M.Phil. B.Arch.	2 only All
4.	Noise	M.Sc. (Physics)	2 only
5.	Underwater acoustics	M.Sc. (Physics) M.Sc. (Marine)	2 only 1 only
6.	Biomedical ultrasonics	B.E.	1 only
7.	Transducers	M.Sc. M.Phil.	2 only 1 only
8.	NDT	B.E.	50
9.	Studio acoustics	B.Sc.	1 only

Annexure 1

Acoustic Experiment at Various Levels

A. Pre-University and graduate level

1. To determine sound velocity using Melde's method
2. To study beats and coupled oscillation
3. To study laws of string using sonometer
4. To determine velocity of sound in air using resonance coloumn.
5. Study of wave properties by Ripple Tank.
6. To determine frequency of tuning fork using Kundt's tube.

B. Postgraduate level (common experiments).

1. To evaluate velocity of sound in different liquids by ultrasonic interferometer.
2. To study light diffraction by ultrasonic waves.

C. Postgraduate level (special experiments).

1. Flaw detection in solids by ultrasonic pulse method.
2. Propagation of waves in diatomic crystals.
3. To study wave fronts in an enclosed room.
4. Case studies: Autitoriums, Cinema Halls, Broadcasting studies.
5. Ultrasonic methods for studying materials.
6. Demonstration of beats, doppler effect, echo, resonance.
7. Measurement of sound pressure level using SPL meter.
8. Measurement of reverbation time.
9. Measurement of Audiometer.
10. Hearing aid measurements.
11. Magnetic recording and reproduction system.
12. Measurements of attenuation characteristics of a room.
13. Demonstration of stereophonic recording.
14. Study of construction of artifical ear, artificial mastoid, pistonphone.
15. Electro acoustic measurement of a hearing aid according to ISI standard.
16. To plot polar characteristics of a loud-speaker at different frequencies.

17. To find polarity of given loud-speaker.
18. To plot frequency response of loud-speaker.
19. To study the variation of output impedance of a microphone at different frequencies.
20. To study public addressing system.
21. Loud speaker Response and Cone Loading.
22. Radiation pattern of woofers and tweetors.
23. Characteristics polar response of moving coil microphone.
24. Characteristics polar response of condenser/ electric condenser microphone.
25. To study transmission loss and transmittivity in various materials.
26. To study the characteristic and response of stylus pick-up.

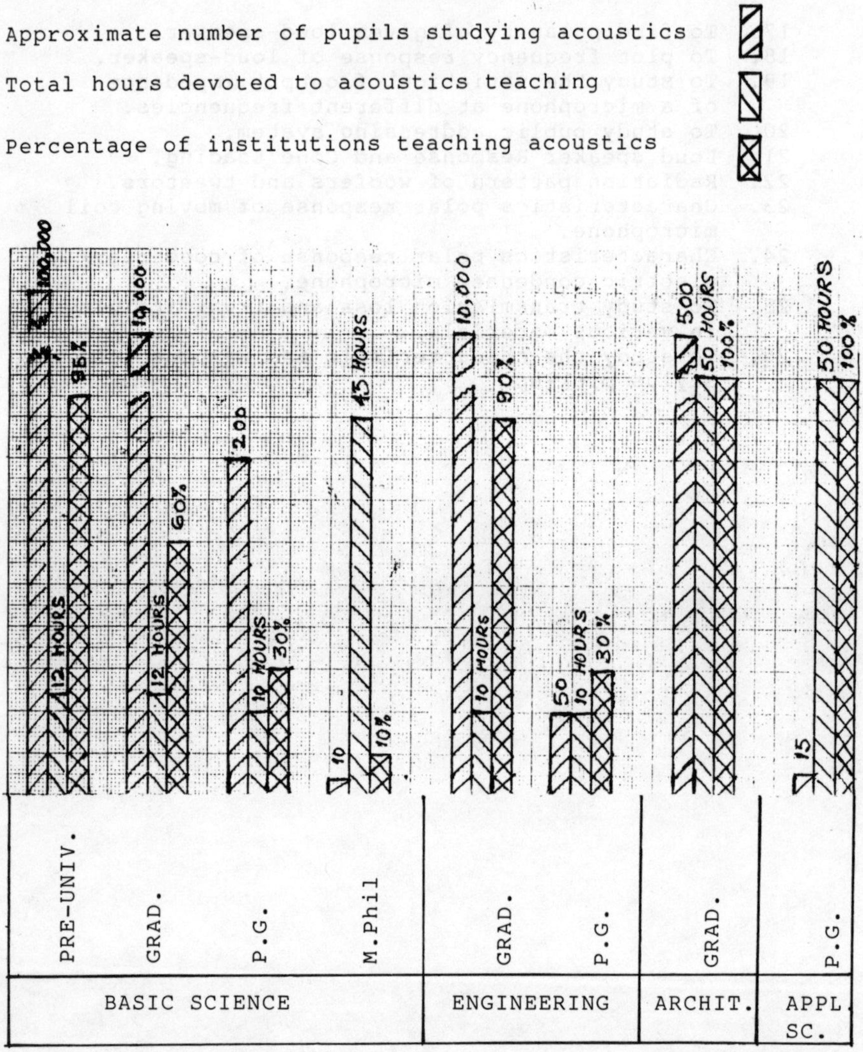

Approximate number of pupils studying acoustics
Total hours devoted to acoustics teaching
Percentage of institutions teaching acoustics

Fig.1 INSTITUTES, STUDENTS AND HOURS FOR ACOUSTICS TEACHING

ADVANCED EDUCATION IN INDUSTRIAL ACOUSTICS IN DENMARK

Professor Leif Bjørnø
Industrial Acoustics Laboratory
Technical University of Denmark
Building 425, DK-2800 Lyngby, Denmark

INTRODUCTION

The interdisciplinary character of acoustics and its central position in modern research and development make necessary a current renewal of topics taught and research subjects taken up in order to keep acoustics in the forefront of development in the community based on advanced technology. The increasing computation and communication possibilities using electronic means form a challenge to education in acoustics and lead to lifelong education to meet the demands from the modern society. New fields are opening up at the frontiers of acoustics research which make necessary, frequently due to lack of ressources, to abandon old and not so fast developing fields of research in acoustics. This dynamical world of acoustics research is reflected back on the teaching of acoustics which, in order to keep pace with the research and development, also has to be dynamic in its production of new scientists, engineers and teachers ready to meet future challenges in acoustics.

This paper will in particular emphasize the significance of and experience obtained by advanced education in industrial acoustics in Denmark. The exposition will comprise a description of the graduate and post-graduate education systems currently being applied in Denmark followed by a detailed description and discussion of the development in industrial acoustics based on current research projects of significance to present and future education in industrial acoustics at the Technical University of Denmark.

THE EDUCATION SYSTEM

The field, industrial acoustics, comprises in principle all production and exploitation of new research results in acoustics, with special emphasis on advanced industrial applications of research results. The Industrial Acoustics Laboratory is related to the Institute for Manufacturing Engineering thus forming a part of the mechanical engineering sector at the Technical University of Denmark. The interdisciplinarity of industrial acoustics is reflected by the fact that the staff at the laboratory have their educational background in physics, mechanical and electrical engineering.

THE FUNDAMENTAL AND GRADUATE EDUCATION

When a young male or female join the University they are normally 18-20 years of age and have just finished their high-school education, most frequently with an education having a strong basis of physics, chemistry, geometry and mathematics. Their first two years of university studies will in particular give them a solid background in the fundamental sciences related to the engineering education, i.e. mechanical, electrical, chemical or civil engineering, they have chosen. After two years of education in general and during subsequent years the young engineering students meet the courses taught by the Industrial Acoustics Laboratory. These courses are primarily a fundamental course in theoretical and applied acoustics for mechanical engineers. This course is, however, also attended by several students in electrical engineering. All courses are structured into a module system which for the fundamental course means two lecture and exercise modules, each of 2 x 35 minutes duration, a week. The next course given by the Laboratory is a course in machine noise in the industry, which aims at giving the students a thorough knowledge of the noise sources by machines in the industry, how to analyse and abate the noise from already existing machines and how to construct low-noise machines. Moreover, this course aims at giving the students a fundamental background for machine diagnosis and process control by the use of noise and vibration. The third more formal course taught by the staff of the Industrial Acoustics Laboratory concerns the planning and performing of small theoretical and experimental research projects related to on-going research in the Industrial Acoustics Laboratory. Several of these research projects shall be treated in later sections of this paper. Later, before graduating with a degree in engineering from the Technical University, the students are confronted with a more comprehensive research and development project having a duration of 5-12 month. This research and development project forms the basis for a thesis to be submitted to the Technical University of Denmark as a partial fulfilment of the demands to be satisfied for obtaining a M.Sc. degree in engineering at the University. The nominal time for a study leading to a M.Sc. degree in one of the four engineering directions is $5\frac{1}{2}$ year, but frequently the studies last one or more years longer.

THE POST-GRADUATE EDUCATION

The post-graduate education in industrial acoustics at the Technical University of Denmark offers several educational lines. These lines are (1) the Danish Ph.D. Education, (2) the Industrial Research Education and (3), most recently, also a Scandinavian Education in Industrial Research. All three educations are strongly research oriented and have a duration of 2-3 years.

1. THE DANISH Ph.D. EDUCATION [1] is a $2\frac{1}{2}$-3 years education at the Technical University of Denmark. It is in principle offered to candidates who have obtained excellent results during their graduate studies

and who have shown abilities for scientific studies. The economical support for the students during their Ph.D. studies comes via the University from the Ministry of Education and is on a level with salaries obtained by engineers working in the industry. Ph.D. students are not permitted to have any other sources of income and are expected to spend their full working capacity on their research. However, their supervisor, originally only a university professor, but now even an experienced lecturer, may decide that the Ph.D. student takes part in the teaching activity in the laboratory related to his research subject in general. The research project, most frequently involving theory as well as experiments, is formulated in a collaboration between the student and his supervisor and a final and detailed plan for the studies also comprising advanced courses to be attended by the Ph.D. student is worked out and are send to the governing bodies at the University for acceptance, normally after half a year of studies.

The studies may also include studies of shorter duration in advanced laboratories in other countries as well as participation in conferences abroad, in particular in relation to presentation of early research results. The Ph.D. studies finish with a thesis produced by the student in consultation with his supervisor. This thesis is submitted for evaluation by one or more examiners from other universities, frequently universities abroad, and is defended orally after a 45 minutes public lecture. During the oral defence the examiners and the superviser ask questions related to the thesis work. The broader basis of knowledge acquired during the studies is controlled either through examinations currently passed by the Ph.D. student during his studies and/or through another lecture of 45 minutes duration given by the Ph.D. student on a subject determined by supervisor and examiner and given to the Ph.D. student one week in advance of his lecture. If these formalities are satisfied in an acceptable way the supervisor and the examiner writes a recommendation to the governing bodies of the University that the Ph.D. degree is conferred on the Ph.D. student. 8 students have hitherto finished Ph.D. studies in industrial acoustics topics at the Technical University of Denmark.

2. THE INDUSTRIAL RESEARCH EDUCATION [2] is a two years education given in a collaboration between a university department and an industrial company doing research and development on an advanced level. The industrial research education is on a scientific level considered to be on a par with the Danish Ph.D. education.

The fundamental idea behind the industrial researh education is that the need in a modern society for qualified scientists to work in industry must be satisfied in a close collaboration between universities and industry because the university environments will not be able to give the student the industrial experience necessary for later succesful research done under industrial conditions. The industrial research education is primarily aiming at direct and practically applicable problem solutions, i.e. it shall lead to the development of fundamental knowledge on which new and advanced products can be based.

This fertile collaboration between university laboratories and industry has led to many valuable new research results and products developed over the years since the start of this education in 1970. This education is in principle offered to candidates from universities and institutes for higher education if the candidate has the right research qualifications. A steering committee appointed by the Danish Academy of Technical Sciences is responsible for the nominal structure of this advanced education and for each individual education a research leader group is formed to guide the student in his research and in his choice of courses to be attended. The research leader group normally consists of the person being responsible for research in the company involved in the education and a university professor or a lecturer having a good background in research. The courses to be attended are selected among advanced courses in research management, research planning and technical courses related to the research subject. These courses are frequently given at universities or business schools abroad. The student is assumed to spend more than half of his study time in the industrial company involved in the education.

The salary received by the student is the salary normally given in industry, but in order to ensure that the student concentrates all his effords on his research and development project, half of his salary and all educational costs involving fees for courses, travel costs etc. and publication costs for the final thesis is paid by the Ministry of Industry via the steering committee. This committee is also responsible for that all industrial research educations are carried out on the same high level over the whole country. The committee also appoints examiners to evaluate the final thesis submitted by the student. This thesis is by the student defended in a public procedure very much like the one leading to the Danish Ph.D. The diploma for the industrial research education is issued by the Academy of Technical Scienses. About 140 candidates have since 1970 received a diploma for an industrial research education and of these candidates 8 have been involved in industrial acoustics research and development.

3. THE SCANDINAVIAN EDUCATION IN INDUSTRIAL RESEARCH [3] involves companies and universities in the five Scandinavian countries, Denmark, Norway, Sweden, Finland and Iceland. This advanced education started in 1986 is particularly characterized by collaboration across the national borders, as it involves a university in one country and a company and a candidate in another country. This post-graduate education has a duration of 2-3 years and during this period the candidate is assumed to spend a considerable part of his study time in the industrial company frequently in another country than his own. This study is aiming at an exploitation of the good experiences related to collaborative education schemes between universities and industry already existing in more of the Scandinavian countries. Like the industrial research education,on which the Scandinavian education in industrial research to a large extend is based and follows in its structure and general aim, individual study planes are worked out for each candidate in a collaboration between the members of the research leader group and the candidate. These study planes must be ready after half a year of studies and has to be sub-

mitted for acceptance by the steering committee responsible for the
studies, for the Scandinavian education in industrial research a committee,
the Scandinavian Industrial Research Committee, consisting of members
appointed on background of their research qualifications by national
research organization in the five countries, in Denmark by the Academy
of Technical Sciences.

The economical support to the education covering about 50% of the salary,
all travel costs, extra living costs related to part time moving to
another Scandinavian country, fees for courses, publication costs etc.
are contributed from the Scandinavian Industrial Foundation.

To the Scandinavian education in industrial research the same demands
to the candidates research qualifications and to the level of research
and development in the industrial company involved are valid as the
demands to the industrial research education, and the procedures used
for evaluation of the thesis produced and the broad background knowledge acquired by the candidate is very much alike in these two advanced
educations. However, the Scandinavian education in industrial research
may, moreover, lead to a Ph.D. degree at one of the national universities if the original study plan for the candidate satisfies certain
demands set by the universities.

The diploma for the Scandinavian education in industrial research is
issued by Scandinavian Industrial Foundation. One student from Norway
has up to now started on a Scandinavian education in industrial research
in a collaboration between a Norwegian company and the Industrial
Acoustics Laboratory at the Technical Universiy of Denmark.

DEVELOPMENT ASPECTS IN INDUSTRIAL ACOUSTICS

Beside the fundamental education in acoustics for students in mechanical
engineering, advanced teaching and research in industrial acoustics at
the Technical University of Denmark comprises three main areas. (1)
Machine noise abatement, machine diagnosis and process control using

noise and vibration. (2) Medical and industrial applications of ultrasound.
(3) Development of transducers and measuring systems.

1. MACHINE NOISE AND VIBRATION studies have internationally developed
very fast during recent years in particular aiming at the creation of
the necessary basis of knowledge for construction of low noise machines
and industrial equipments. This development will most probably continue
and will be amplified in the years to come as new areas of research in
machine noise and vibration are opening up and are becoming more widespread in technologically developed countries. These areas include for
instance machine diagnosis and process control by the use of noise and
vibration. The experiences obtained by the use of multi-sensor systems
for machine control and the use of intensity (sound and vibration)
measurements by identification and ranking of sources of noise and
vibration, together with the exploitation of new analytical and numerical
procedures, modal analysis, finite element methods, statistical energy
analysis etc., developed over recent years parallel with the development
of the capacity of available computers, have shown that a considerable
improvement in the economy related to production in industry and the
maintenance of the production apparatus may be obtained and that more
identical, high quality products may be produced if modern production
methods involving control based for instance on sound and vibration are
introduced. A step to take in the very near future is the introduction of
artificial intelligence, in particular expert systems, in the signal
processing, evaluation and steering of the industrial processes. In this
context sound and vibration are expected to play a very dominating role
[4].
Several research projects are currently being performed in the Industrial
Acoustics Laboratory related to machine noise and vibration.

2. MEDICAL AND INDUSTRIAL APPLICÁTIONS OF ULTRASOUND is a field in fast
development. Ultrasonic equipment used for medical and biological measurements, including high resolution scanners which permit several

different medical diagnostical investigations to be performed using the same main instrument, and therapeutic equipment based on high-intensity ultrasound are devices with a high development potential. The disintegration of kidney stones and other body stones by focussing of shock waves or high-intensity ultrasonic waves, characterization of biological media including human tissue by means of nonlinear ultrasonic methods, the study of biomass in the sea by the use of back scattering spectra in the frequency range of 15-200 kHz and measurements of the growth of bacteries of various types are a few biological ultrasound studies currently performed by the Industrial Acoustics Laboratory. Also the industrial application of ultrasound is strongly increasing. The studies of ultrasonically based processes involving deformation, welding, cutting, mixing etc. of materials and the studies of industrial control and measuring processes based on ultrasound are numerous in countries with an industry based on high-technology. Localization of defects in structures in air and under water, the search for tubes and cables in the sea bed and in particular ultrasonic NDT studies of bonded structures, fiber reinforced and composite materials are areas of research of current interest to research workers in industrial acoustics.

3. DEVELOPMENTS OF TRANSDUCERS AND MEASURING SYSTEMS are necessary for reliable registrations of sound and vibration signals. Mostly in Japan, but also in the U.S. a very fast development in transducers and measuring systems has taken place during recent years in particular based on the development of new transducer materials, principles and constructions. The development has shown that the company in possession of the most advanced and best suited transducer most frequently will be the winner of the competition among manufacturers of measuring systems. Among the many new areas of transducer technology where a strong and continuous development shall be expected in the nearest future are transducers based on: ceramics, polymers, biological and chemical processes, silicon and fiber optics [5].

Currently transducer research is being done at the Industrial Acoustics Laboratory related to: New calculation procedures for transducers involving among other things finite element procedures, new transducer materials including polymers, fiber optics and piezoelectric composites, new applications as for instance robotics and deep sea applications.

Most research and development projects in industrial acoustics mentioned above are carried out in a collaboration with industrial companies or research institutes in Denmark and abroad. Some of the projects have received financial support from the EC under the comprehensive research programming financed fully or in part by the European Communities. Several research students and visiting professors from countries abroad have worked for shorter or longer periods in the Industrial Acoustics Laboratory bringing with them research traditions and fundamental and specialized knowledge from their own home countries. This contribution of cross-fertilization is extremely useful for the advanced research and education in Industrial Acoustics. Moreover, a contract research division of the Institute for Product Development, a non-profit research and development institute situated at the Technical University of Denmark, but not being controlled by the governing bodies of the University, was established some years ago in the Industrial Acoustics Laboratory and has contributed essentially to the strong contacts between the laboratory and industry in Denmark and abroad.

The close relations between research and education in industrial acoustics in Denmark is very fruitful for both parts. Early transfer of interesting research results to the teaching of engineering students at various levels is a must in order to give an up-to-date teaching which is lively, inspiring and oriented towards future applications by giving the students the best tools possible for their future work in the Danish industry. The students own participation in small or larger research projects related to on-going research in the Laboratory is an offer being well received by the students as it leads to the education of scientists with a background in industrial acoustics who are going to form the backbone of the scientific staff in industry in the years to come. Good employment possibilities in industry, exciting research and development projects and in particular an excellent basis for personal development are offered to students who decide to study industrial acoustics in a graduate or post-graduate study. It is my general opinion, among other things based on the development going on in acoustics laboratories in the other Scandinavian countries and in many other technologically developed countries, that industrial acoustics will take its great share of education and development in modern acoustics in the years to come.

REFERENCES

1. Guidance concerning the Ph.D. study at the Technical University of Denmark. April 1982. (In Danish).

2. The Industrial Research Education. ATV, Erhvervsforskerudvalget, March 1987. (In Danish).

3. Scandinavian Education in Industrial Research, Nordisk Industrifond and ATV, November 1986. (In Danish).

4. Bjørnø, L., Trends of developments in Industrial Acoustics. Proc. Nordisk Samarbejdsmøde, Ustaoset, February 1986. (In press) (In Danish).

5. Bjørnø, L. et al., New Sensor Technology. Proposals on future research activities under the second part of the European Community research programme BRITE. January 1987.

HOW TO TEACH SOUND-ENGINEERING ?

Gustaw Budzyński, Marianna Sankiewicz

Sound Engineering Department
Institute of Telecommunication
Technical University of Gdańsk
80-952, Gdańsk, Poland

Giorgos Papanikolaou

Institute of Telecommunication
University of Thessaloniki
Thessaloniki, Greece

ABSTRACT

Specific educational problems arising at the formation of sound-engineers are discussed. Discrepancies in existing educational programs are characterized. New aspects in teaching sound-engineers are presented, from both theoretical and practical points of view. The role of human factors is stressed. Concluding remarks show some practical means, which would help to improve the teaching process of sound-engineering. First of all, the importance of an information exchange and of a broad discussion among the experienced teachers and scientists in acoustics, is emphasized.

INTRODUCTION

Development of every branch of science depends obviously on appropriate education. The term 'appropriate' denotes here many requirements concerning educational programs, their contents, their levels, organization of studies, etc. Many branches of science have traditionally established forms of education, e.g. architecture, astronomy, civil engineering, chemistry, electronics, etc., while many others have not.

Acoustics is situated among those latter and the actual ICA Conference is needed due to that unfavourable situation. It stems from the diversity of acoustical subject-matters and from their strong attachemennt to various branches of science. Thus, acousticians, namely those who teach acoustics, are mostly dispersed at various university-faculties, -schools, or -departments, being submitted there to influence and decisions issued by people specialized in various domains, yet almost never in acoustics. It is obvious that the mentioned decisions may influence negatively the development in acoustical teaching, e.g. by limitations of funds, laboratory equipment, etc., as otherwise acoustics-schools might turn to be competitive with well established branches of science. Moreover, acoustics as a subject is a peculiar one and it needs specific methods in teaching.

Acoustics is one of the most ancient sciences of the human civilisation. However, its great development is so recent, so diversified and so distributed, that the scientists and teachers working on acoustics could not, so far, concentrate and create acoustical faculties or departments, or, generally speaking, independent units of higher education. In those, quite a new modern manner of teaching might easily be introduced.

Acoustics need not to be taught by the use of analogies only, based traditionally on mechanical or electrical concepts. It should be presented from its foundations up to the large spectrum of applications, as an entity. Such a teaching course ought to be aimed at the formation of acoustical engineers, able to replace e.g. architects, mechanical-, electrical-, electronics- and other engineers, being only superficially adapted to work on acoustical problems. However, solving such problems in practice nowadays, is more or less successful. The actual industrialized civilization brings those important problems more and more frequently, hence stems the growing need for modern teaching of professional acousticians-engineers.

The above mentioned concepts concern the main idea how to teach acoustics. However, before a required general solution is found, particular questions connected with some specialized fields of acoustical teaching should be answered. Such a distinctly separated field of acoustics, very important in practical applications is the Sound-Engineering. Then, how to teach Sound Engineering, how to form sound-engineers ?

FORMATION OF SOUND-ENGINEERS

Before entering into the teaching problems the definition of sound-engineering should be recalled and commented. The term 'sound-engineering' is frequently used in a nearly the same meaning as 'audio-engineering'. Both denote the domain of engineering connected directly or indirectly to audible sounds. Similarly, yet in a narrower sense, the term 'noise-engineering' is being used. Contrary to it the term 'acoustical engineering' is very broad and it contains all engineering in the domain of acoustics.

However, the term 'sound-engineering', besides its general meaning is being used in a more specialized form.

So the name of 'sound-engineer' is being employed in parallel with 'recording engineer' or 'balance engineer' to denominate a highly specialized person, who knows very well the technology of recording, broadcasting, film, video post-production etc., one who is educated in music, and who is able to fulfill all the duties in a sound-control-room in order to achieve an acceptable final effect of the sound transmission in space or time.

Due to the ambiguity of the above terms, a German specialized professional name 'Tonmeister' is being frequently used in English. It has been even officially adopted as a professional title by the Surrey University in Guildford [4]. There is, however, another German professional denomination 'Toningenieur', which has a very close, yet slightly different meaning. Both terms have their counterparts e.g. in Russian [15], as well as in Polish, where 'Tonmeister' means 'Rezyser Dzwieku' and 'Toningeniieur' means 'Inzynier Dzwieku'. However in other languages generally no adequate counterparts are in use.

It is not a question of a lacking term only. It results rather from the absence of appropriate teaching on a higher level and from the lack of relevant professional titles in many well developed countries. E.g. in France, although the term 'ingenieur du son' exists, which means in translation exactly the same as 'sound engineer', sound engineers for broadcasting, television, recording and similar professional activities are being trained practically to their future jobs, without need of any formal higher education in the appropriate domain. One must admit, however, that even in France, which is rather reluctant in accepting new teaching concepts, an attempt to organize teaching for sound engineers at a university level has been recently undertaken [11].

The need of a highly specialized, high level education for sound-engineers is now generally understood and accepted. Several universities in Europe and in America have undertaken appropriate teaching [5]. The 1984 edition of the 'AES Directory of Educational Programs' contains a list of 38 such universities. Very often a combined process of teaching made by two faculties in common is applied: a musical one plus a technical one. As there is no traditional way of teaching sound-engineers, so the existing programs vary in many aspects.

It is impossible to analyze here all available programs in details, to compare them critically and assess their merits or imperfections. Some general comparisons are presented in Table 1, rather in order to exemplify the discrepancy of concepts, as well as to characterize peculiarities of sound-engineering educational programs.

The figures given in Table 1 have been evaluated, basing on accessible publications ([7][13][14][16]). As particular programs are uncomparable, so several simplifying assumptions had to be made in order to present a general characteristics uniformly.

Table 1   Percentage of three subject-groups in educational programs for sound-engineers.

University:	Subjects % concerning:		
	Music	Gen.Science	Engineering
1. Berlin	62	18	20
2. Detmold	48	34	18
3. Dusseldorf	30	8	62
4. Gdańsk	8	19	73
5. The Hague	29	15	56
6. Warszawa	66	13	21

Remarks:
1 - Staatliche Hochschule fur Musik und darstellende Kunst, in Verbindung mit der Technischen Universität Berlin. Tonmeister course.
2 - Staatliche Hochschule fur Musik Westfalen-Lippe, Detmold Tonmeister Institut.
3 - Fachhochschule Düsseldorf, in Verbindung mit dem Robert Schumann-Institut der Musikhochschule Rheinland. Sound and picture technique.
4 - Technical University of Gdańsk, Faculty of Electronics, Sound Engineering Department.
5 - Royal Conservatory, The Hague, Sound Engineering Course.
6 - Frederic Chopin Music Academy, Warsaw, Sound Recording Faculty.

Two distinct tendencies are noticeable among the above quoted percentages: that of prevailing music subjects (e.g. Nos 1 and 6) and that of prevailing engineering subjects (e.g. Nos 3 and 4). This differentiation of educational programs is typical and it founds its reflection in professional profiles of sound-engineers. Just because of its existence the two formerly mentioned German professional titles are unavoidable and in common use.

Thus, the problem of teaching sound-engineers is complex from its very beginning. However, its complexity is the more intricated, the more modern are the engineering systems, instrumentation and methods, as well as, the deeper is the insight into the general foundations of acoustics and, especially, into the psychophysiology of perception.

On the other hand, teaching sound-engineering requires a great deal of parallel practical training and acquiring many skills, the importance of which surpasses any requirements met in other engineering professions. Those aspects are worth to be mentioned here, even superficially.

NEW ASPECTS

The theoretical aspects are similar to those encountered in teaching different acoustical professions, i.e. subjects like the foundations of acoustics, acoustic measurements, electroacoustic transducers and other general ubjects. An especially important theoretical background is necessary in psychology and physiology of hearing, as well as, in musical acoustics.

New theoretical knowledge is necessary concerning the modern signal theory, the information theory, the digital communication systems, and the theory of computer programming. However the most characteristic feature of teaching in sound-engineering, is the decisive role of practical training and laboratory works, and least but not last, a necessary portion of multidirectional abilities, which are not so important in the remaining fields of acoustical engineering.

The practical aspects require familiarity with particular subjects, such as: sound systems, sound sources, transducers, converters, sound mixing, sound recording and reproduction, sound synthesizers, room acoustics, sound reinforcement, computer-aided techniques, and other subsidiary skills. Playing some traditional music instruments, or performing in an orchestra is a significant example among those skills. It should be stressed here, that the above-said familiarity means a real capacity to perform the mentioned functions better or worse, at any rate independently. Thus, the role of education in appropriately equipped laboratories could not be overestimated.

A problem, which should not be overlooked here, is to consider the so-called human factors. Subjective factors play an important role at almost all types of sound transmissions. A sound-engineer, responsible for the quality of sound productions, must be able to take into cosiderations all possible influence of human factors. E.g. he must know the stability of his own senses and their response influenced by the fatigue of a prolongued work; he must be able to appreciate the mood of the musicians or singers, which may favour the success of a recording; he must be capable to imagine subjective responses of his future listeners assessing his work.

It is relatively easy to mention all above-quoted factors influencing the job of a sound-engineer. It is however much more difficult to create an educational system matched to the demands of the future sound-engineers. The rapid progress in sound techniques must be taken into account. On the other hand, the possibilities of getting the modern equipment into educational laboratories are limited by economical respects. Thus, practical solutions must be sought to find full value substitutions of modern expensive professional equipment, thus creating, appropriate conditions for teaching all necessary subjects and for students practicing all the required skills.

Practical solutions depend on a cooperation between universities, generally unable to spend money on a very expensive equipment for sound-engineering laboratories, especially for a professional recording studio, and other institutions involved in sound technique. In West countries these are usually the industrial enterprises. E.g. the equipment at Surrey University worth about 200.000 is the result of generous cooperation of Solid State Logic, Bruel & Kjaer, Calrec, Sony, Dolby and other companies [6]. On the other hand, in Poland, there is a well equipped professional recording studio at the Academy of Music in Warsaw. The modern equipment there is payed by the Ministry of Culture. However, Polish universities depending on Ministry of Higher Education have not such possibilities. Only thanks to a close cooperation between local Brodcasting Centre in Gdansk and the Technical University of Gdansk, moreover, thanks to the support of the Polish Radio and Television it was possible to equip a small professional studio, which serves now as a main educational means for teaching sound-engineers in Gdansk. More detailed information about that studio outfit, as well as, about its activities and laboratory subjects, is contained in a separate contribution to the Conference [3].

A recording studio is a necessary educational instrument for the formation of sound-engineers. Other instruments are, however, indispensable too. Students must be trained in applications of computer-techniques to sound-engineering problems. Examples showing parts of teaching programs applied at the Sound Engineering Department of the Technical University of Gdansk, are presented in a special contribution to this Conference [10]. Other applications of computer-techniques to sound analysis and to bowed string vibrations are described separately [9][12]. Acoustical modelling by analogue technique remains an efficient method of solving many sound-engineering problems. Possible applications of that laboratory technique are then described in another contribution [17].

CONCLUDING REMARKS

It is obvious, that the question posed on title cannot be answered explicitly. Yet, some partials of the future answer have been sketched above. In that situation any rigorous conclusions would be premature. Possibly other contributions to this Conference and subsequent discussions during debates will approximate a demanded answer. However, some remarks made here seem to be actual and helpful for expected discussion.

So far, nothing was said about the mixed media of sound- and vision-engineering. The revolutionary development of vision-techniques and their intricate connexion with sound-techniques renders the discussed problem of teaching more complex and difficult. The actually graduated sound-engineers will function sooner or later as sound- and vision-engineers. Thus, appropriate modernization of teaching programs should be already undertaken. E.g. at the Sound Engineering Department in Gdansk the course on Sound Systems is already changed into Sound and Vision Systems. The course on Psychophysiology of Vision is being prepared. The Laboratory on Sound Studio Technique comprises now several items on Television Studio Technique. This technique becomes now a basic solution for all problems of synchronisation of image and sound. Of course, professional video-recorders are too expensive; however, semiprofessional or general use equipment permits to fulfill the necessary training tasks. Nevertheless, parts of the television equipment at the Gdansk laboratory e.g. vision amplifiers, racks, monitors, cameras, mixers etc., are of professional, although not modern types.

The progress in teaching sound-engineering should not, of course, be limited to practical problems exclusively. Investigation of various fundamental scientiific questions is very important in order to develop independent thinking, criticism and creativity in future engineers. Many problems important for sound-engineers still require a deeper insight and a better understanding. E.g. inspite of the discovery of an important influence of the so called 'lateral reflections'[1] on the quality of sound in rooms, some essential properties of the directional hearing remain, so far, unclear, which prevents rational solutions in some sound-systems [2]. Such problems need further theoretical and experimental investigation.

A special attention should be paid to a division in acoustics created by independent activities of acoustical engineers and of sound-engineers. Looking from another point of view this is a division due to activities of the two independent organizations: the Acoustical Society of America and the Audio Engineering Society, as well as their counterparts elsewhere. There is a lack of information between those two big professional groups of acousticians, while both groups ought to be tending to a certain degree of integration, at least in the field of education. The problem of divisions among acousticians is known and its harmful effects were discussed [8].

Everybody wishes the process of education of the sound-engineers, and more generally, of the acousticians to be designed and executed as well as possible. The forthcoming discussion will most probably show that appropriate ways are known, at least partially. However, the results of education depend not only on the knowledge of teachers, but also on adequate funds provided for the teaching process. Apart from economic questions, one may conclude, that the teachers should augment their knowledge on educational programs and on organization of the teaching process in acoustics, in order to meet growing demands of the human society.

The exchange of information and the discussion among the experienced university teachers and scientists is the best and unique method to obtain the desired final results. The authors of this contribution hope their passion for teaching sound-engineers both in Poland and in Greece will arouse a vivid discussion among the participants of this Conference. Such a discussion should become an efficient instrument of disseminating useful ideas on the education and development in acoustics.

REFERENCES

[1] M. Barron: "The subjective effects of first reflections in concert halls – the need for lateral reflections", J. of Sound & Vibrations, Vol.15, No.4, 1971.

[2] G. Budzyński: "Theory of the reflective localization of sound sources", Archives of Acoustics, Vol.11, No.1, 1986.

[3] G. Cypukow, J. Dyżewski: "Educational Recording Studio", ICA Conference 'Prospects in Modern Acoustics – Education and Development', Gdańsk, 1987.

[4] J. Eason: "Tonmeister Training – UK", Studio Sound, p. 44-46, June, 1983.

[5] A.E.S. Education Committee: "Directory of Educational Programs", Audio Eng. Society, U.S.A., 1984.

[6] P. Evans: "Surrey Course – Full Report", APRS News, January, 1985.

[7] H.L. Feldgen: "Tonmeister", Blatter zur Berufskunde, Bd.3, S. 1-22, Bertelsmann, Bielefeld, 1974.

[8] G.L. Fuchs: "L'enseignement de l'acoustique", Impact, Vol. 35, No.2/3, p. 139-153, 1985

[9] A. Kaczmarek: "Minicomputer application to organ-sound analysis", ICA Conference 'Prospects in Modern Acoustics – Education and Development', Gdańsk, 1987.

[10] A. Kulowski: "The ray method as a teaching tool", ICA Conf.'Prospects in Modern Acoustics – Education and Development', Gdańsk, 1987.

[11] Panel discussion on Education of Recording Engineers, 77th AES Convention, Hamburg, 1985.

[12] Z. Perucki, B. Kostek: "Computer modelling of the bowed string oscillations", ICA Conf. 'Prospects in Modern Acoustics – Education and Development, Gdańsk, 1987.

[13] M. Sankiewicz; "Course Program – Sound

Engineering" (in Polish), Technical University of Gdańsk, Faculty of Electronics, 1985.

[14] M. Sankiewicz: "Sound-engineers educational problems"(in Polish), First Symposium on Sound Engineering, p. 123-128, (ed.) M. Sankiewicz, G. Papanikolaou, Gdańsk-Thessaloniki, 1985-1986.

[15] Technical Commission O.I.R.T.: "Requirements on formation of sound engineers in broadcasting and television" (in Russian), TK-II-833, Prague, 1971.

[16] J. Urbański: "Teaching program on the Faculty of Sound Recording of the Music Academy in Warsaw" (in Polish) Wyd. Radia i TV, Warszawa, 1972.

[17] A. Witkowski, M. Cwieczkowski: "Acoustical modelling laboratory", ICA Conf. 'Prospects in Modern Acoustics - Education and Development', Gdańsk, 1987.

Education in vibroacoustics

Z. ENGEL

Institute of Mechanics and Vibroacoustics
University of Mining and Metallurgy
Al. Mickiewicza 30, 30–059 Kraków

ABSTRACT

Vibroacoustics is the science of all vibrational and acoustic phenomena occurring in nature and industry, in machines and appliances. This paper presents the utilitarian aim of the vibroacoustics of machines and the main tasks of this new branch of science. Some problems connected with the training of specialists – vibroacousticians – are also mentioned.

A new branch of science, vibroacoustics, was started about 20 years ago. It includes all vibrational and acoustic phenomena occurring in nature and industry, in machines and appliances. The term "vibroacoustics" was used for the first time in Cracow. The authors of this new field of knowledge made a point of consolidating two disciplines of science, known and well developing in Poland, namely, the theory of vibration and acoustics. The creators of vibroacoustics aimed mainly at a joint considering of the vibrational and acoustic phenomena taking place in machines and appliances.

The practical aim of the vibroacoustics of machines can be formulated as follows: the lowering of the level of vibroacoustic disturbances in machines, appliances and their surroundings to a minimum possible at the given stage of knowledge and technology, and the application of information comprised in vibroacoustic processes for the evaluation of the quality of products and manufacturing processes.

Against the background of a thus formulated purpose we can define the main tasks of vibroacoustics, which are /Fig. 1/

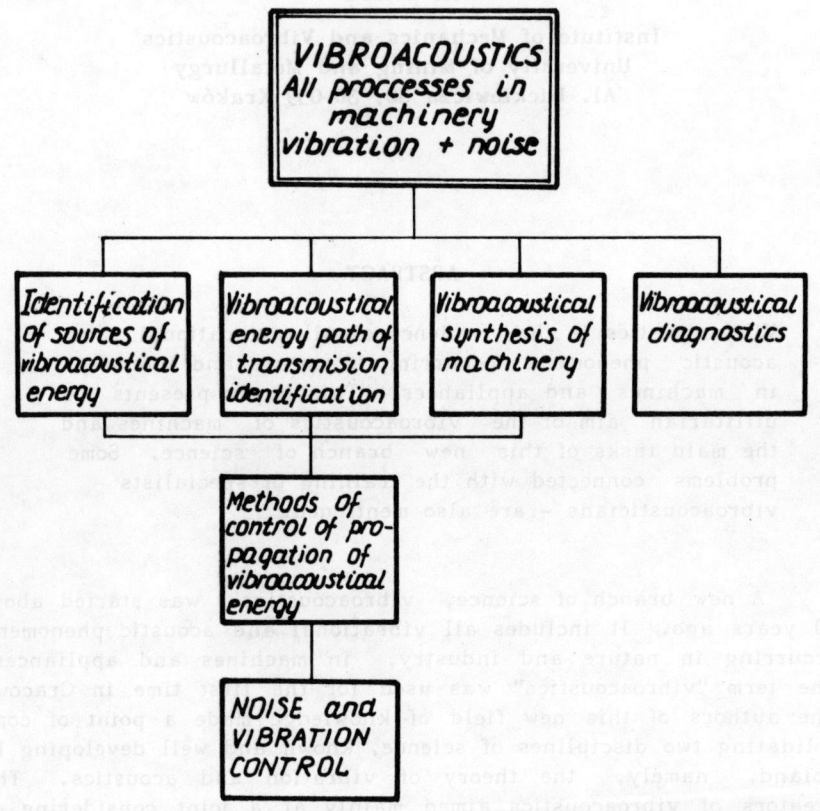

a/ the identification of the sources of vibration and noise; it consists in the localization of particular sources within a plant or machine, the determination of the characteristics and interdependence between individual sources, the determination of vibroacoustic power and of the nature and causes of the generation of vibration and sounds, b/ the identification of the paths of propagation of vibroacoustic energy in machines and constructions, the working-out of the methods for the preventive and remedial inspections of these phenomena, the working-out of new methods of analysis, intermediate between the undulatory and the discrete approach to these phenomena, the elaboration of the methods for the inspection of the emission and propagation of vibroacoustic energy in industrial environment and the working-out of the methods for the controlling of these phenomena in the industrial environment,

c/ vibroacoustic synthesis of machines, the structural, kinematic and dynamic synthesis leading to the attainment of the optimal vibroacoustic activity when the machines are in operation,

d/ the utilization of vibroacoustic signals for the purposes of technical diagnostics, as those signals bear information about the condition of machines and devices.

Parallel to the development of vibroacoustics, education of specialists in this field was initiated. At the Institute of Mechanics and Vibroacoustics, Technical University of Mining and Metallurgy in Cracow, specialists in vibroacoustics are educated on the basis of a general course in mechanics.

The course in vibroacoustics, including the completion of the diploma thesis, takes 5 years /10 semesters/. The students have some subjects in common with the students of the course in mechanics.

Table 1 shows the structure of didactic classes

	Division of subjects	Number of class and laboratory hours
A	General subjects	1410
B	Basic subjects of the course in mechanics	2235
C	Subjects of specialization	630
	Total	4275 hrs

The basic disciplines in the field of mechanics include, among others, courses in the following subjects:
- mechanics   165 hrs
- strength of materials   180 hrs
- theory of vibration   60 hrs
- fluid mechanics   75 hrs
- foundations of machine construction   240 hrs
- materials science   105 hrs
- metrology   105 hrs
- foundations of informatics   75 hrs
- theory of machines   45 hrs
- machine dynamics   60 hrs
- anti-vibration insulation   60 hrs

After the completion of the course in the basic disciplines, the students who specialize in vibroacoustics, study the following subjects:
- dynamics of continuous material systems
- identification of material systems
- signal processing
- foundations of acoustics
- foundations of vibroacoustics of machines
- stochastic processes
- electro-acoustics

The students specializing in vibroacoustics may graduate in three branches of this science:
1. Noise and vibration control
2. Vibration technique
3. Vibroacoustic diagnostics

In accordance with the branch in which they intend to graduate the students choose from the below listed subjects:
- vibroacoustics of machines
- vibroacoustic measurements
- flow acoustics
- room acoustics
- vibrotechnique
- vibrational processes
- diagnostic identification
- diagnostic metrology
- diagnostic systems
- protection from noise and vibrations

In the last semester of study the students complete their theses for a masters degree. Some subjects of the theses are given below by way of example:
1. Vibroacoustic discrimination of state characteristics and classification in diagnostic experiment.
2. An analysis of the possibilities to lower the level of vibroacoustic energy by a change in the rolling programme.
3. Localization of the sources of vibroacoustic energy in the operating lines of gas stations.
4. Numerical analysis of vibroacoustic signals.
5. The effect of traffic parameters on the noise level of the crossroads selected.
6. The effect of vibration parameters on the quality of work of concrete thickeners.
7. Methods for the lowering of vibration and noise level in portable grinding tools.
8. A selection of elastic systems of suppression in large acoustic chambers.
9. Acoustic adaptation of the pouring shop.
10. A desing for anti-noise and vibration protection of natural gas reducers.
11. The evaluation of radiating coefficient for sound surface sources.
12. Model testing of the generation of impact noise.
13. Identification of the moulding machine-exemplified mechanical system for purposes of the minimization of noise and vibration.
14. A dynamic analysis of the vibrating screen.
15. An analysis of vibration of sandwich beams induced by a moving load.
16. Digital simulation experiment in the diagnostics of machines.
17. Emission of vibroacoustic energy by flat-surface sources of sound in the case of a variable distribution of vibration speeds.
18 An analysis of vibrations of the machine-framed structure system by the method of deformable finite elements.
19. Methods for lowering the noise level of circular saws.
20. The protection zone of the complex of "Lenin" steelworks in Nowa Huta, Cracow.
21. Noise hazard in Cracow, with special attention given to the Old City.
22. Sound-absorbing and sound-insulating screens.

23. Dynamic vibration damper with the automatic control of tuning
24. Infrasonic noise, as exemplified by compressors.

In addition to fundamental education, the Institute of Mechanics and Vibroacoustics offers postgraduate and doctorate studies. Lectures on "Protection agains vibration and noise" are also delivered for students of other Faculties of the Technical University of Mining and Metallurgy.

The Institute provides various facilities for carrying on both research and didactic work. Only some of its laboratories for studying vibroacoustic processes are listed below:
1. Anechoic chamber, 1000 m^3 in volume, chamber mass - 600 tons length of acoustic wedges of mineral wool - 100 cm, working surface area designed for machines - 12 m^2, load capacity of testing platform - 80 kN.
2. Anechoic chamber, 220 m^3 in volume.
3. Echo room, 186 m^3 in volume, built of non-parallel concrete walls /in accordance with ISO_3 standards/.
4. A set of 2 echo rooms, 186 m^3 in volume each, for testing the insulating power of partitions.
5. Measurement room for the chambers.
6. Laboratory for studying sound-absorbing and sound-insulating materials.
7. Laboratory for testing the dynamic properties of vibro-insulators.
8. Metrological laboratory.
9. Laboratory of machine dynamics.
10. Laboratory of theory of machines.
11. Workroom for analyses of results of measurements.

The laboratories, listed above, furnished with appropriate test stands, computers and measuring apparatuses, as well as a set of competent workers, make further development of the new branch of science, vibroacoustics, possible.

ULTRASOUND AND MEDICINE

L.FILIPCZYNSKI
Department of Ultrasonics, Institute of Fundamental
Technological Research, Polish Academy of Sciences,
00-049 Warsaw, POLAND

A.KOWALSKI
Department of Biophysics, Institute of Physiological
Sciences, Medical Academy, 02-004 Warsaw, POLAND

A rapid development of ultrasonic methods and their application in medicine has been observed during the last three decades. Nowadays ultrasonic methods are used in diagnostics, therapy, as surgerical tools [7,19] and in the last times as noninvasive method of destroying kidney stones [2,9]. However, one can observe the most rapid development in ultrasonic diagnostic methods. Nowadays these methods are as widely used in medicine as the X-ray techniques.

Many techniques for obtaining images of interior organs of the human body by means of ultrasonic waves have been developed. They are widely used in obstetrics, gynecology, cardiology, internal medicine, ophtalmology, neurology, surgery and so on.

This unique ability of ultrasound waves is due to exceptionally advantageous conditions of their propagation in soft tissues. Attenuation in soft tissues — almost directly proportional to frequency up to 100 MHz — is relatively low, taking into account the size of the organs under examination. Next, acoustic impedances of various soft tissues differ from each other only a little due to differing elastic properties, their densities being practically constant [3]. The acoustical mismatch is so small that only 1% of intensity is reflected when the wave is incident normally to the plane boundary of two different tissues. Therefore ultrasonic waves penetrate deeper into successive tissues and the small reflections at every boundary produce echoes which can be transformed into an image of the whole organ with its many internal details [4].

It is a unique situation which does not exist in sonar or in radar, where the wave is completely reflected from the submarine's or airplane's surface, thus producing only information about their outside size, or occasionally their outside shape.

Many techniques for obtaining images of the human body by means of ultrasonic waves have been developed, for instance, static and dynamic (real time) B-mode presentation with linear, sector, compound and automated ultrasonic scanning. The information obtained by means of the ultrasonic wave penetrating the human body is coded in its time of passage, in amplitude, frequency and phase. It is decoded and displayed in the corresponding echo method on a CR-tube as the A-, B- or TM presentation (amplitude, brightness and time-motion display, respectively) or by the methods based on the Doppler efect. Also digital reconstruction, similar to that one used in computerised X-ray tomography, is possible. Interesting results obtained by means of the ultrasonic transmission and reflection tomography could be already demonstrated, showing also scatter and absorption images [16,17].

Improvements like the grey scale, based on the information enclosed in amplitudes of echoes caused by specular reflections from tissue boundaries and by diffuse reflections from tissue interior organization made it possible to image the tissue texture, thus creating fundamental basis for tissue characterization. Succesful results in this research area may dramatically increase the value and the range of application of ultrasonic medical diagnosis. This may enable us to identify certain pathological structures such as, for instance, malignant tumors.

The problem is complicated as the interaction of ultrasonic waves with tissue is very complex. Absorption, scattering, wave velocity, impedance, their spatial distribution and dependence on frequency affect the ultrasonic wave in various ways. The information obtained in vivo is very complex; the problem requires futher reserch in spite of the fact that it was initiated already many years ago showing great promise. The equipment for tissue characterization is still not sufficiently developed to be used in clinical conditions, and only a small base of clinical data is available for this purpose [15]. In cardiology, by means of two-dimensional imaging systems, backscatter changes caused by ischemia could be visualised [1]. Several types of tissue characterization parameters can be imaged (spectral, discriminant, cepstral, scatter images) [8,11,18].

Dynamic focusing increased the image resolution to a great extend. A microprocessor controlled digital memory capable of storing a great number of ultrasound data has created new capabilities for application of various signal processing techniques, for image reconstruction, data extraction and quantification and for computation of characteristic values of visualized organs (e.g. in cardiology heart volumes, velocity and displacement of valves, blood velocity, cardiac output, etc.).

The resolution of ultrasonic echo methods - an important factor in diagnosing various pathologies - depends to a great extent upon soundfield formation. To obtain a proper axial resolution the generation of ultrasonic pulses of short duration is necessary. This is in principle possible by a careful design and coupling of the piezoelectric transducer to electric transmitter and receiver circuits and to biological medium. As the acoustic impedance of the human body tissue is by one order of magnitude smaller than the impedance of the piezoelectric transducer, matching layers have to be used to compensate this mismatch for a wide bandwidth of frequencies. To design an optimum matching system one uses a computer-aided step by step procedure as no satisfactory method has been found to express the performance of the system by analytic formulae. New types of plastic transducers made of PVDF are here very promising due to their low acoustic impedance comparable with the impedance of biological medium, in spite of the small electromechanical coupling coefficient.

A dramatic increase of the lateral resolution of diagnostic devices was possible by means of the dynamic focusing system. In this system a transducer array is used in which all piezoelectric elements are separately excited by electrical pulses. By proper delay of these pulses one can focus the beam at various points on the ultrasonic beam axis. In the dynamic focusing system the focus position is shifted, or propagates with the wave velocity, thus enabling us to obtain maximum lateral resolution just at the points where the echo originates. Additionally a dynamic aperture controlling system maintans the optimum active aperture of the transducer from the near field to the far field.

To supress side lobes in the ultrasonic beam one uses nonuniform velocity distribution on the transducer's surface. The Gaussian velocity distribution eliminates the side lobes completely, however, it makes the beam a little wider [5]. In this case one chooses modified velocity distribution, as for instance the Henning distribution used in electromagnetic array antennas.

In this manner one can minimize the sensitivity of the ultrasonic device to spurious echoes from "of-axis" targets, thus improving the signal to noise ratio.

In principle, it is possible to apply all the described electronic solutions in one transducer array system together with linear, sector or even compound system, scanning in real time the human body.

To meet all the above mentioned requirements enabling

us dramatically to improve the diagnostic ultrasonic image, it is necessary to transform the ultrasonic device into a sophisticated computerized electronic system in which ultrasound represents only a small part of the whole.

The newest trend in development of ultrasonic diagnostic devices is to construct an array transducer with several hundred piezoelectric elements, every one of them having its separate programmable transmitter, receiver, delay line etc. which are soft-ware controlled. In other words, one can say that this solution represents several hundred separate diagnostic devices which are coupled mechanically into one ultrasonic probe and their parallel work is controlled by a common computer [10].

The most crucial problem facing the present development of ultrasonic diagnostic methods consists of further miniaturization of ultrasonic piezoelectric array transducers. The piezoelectric ceramic used for this purpose should have a high ratio of electromechanical coupling coefficients for thickness and transverse vibrations k_{33}/k_{31} to minimize cross-talk between neighbouring piezoelectric elements. Also a proper acoustic isolation between these elements and an adequate acoustic matching to the body are necessary. To achieve this purpose advanced microelectronic technology has to be used, especially if one takes into consideration the possibility of connecting piezoelectric and electronic (e.g. amplifying) function of array elements. An interesting imaging method with parametric generation of ultrasonic waves was recently showed to be possible in medical diagnostics. In this system, originally used in sonar, nonlinear properties of biological media are utilized. In this way possibilities are created for obtaining new clinical information from images showing B/A nonlinear parameter distribution in tissues.

An important part in ultrasonic diagnostics is played by Doppler methods [12]. They allow us noninvasively to follow the blood flow, to measure its velocity, velocity profiles in the vessel cross-section, to detect disturbances (e.g. turbulence) of the blood flow and to localize stenosis and oclusions in various vessels.

In cardiology, when connected with visualization echo methods one can obtain in real time the image of moving heart structures, as well as the color image of blood flow streams in the heart chambers. The colours corresponding to different blood velocities allow us to observe the effects of the heart function, to recognize pathological turbulent or jet flows and even to determine the pressure gradients in the heart using for this purpose the simplified Bernoulli formula.

As a future development one can expect whole new classes of B-mode ultrasonographs combining conventional pulse echo and the Doppler imaging methods in such a way that by switching over automatically the stationary echo cancellation system it will be possible to obtain the image of tissue interfaces and texture as well as the blood flow map coded in different colours showing simultaneously the blood velocity.

This technique may be very useful for many purposes. For instance, one could show the changes in the blood flow in the case of malignant and nonmalignant breast cancers. This information may increase considerably the diagnostic value of ultrasonic methods for tumor examination. Various Doppler methods were developed by many authors during the 2 last decades. We can mention here the continous wave (c.w.) and pulse methods, c.w. mapping method, multigate method, automatic measuring method eliminating the angle dependence, stationary echo cancellation method, multi-pseudorandom Doppler and so on [6,14].

Two ultrasonic methods are very hopeful for neurology. The first of them is the transcranial Doppler method which enables us to measure the blood flow in brain arteries. Due to a high absorption in the skull bone, one uses here as low frequencies as 2 MHz. Nevertheless, the sensitivity of the method is still too small, as limiting the measurements to great vessels only. Even the application of the real time FFT signal processing necessary in this case, does not solve the problem satisfactory.

The second interesting method for examination of the cerebro-vascular system is the input vessel impedance (IVI) method [13]. The IVI is defined as the ratio of blood pressure to the blood flow for successive harmonics of the heart beat frequency. The instantaneous blood pressure can be determined from vessel diameter variations which are measured by an echo tracking system. From these measurements and from blood velocity measurements the blood flow can be found. The measurements are performed by a special developed device with a double ultrasonic probe for echo and Doppler measurements and the results are computed on line. When applied in carotid arteries they are able to characterize the celebro-vascular system showing also the practical value for detecting stenosis and oclusions.

The ultrasonic quantitative blood flow measurement method makes it perhaps possible to evaluate quantitatively the collateral circulation in ischemic extremities, if it is used simultaneously with the rheographic method. This could be shown by measuring the blood volumetric flow in the femoral arthery with the ultrasonic pulse Doppler method, and in the femoral segment by means of the tetrapolar rheography. The difference between these two values corresponds to the collateral circulation in the tight. Then one can introduce the Collateral Circulation Index which may give a new insight into the quantitative characterization of the blood flow pathology in ischaemic extremities [6].

One can say that nowadays technical possibilities exist to transform Doppler methods from qualitative into quantitative ones, although they are still not used quantitatively in routine clinical examinations. However, it seems that this trend in the development of Doppler methods is natural and reasonable, as it can provide us with a new information on the blood circulation. Moreover, quantification of Doppler data is necessary for solving the problem of vascular impedance - the very perspective and attractive fundamental problem in diagnosis of cardiovascular system.

During the last 20 years many attemps were made to apply the holography to ultrasonic imaging. In spite of impressive holographic reconstructions of this imaging methods, even best holographic reconstructions are inferior than the images obtained by B-mode scanners.

In conclusion one can say that the unique ability of ultrasonic waves providing us with valuable information, which would be difficult or even impossible to obtain otherwise - is due to exceptionally advantageous conditions of wave propagation in the soft tissue. The resulting effects to a great extent depend on the generation of very complex ultrasonic fields which are a function of time, space and frequency. The effects depend as well on a reasonable and clever extraction of significant information from fields distorted by frequency dependent absorption, scattering, reflectivity and their spatial distributions in nonhomogeneous and nonlinear biological media.

To maximize the medical usefulness of ultrasonic data and especially images, various signal-processing approaches which were developed in connection with other imaging systems (e.g. smoothing, contrast and edge enhancement, filtering, deconvolution etc.) should be applied to any imaging system.

In the research and development of these methods the contribution of highly educated, ultrasonic specialists is indispensable. How to educate specialists of this type and how to bring acoustics closer to physicians is a problem which is solved in every country in a separate way. I am able to present here mainly the state of the art existing in Poland.

EDUCATION

Ultrasonic methods used in medicine represent a wide field of interdisciplinary science and technology owing to the close cooperation of medicine, physiology, acoustics, hydromechanics, electronics and informatics. Ultrasonic specialists are trained at universities and at technical universities. At the Warsaw Technical University postgraduate courses on ultrasonic medical instrumentation are organized every second year.

Since the ultrasonic medical instrumentation is becoming more and more complicated every year, a new type of operating personnel - ultrasonographers - are employed in hospitals of some countries. This personnel consists of highly educated physicists and electronic specialists with good training in ultrasonic acoustic who also understand the language of medicine. However, in the majority of coutries, the physicians themselves examine the patient by means of the ultrasonic instrumentation and a special technical personnel is used only for the service and regulation of the instrumentaion.

The education of physicians in acoustics is performed in our country on two levels; on the lower level in Medical Academies (medical schools) and on the higher level in the Medical Center of Postgraduate Education in Warsaw.

Foundations of acoustics, physiological acoustics and ultrasonic physics - including principles of diagnostics and therapy - are represented in a course of Biophysics during the 2-nd year. A specialized course on ultrasonic diagnostics is taught during the 4-th year in the framework of Radiology. It is expected that this lecture will soon expand into Medical Imaging enclosing ultrasonic, X-ray, NMR and nuclear visualisation techniques.

On the higher level, in the framework of the postgraduate education every year 6 courses are organized for 40-60 physicians on ultrasonography. The problems of ultrasonic therapy and surgery are represented in courses on physico-therapy, and the problem of ultrasonic hyperthermia partially in the course on oncology.

If the look at this education scheme we can notice the expanding role of acoustics which is played in the modern medicine.

REFERENCES

1. Chandrasekaran K. et al., Echocardiographic visualisation of acute myocardial ischemia - in vitro study. UMB, 12, 785-793, (1983).
2. Chaussy C., Brendel W., Schmiedt E., Lancet, 2, 1265--1268, (1980).
3. Dunn F., Ultrasonic properties of tissues. In: Ultrasonic differential diagnosis of tumors. Kossoff G. et al. (Eds), Tokyo 1984, 3-13.
4. Filipczyński L., Ultrasonic medical diagnostic methods. In: Acoustics 1974, Stephens R.W.B. (Ed.), Chapman and Hall, London 1975, 71-85.
5. Filipczyński L., Etienne J., Acustica, 28, 21, (1973).
6. Filipczyński L., Nowicki A., Powałowski T., Perspectives of Doppler ultrasound techniques in the study of arterial circulation. In: Informatics and Bioenginee-

ring in Medicine, Stipa S. et al. (Ed.) Serono Symposia, Rome 1984, 105-115.
7. Hill C.R. Ed. Physical principles of medical ultrasonics, Wiley, London, 1986.
8. Lizzi F., Feleppa E., Yaremko M., Tissue characterization imaging, WFUMB, Sydney 1985, 497-498.
9. Martin X., Mestas J., Cathignol D., Duberand J., Lancet, 2, 1005, (1985).
10. Maslak S., Computed sonography. In: Ultrasound Annual 1985. Sanders R., Hill., (Eds), Raven Press, New York 1985, 1-16.
11. Nicholas D., Nassiri D., Garbutt P., Hill C., Tissue characterization from ultrasound B-scan data, UMB 12, 135-143 (1986).
12. Peronneau P., Diebold B., (Eds), Application cardiovasculaires de l'echographie Doppler, INSERM, Paris 1983.
13. Powałowski T., Peńsko B., Non-invasive ultrasonic method for the blood flow and pressure measurements to evaluate hemodynamic properties of the cerebro-vascular system, Archives of Acoustics, 10, 3, 303-314, (1985).
14. Nowicki A., Klepper J., Reid J., Spencer M., An imaging gate pulse Doppler for examination of coronary bypass graft patency. In: Cardiac Doppler Diagnosis, Spencer M. (Ed.) Martinus Nijhorff, Hague 1983, 51-60.
15. Robinson D., Gill R., Kossoff G., Quantitative Sonography. UMB 12, 555-565.
16. Roehlein G., Ermert H., Schmolke J., Advanced in reflection - mode computerized tomography. Symposium on Ultrasonic Imaging and Tissue Characterization. Washington 1985, Abstracts, 100-101.
17. Sehgal C., Greenleaf J., Derivation of scatter and absorption images from speed and attenuation reconstruction. Symposium on Ultrasonic Imaging and Tissue Characterization, Washington 1985, Abstracts, 98-99.
18. Thijssen J., Nicholas D., (Eds), Ultrasonic Tissue Characterization, Martinus Nijhorf, Hague 1982.
19. Wells P.N.T., Biomedical Ultrasonic, Academic Press, London 1977.

Towards forming of a Data Bank about didactic Materials and Systems, as a means to increase the efficiency of Education in Acoustics.*

Paul FRANÇOIS

Convener of the Working Group 2
"Survey of Educational Tools"
Federation of the Acoustical Societies of Europe (FASE)
134bis Rue Lasègue, F-92320 CHATILLON, France

ABSTRACT

This paper was prepared by a specialized working group within the Federation of the Acoustical Societies of Europe (FASE). After reporting on a rapid and non exhaustive enquiry, FASE gives some suggestions in order to improve and systematize the circulation of information about didactic materials and systems (Educational Tools). Mainly interested: national Societies of Acoustics, organisators of scientific metings on Acoustics and Acoustical reviews and journals.

1)- Introduction

The interest of didactic materials and systems, such as audiovisuals, software, panels, demonstration gadgets, films, etc..., needs no further demonstration.

With GORDON[1], we have to point out that if modern equipment is very desirable for serious measurements, it is a serious danger for the students that operating the apparatus becomes more significant than understanding the experiment. So, let us not sneer at somme of

─────────
* Talk presented at the International Conference on "Prospects in Modern Acoustics Education and Development", GDANSK (Poland), May 19-21 1987.

the excellent experiments that were performed by TYNDALL, Lord RAYLEIGH, Sit William BRAGG, BOUASSE, and many others who lacked the benefits of electronic. We have to discover again lecture demonstrations that are stimulating and do not rely on expensive equipment.

The comunity of the Acousticians has to give itself the means to know what exists in this area and where it is possible to find it. The aim of this paper is to deliver the reflexions of F.A.S.E. and to propose some solutions to achieve the target defined in its title.

2)- General action of FASE in Acoustics Education

The main object of the Federation of the Acoustical Societies of Europe (F.A.S.E.), beside helping a better coordination of the activities of its members (26 national Societies of Acoustics), is to encourage progress in Acoustics by all kinds of action, particularely in the double area of Education and Research (See article 2 of its Constitution, signed in 1972).

The action of F.A.S.E. in the area of Research is covered by organisation, at regular intervals, of scientific meetings: Congresses (limited to 3 topics) and Symposia (limited to 1 topic), that in coordination with I.C.A.. Description of this activity is not in question in the present paper.

In the area of Education, one of the first steps of F.A.S.E. (1st resolution taken in LONDON, in 1976) was to decide "to encourage all Acoustical Societies in their effort to increase international exchanges in the field of Acoustics and to urge them to disseminate information on scholarships available and, generally, to help them in this work". It is why, as tool of this action, every year since its foundation, F.A.S.E. circulates to its members a document entitled "Education in Acoustics" with all details on place and content of the studies which are organised.

By the same way, in 1982, in order to increase the efficiency of its action (to enlarge it), a memo of agreement was signed between A.S.A. and F.A.S.E., with the following indication concerning the major themes of common interest: "Exchanges of informations concerning didactic acoustics (teaching), and possibly also the preparation of suitable documents for the promotion of devlopment in this field".

3)- Towards a survey on Educationnal Tools in Acoustics

Carrying on with its reflexion on how to increase the efficiency of Education, F.A.S.E. takes into account the problem of lectures demonstrations as the means to stimulate the interest of students, as well as the means to get a better understanding of the physical phenomena.

F.A.S.E. notes that many efforts are made all over the world in this way, not only concerning lectures demonstrations, but including this particular aspect of the topic "Education Methods". As example, taken in our not exhaustive enquiry, we would like to give under, some typical information:

- Meetings, Conferences, Congresses (interesting different levels of teaching):

    . higher level: British Conference on the teaching of Vibration and Noise in SHEFFIELD (1st: 1975, 2nd: 1985);

    . second degree: annual french conference of an association of physicians

    . technical level: FORMINTEC, Paris, October 1986 (Problems of adaptation between Education and the need of the Industry).

- Exhibitions:

    . EDUCATEC 86, Paris, October 1986; regarding didactic material
                and systems (technical level).

- Books and Papers:

    . Demonstration descriptions of higher level:

                - SCHROEDER[2], during FASE 82
                - BAADE[3], for ASEE 83
                - ALBERS[4], for ASA

    . Demonstration descriptions of elementary level:

                - New UNESCO Handbook[5]
                - American Science[6]
                - RUMEDE, for "Palais de la Découverte"

    . Audiovisuals:

                - STUMPF[7]
                - Penstate University[8]
                - "La Villette" guide[9]

    . Publicity of Manufacturers of gadgets or facilities:

                - RIVERBANK (USA): forks
                - SEREME (France): small anechoic enclosures

- Museums (especially interactive):

                - Exploratorium, S.Francisco (USA)
                - City of Sciences and Industries,
                    Paris-La villette (France).

- Specialized Organisations:

     - Section "Education" in ASA,
     - ASEE (Engineering Education),
     - France-DIDAC (French Association of Manufacturers of didactic material and systems),
     - EFETA: European Federation of Education and Training Associations (of Manufacturers, as listed in Appendix A),
     - CIME (International Council in Teaching means).

Taking into consideration this amount of effort (of realizations), and noting the necessity to systematize these efforts, the Council of F.A.S.E. took the resolution in 1982 (N° 5) by which it is to be considered that a survey on Educational Tools used for the instruction and popularization of Acoustics would be useful.

A Working Group (N° 2) was established in order to examine the feasability of such a survey, which should cover the following sources or informations:

a)- Museums (Colections, interactive demonstrations, etc...)

b)- Travelling Exhibits (Commercial, Historicals, etc...)

c)- Wall Panels   (ditto)

d)- Films (Scientific, general, availibility, etc...)

e)- Demonstration Materials (Objective, Volume, origine, etc...)

f)- Documents describing practical works in Acoustics.

4)- <u>Transfer of the Task to more appropriate Organizations or People</u>

On the way, the F.A.S.E. appeared not structured (no personal and financial means) to follow with the work and to solve itself the practical problems arising from such a survey. It is why the Council asks the Working Group 2, in charge of this survey, to present its conclusions during the International Conference on "Prospect in Modern Acoustics Education and Development" hold in Gdansk (Poland), 19-21 May 1987, in order to get reactions from appropriate and interested Peoples or Organization.

5)- <u>Definition of the task which could be carried out.</u>

Acoustical Societies and their specialized Working Group or Committee, Technical and Scientific Reviews, Congresses, Organizations, etc... are in the best place for doing something, systematically, to get the above defined target.

The suggestions of F.A.S.E. would be the following:

- in Conferences on Acoustics:

    . inside the topic "Education", which is more and more introduced in programs, one place must be found to present lecture demonstrations (See the case of FASE 82 with the presentation of SCHROEDER [2]),

    . posters sessions and exhibitions are also appropriate places to present such demonstrations and the associated equipment.

- in Acousticals Journals and Reviews:

    . specialized Sections could be open in order to present didactic equipments and systems by their authors,

. a call for answering the enquiry of F.A.S.E. could be
published,

. a call for announcments in publicity section or pages
could be made by the manufacturers (or equivalent)
of didactic equipment and systems.

- in Acoustical Societies:

. they would have to support the action of their members
in this area and to take care that the propositions made
above will be taken into consideration,

. in that way, they would have to encourage the edition
of specialized documents (books, guides, listings, etc...)
and to take care that appropriate announcements will be made
in acoustical reviews and journals,

. another action could be to interest the authorities of their
country in charge of noise abatement, to grant such an
inventory as a means of increasing the efficiency of their
action,

. to cooperate with the specialised team of an interactive
museum and to report on the results, would be also one of
the ways to be explored.

6)- Conclusions

Finally, F.A.S.E. thinks that a "step by step" action is the only way to solve the problem, with the background idea that all the information so collected could constitute a guide which will be enriched year after year. The present problem is to begin.

7)- Bibliography

[1] - D.S. GORDON
He who can, does
Introduction to 5th British Conference on the teaching
of vibration and noise, held in Sheffielf (U.K.),
2-4 July 1985.

[2] - M.R. SCHROEDER
Eight descriptions of lecture demonstrations made in his
laboratory during FASE 82 in Goettingen (Germany),
13-17 September 1982.

[3] - P.K. BAADE
Lecture demonstrations (three) of Acoustics resonance
and feedback
ASEE Conference (U.S.A.), 1983

[4] - V.M. ALBERS
Suggested experiments (seven) for laboratory courses
in Acoustics and Vibrations
ASA and Penn State University Press (2nd Edition), 1972.

[5] - Associate Authors
New Handbook of the U.N.E.S.C.O. for Science Education
1974.

[6] - J. WALKER
Amateur's experiments
in french edition of "American Science", March-April 1983.

[7] - F.B. STUMPF
Audiovisual materials and microcomputer softwares for
teaching Vibration and Sound
J.A.S.A., Vol. 77, N° 6, June 1985.

[8] - Penn State University
Description or four movies related to Acoustics
J.A.S.A., Vol. 65, NO 5, May 1979.

[9] - G. RUMEBE
L'onde sonore (demonstrations)
Rev. "Palais de la Découverte", Spécial N° 10, Avril 1977.

8)- Appendix A - List of Associations members of the European
Federation of Education and Training Associations):

. Associazione Didattica Italiana (A.S.D.I.)
. British Educational Equipment Association (B.E.E.A.)
. Deutscher Didacta Verband (D.D.V.)
. FRANCE DIDAC

The use of films in teaching acoustics

M. HECKL
Institut für Technische Akustik
Technische Universität Berlin
Einsteinufer 25
1000 Berlin 10

ABSTRACT

The main subject of acoustics is to understand and
to make use of the excitation, propagation and dissipation of waves in gases, fluids and solids. The
mathematical methods for describing these phenomena
are rather advanced and can be used to solve many
radiation, scattering, reflection, transmission and
absorption problems. The presentation of the results
of the calculations, however, poses some problems at
least in the first year of an acoustics course,
because the usual way of presenting results by graphs
and tables does not convey the student the very
essence of a wave field which is a one- two- or
threedimensional <u>timevarying</u> phenomenon.
A good help and useful teaching aid in this respect
is the use of films. The films are made by calculating the positions of many volume elements and dis-
playing these positions as dots or lines on a screen.
If such pictures are calculated for many different
times within one period and photographed, a moving
picture can be obtained. Examples of films made
this way are shown. The limitations of the method
are discussed.

1. INTRODUCTION

Films and video tapes are very useful teaching aids.
They allow the repeated presentation of interesting and
complicated experiments; they give the possibility to show
what real situations in a factory, in a power plant, in the
engine room of a ship look like and how difficult noise

control in such an environment can be. In this and many other applications of films as a teaching aid one may also make use of the fact that a camera can be installed in a very rough environment (e.g. high temperature), that the optical system allows large amplifications, and that time can be compressed or expanded. The latter two effects are especially important in acoustics, because usually we have to deal with motions in the range of 10^{-6} m or less and with times of 10^{-3} sec or less.

In this paper the type of films, which give a very detailed picture of real situations shall not be treated. Instead we shall deal with films as an aid for teaching theoretical acoustics. The hope is that this is of some help for students to get an easier access and a better understanding of theoretical problems.

2. USE OF COMPUTERS TO MAKE FILMS

The principle of making films about theoretical problems is very simple. It consists of the following steps:

a) select the problem that should be presented in a film;
b) idealize the situation in such a way that the problem can be presented by mathematical expressions, that are used in theoretical acoustics;
c) write a computer program, that allows to calculate the motion (or any other quantity such as pressure or temperature) of each individual area element in a given area (the screen);
d) calculate the motion (or other quantity) for different times with time intervals that are a fraction of the period of the event;
e) choose the parameters that go into the equations in such a way, that the effect that should be shown, becomes easily visible;
f) make a picture of the whole field of area elements for

each time frame.

Obviously this type of procedure is possible only with the help of a computer, because one typically has to split up a screen into at least 1500 area elements (represented by dots on the screen) and one has to make a minimum of 6, and on average of 12 time frames per period; sometimes (e.g. when an impulsive source has to be presented) this number may go up to 100. It is, however, not neccessary to have a large computer. If one is willing to use the "mickey-mouse-method" (i.e. make each picture individually) a core memory of 64 kByte and two floppies are sufficient. If one wants to see the moving pictures on the computer display directly a core memory in the megabyte range is neccessary. In any case a good display is needed; with the use of colours the author has no experience.

3. EXAMPLES

In the following a few examples of acoustic phenomena and their visual presentation are described. They all deal with sound radiation. This restriction is only due to lock of space and time. Other problems such as scattering or dissipation or sound propagation in inhomogeneous, liquids or solids might as well have been considered. In all cases shown here the wave fields are two-dimensional. One obvious reason for that is the two-dimensionality of the screen. Another reason is that in two dimensions the decay of sound waves due to geometrical spreading is less than in three dimensions; this helps to keep the dynamic range of the signals low, this is important in film making, because - as will be shown later - the visible dynamic range is only 20 dB.

3.1 Radiation From Vibrating Cylinders

The radiation of sound from a cylinder vibrating with angular frequency ω is given by

$$p(r,\phi,t) = \text{Re}\left\{\sum_{n=-\infty}^{+\infty} p_n e^{jn\phi} H_n^{(2)}(kr) e^{j\omega t}\right\}. \tag{1}$$

Here r,ϕ are cylindrical coordinates, p is the sound pressure, $k = \omega/c$ the wave number, c the speed of sound, $H_n^{(2)}(...)$ the Hankel function of the second kind of order n. We assume that the medium is isotropic and unbounded. The mode number n describes the type of vibration, if $p_n = \pm p_{-n}$, the cylinder has 2 n stationary nodal points along a circumference. The pressure amplitudes p_n are obtained by applying the boundary condition that on the cylinder surface r = a the normal component of the velocity in the medium has to be the same as that of the vibrating cylinder. This leads to

$$p_n = \frac{-j\rho c}{2\pi H_n^{(2)'}(ka)} \int_0^{2\pi} v_c(\phi) e^{-jn\phi} d\phi, \tag{2}$$

where $v_c(\phi)$ is the velocity distribution on the cylinder surface and $H_n^{(2)'}(...)$ the derivative of the Hankel function with respect to the argument. The motion of the individual area elements with coordinates x and y can be obtained from eq.(2) by applying Newton's second law which gives

$$v_x = \frac{-1}{j\omega\rho}\left[\frac{\partial p}{\partial r}\cdot\frac{\partial r}{\partial x} + \frac{\partial p}{\partial \phi}\frac{\partial \phi}{\partial x}\right]; \quad v_y = \frac{-1}{j\omega\rho}\left[\frac{\partial p}{\partial r}\frac{\partial r}{\partial y} + \frac{\partial p}{\partial \phi}\frac{\partial \phi}{\partial y}\right] \tag{3}$$

Fig. 1 shows some pictures of wave fields that were obtained this way. In these cases it was assumed that $v_c(\phi) = v_n \cos n\phi$; i.e. the cylinder has 2n stationary nodal lines.

Fig. 2 shows examples when the cylinder motion was given by $v_c(\phi) = v_n e^{jn\phi}$. In these cases the nodal lines rotate and one gets the picture of a rotating body with a tip speed of

$$U = \frac{ka}{n} c \ . \qquad (4)$$

The wave field consits of outwards spiralling waves, as can be seen in Fig. 2.

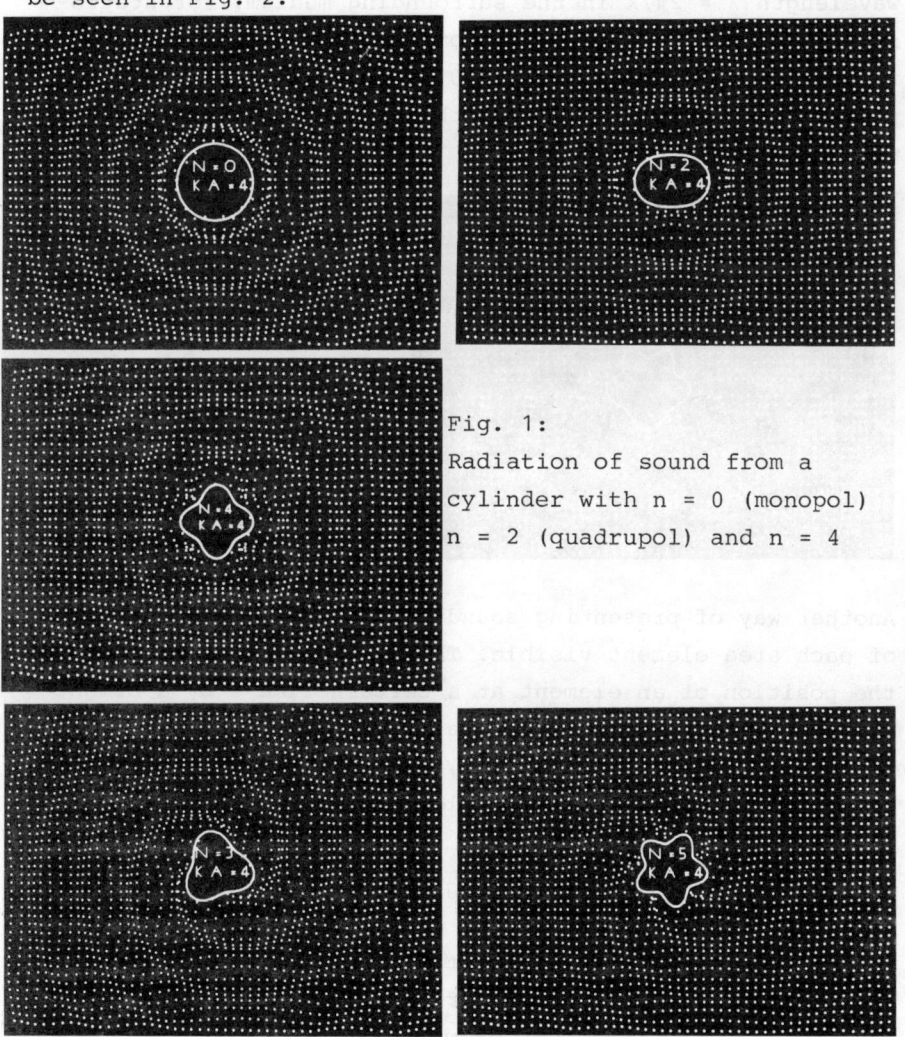

Fig. 1: Radiation of sound from a cylinder with n = 0 (monopol) n = 2 (quadrupol) and n = 4

Fig. 2: Radiation from a cylinder with counter clock-wise rotating nodal lines.

An important effect that can be seen from the wave fields radiated from a cylinder is the "hydrodynamic short circuiting"; i.e. the very low radiation when the wavelength λ_c of the vibration on the sound source is less than the wavelength $\lambda = 2\pi/k$ in the surrounding medium. For stationary nodal points this condition corresponds to $ka < n$ and for rotationg nodal points to $U < u$. In Fig. 3 a "split picture" is shown which makes this phenomenon rather obvious.

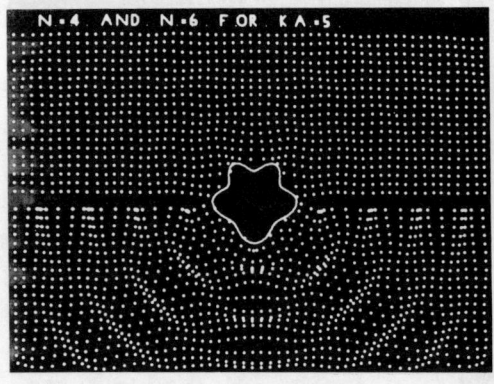

Fig. 3: Radiation from a cylinder with
$n = 6$, $ka = 5$ (upper part) and
$n = 4$, $ka = 5$ (lower part)
short circuiting for $n > ka$
strong radiation for $n < ka$

Another way of presenting sound fields is to make the path of each area element visible. This is done by respresenting the position of an element at a certain time t by a bright dot and the earlier positions at time $t-\nu\tau$ by dots of reduced brightness ($\nu = 0,1,2...$, τ is a time which is typically one eighth of a full cycle). Examples are shown in Fig. 4.

It is seen, that for higher order vibrations the tangential motion becomes more and more important first giving elliptical paths and later (e.g. $n = 6$) a purely tangential motion near the source and almost no sound radiation; i.e. no transport of wave energy.

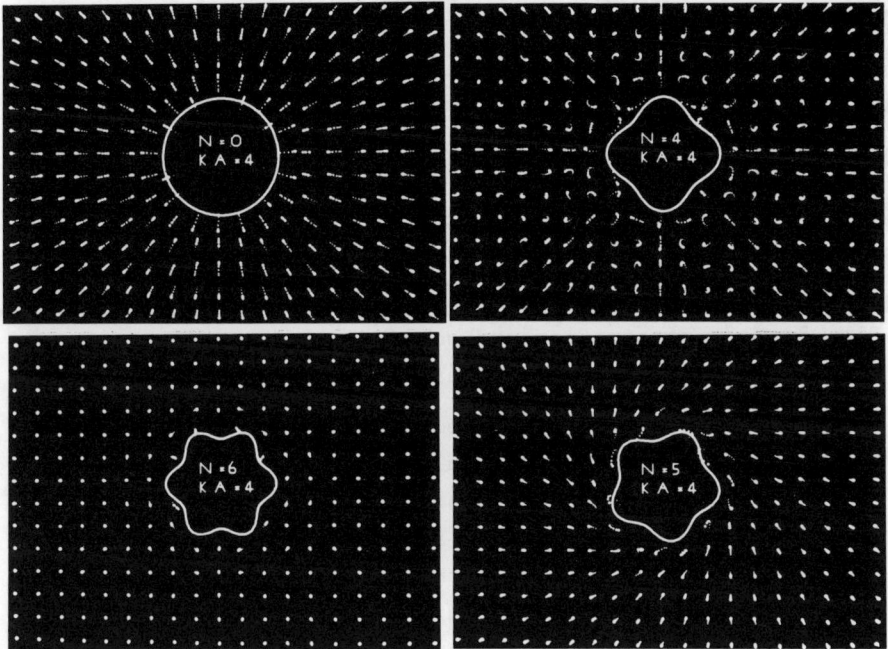

Fig. 4: Particle paths in the wave field near a vibrating cylinder for n = 0, n = 4 and n = 6 when the nodal points are stationary and ka = 4. The lower left photograph shows the result for rotating nodal points and n = 5, ka = 4.

The radiation from a less simple source (i.e. from one where the sum in eq.(1) consists of more than 10 terms) is shown in Fig. 5. On the left side the combination of the different simple mode shapes is given. This is done by showing in the first row the modes number 0,1 and their sum, in the second row the modes 0,1,2 and their sum etc. Since the amplitude of all modes is taken to be equal the vibration pattern becomes more and more localized and would in the limit for n → ∞ go towards a delta function. The sound field generated by the combination of the first seven modes is shown on the right of Fig. 5.

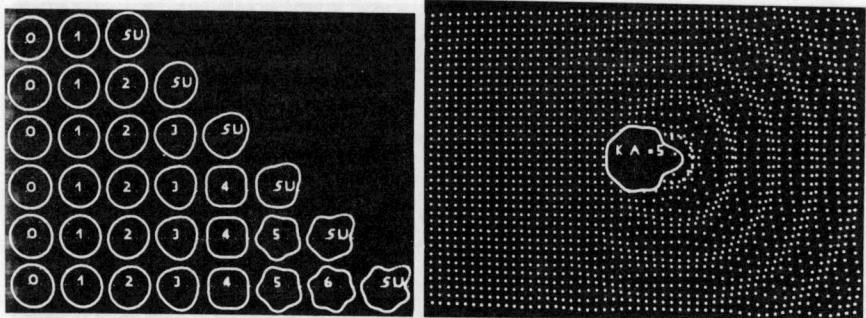

Fig. 5: Mode combination and sound radiation from a cylindrical source with equal amplitudes of mode 0...6.

Since the wavelength in the medium is $\pi/5$ of the cylinder diameter the radiation is mainly in "forward direction"; for the rear the cylinder acts as its own shield (for $ka < 1$ the radiation would be almost omnidirectional). Fig. 5 is only an example of the infinite number of mode shapes that can be produced this way. It is hoped that this type of presentation helps the student to get an understanding of the eigen function expansion methods and its applications.

3.2 Plane Radiator With Given Velocity

If there is a velocity distribution $v_p(x)$ on a plane surface ($y = 0$), the sound pressure is given by the two dimensional version of Rayleith's radiation formula

$$p(x,y,t) = \frac{\omega\rho}{2} \int v_p(x_p) H_0^{(2)}(k\, r_p)\, dx_p e^{j\omega t} \quad (5)$$

with $r_p = \sqrt{(x-x_p)^2 + y^2}$,

where the integration has to be taken over the radiator seize.

Results obtained by using this formula are shown in Fig. 6,7,8. Some visible effects (on a film much more than on a photograph) are: A small piston radiates equally well in all directions, whereas a large piston has a strong forward radiation but also some side lobes (Fig. 6)

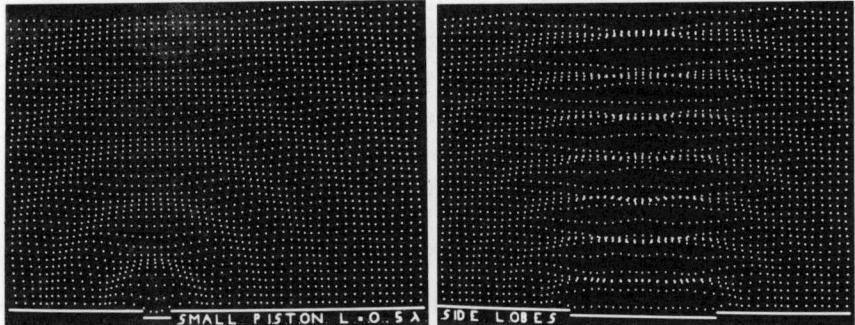
Fig. 6: Radiation from a small and a large piston

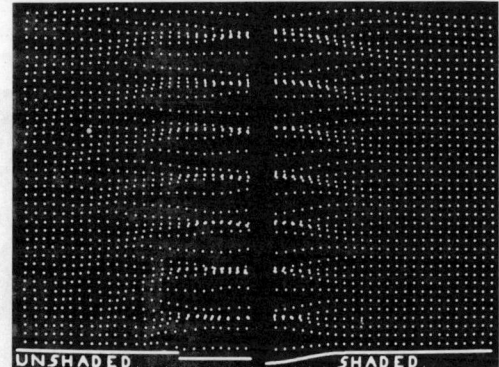

Fig. 7: Influence of shading on the strength of the side lobes

A shaded radiator, i.e. one with a gradual transition from the maximum amplitude at the center to zero amplitude at the edges, has more or less the same radiation in forward direction but less to the sides (almost no side lobes). See Fig. 7.

When a plane radiator has the velocity distribution

$$v_p(x_p) = \begin{cases} v_n \sin \frac{n\pi x_p}{L} & \text{for } 0 < x < L \\ 0 & \text{outside} \end{cases} \qquad (6)$$

the direction of main propagation is determinde by

$$\sin \theta = \frac{n\pi}{kL} = n \frac{\lambda}{2L} \ . \tag{7}$$

Here L is the width of the radiator and λ the wavelength in the surrounding medium. Fig. 8 shows this effect because for n = 1 there is mainly forward radiation, for n = 7 the direction is 31.5°, for n = 12 it is 64°, for n = 14 it would be grazing and for higher values there is no sound beam any more; there is instead a near field above the source and two fairly weak cylindrical waves which seem to originate from the edges. A closer examination reveals that this "edge radiation" depends strongly on the boundary condition there.

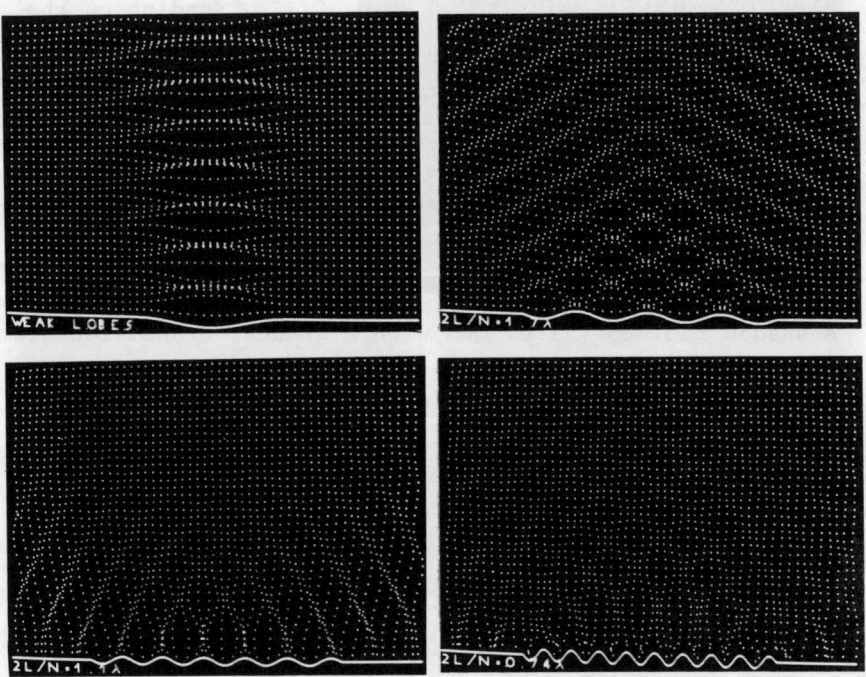

Fig. 8: Radiation from a plane source with a velocity distribution given by eq.(6).

3.3 Point Driven Plates

The bending vibrations of a plate with bending stiffness B, mass per unit area m and loss factor η are described by

$$\Delta\Delta\, v_p(x_p) - \omega^2 \frac{m}{B}(1-j\eta)\, v_p(x_p) = \frac{j\omega}{B}(p_a - p_s). \qquad (8)$$

Here p_a is the driving pressure acting from outside, p_s is the sound pressure at the plate surface. If Fourier transforms are taken and if p_a is assumed to be a point force of strength F_A acting at $x_p = 0$, eq.(8) becomes

$$v_p(x_p) = \frac{j\omega F_A}{2\pi B}\int_{-\infty}^{+\infty}\frac{e^{jk_x x_p}\,dk_x}{k_x^4 - \frac{\omega^2 m}{B}(1-j\eta) + jZ} \qquad (9)$$

with

$$Z = \begin{cases} \dfrac{\omega^2 \rho}{B}\dfrac{1}{\sqrt{k^2-k_x^2}} & \text{for } k^2 > k_x^2 \\[2ex] j\,\dfrac{\omega^2 \rho}{B}\dfrac{1}{\sqrt{k_x^2-k^2}} & \text{for } k^2 < k_x^2 \end{cases} \qquad (10)$$

ρ is the density of the surrounding medium and k its free wave number. In equation (9) the radiation loading is already included, thus one can insert (9) into (7) and calculate the radiated sound.

Fig. 9 shows results obained by programming eq.(9) for a heavy medium (ρ/km = ρλ/2π m = 0,5) and a light medium (ρ/k m = 0,001) when the bending wave length is larger (upper pictures with F > FG) or smaller (lower pictures with F < FG) than the wave length in the surrounding medium.

One clearly can see, that the fluid loading reduces the initial amplitude, that for F > FG there are Mach waves under an angle given by

$$\sin\theta = \frac{c}{c_B} = \sqrt[4]{\omega^2 m/B}/k \qquad (11)$$

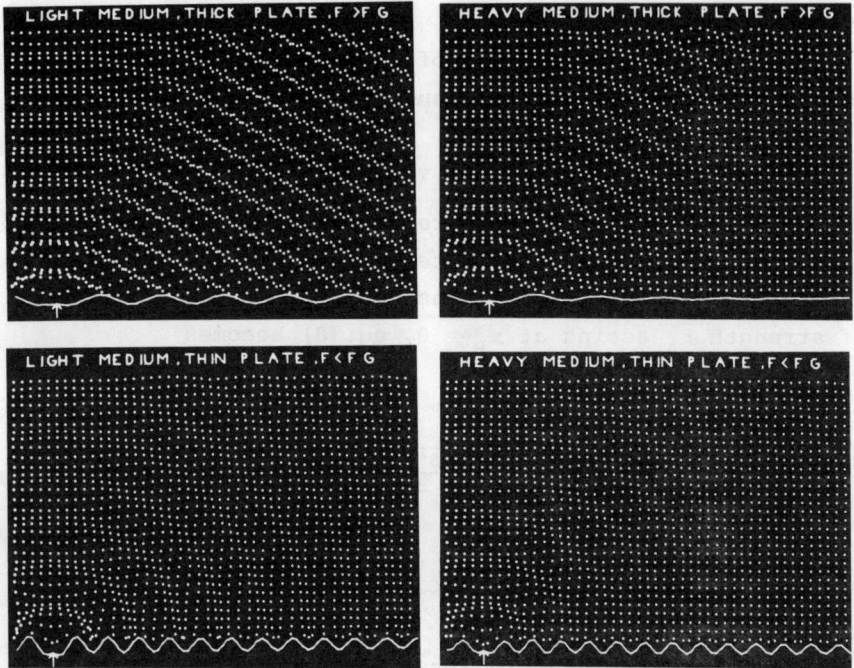

Fig. 9: Radiation from point driven plates in bending

(c_B = bending wave speed), and that these Mach waves generate a strong radiation damping in the heavy medium leading to a fast decay of the bending waves in the plate. In the range F < FG the radiation is much weaker, it comes only from the excitation point; since it is weaker it causes much less radiation damping. Furthermore it is seen that it is more or less of monopole type in the light medium and of dipole type in the heavy medium.

Eq.(9) can also be extended to include the effect of fasteners on the plate. To achive this one assumes auciliary forces F_ν at the positions x_ν of the fasteners. In essence this means that the right hand side of eq.(8) is taken as

$$\frac{j\omega}{B}\left[F_A\delta(x_p) + \Sigma\, F_\nu\, \delta(x_p-x_\nu) - p_s\right] \quad , \tag{12}$$

$\delta(\ldots)$ is the delta function. One now proceeds as before; i.e. taking Fourier transforms and calculating $v_p(x_p)$, which still depends on the unknown values of F_ν. They are obtained by calculating $v_p(x_p)$ at the position x_ν, where the velocity is known (usually it is taken zero). This way a system of linear equations for F_ν is obtained. It easily can be solved and from there on the procedure is as before, the only difference being that now several forces are acting.

Fig. 10 shows two examples for the short bending wave range; i.e. F < FG. In these cases a discontinuity on the plate was represented by two closely spaced points with $v(x_p) = 0$ (indicated by the double arrow). This was done because a single point with $v(x_p) = 0$ allows angular motion and consequently a fairly strong transmission of bending waves.

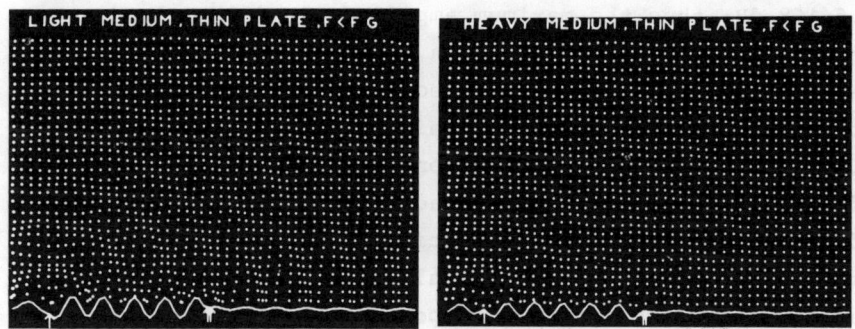

Fig. 10: Radiation from a point driven plate with discontinuity (fasteners) when the bending wave length is short.

The calculation and the film show, that there are standing waves on the plate (not recognisable on the still pictures), and that the radiation seems to come from the excitation point as well as from the discontinuity giving rise to a weak interference pattern.

4. LIMITATIONS OF THE METHOD

The computer made films have many advantages and can be used to clearly present many phenomena, such as radiation, scattering, details of dissipation in the acoustic boundary layer (Stokes layer), non linear effects, wave propagation in solids with all the combinations of compressional and shear waves etc etc. The method has also its limits and drawbacks. The main reason for that is the very limited dynamic range that can be made visible. According to the author's experience the range is limited to 20-25 dB. This figure can be calculated as follows: A typical screen has 500 distinguishable positions in one direction; it one wants to show n wavelengths we find $\lambda = 500/n$; the highest amplitude, which is reasonable is $\pm 0,2 \lambda$, giving a particle velocity of 0,03 of the sound speed (i.e. $M = v/c = 1/30$) and in air a pressure level of 168 dB; thus the maximum amplitude is $100/n$, leading to a maximum value of 20 and a dynamic range of 20:1=26 dB. As a consequence of that, effects that differ by more than 20 dB cannot be shown. This rules out all good sound isolation, phenomena with strong near fields and weak far fields, most of the aero-acoustic sources (where the flow speeds and the sound particle velocities differ by more than a factor of 10), small non-linear effects such as acoustic streaming superimposed on a strong sound field, and other phenomena.

Another problem with such films is, that they might give the wrong impression with respect to the amplitude-wavelength ratio. For a 1000 Hz wave in air this ratio is for a 90 dB level roughly 10^{-6}, in the pictures shown here it is $10^{-1} - 10^{-2}$.

A third restriction, the limitation to two dimensions, is not felt to be very serious, because the important points can be shown on a screen and the use of perspective pictures does not seem to be worth the trouble.

5. CONCLUSIONS

The use of films as an aid in teaching theoretical acoustic is recommended. The precedure is time consuming, because it usually takes many trial runs to find out those parameters that give the most convincing results. When a large core memory is available films can be made in real time.

LITERATURE

Morse, P.M., Vibration and Sound, McGrawHill, New York 1948.

Lord Rayleigh, The theory of sound, Vol. II, Dover Publ., New York 1978.

Cremer, L., Heckl, M., Körperschall, Springer, Berlin 1984.

Heckl, M. and M., Darstellung von Schallvorgängen im Film, DAGA '76, S. 165, VDI Verlag Düsseldorf 1976.

Heckl, M. and M., Einfluß von freien Wellen auf die Schalldämmung von Wänden, ICA 1977, Madrid.

A NEW SET OF RECORDED AUDITORY DEMONSTRATIONS

Adrianus J. M. HOUTSMA
Institute for Perception Research, 5600 MB, Eindhoven,
The Netherlands

Thomas D. ROSSING
Northern Illinois University, DeKalb, Illinois, 60115,
U S A

ABSTRACT

A new set of recorded demonstrations of auditory phenomena is in the final stages of preparation. The 40 demonstrations, grouped into 8 subjects, will be available on a Compact Disc or on a cassette audio tape, and a text booklet will accompany the recording.

INTRODUCTION

In 1978, a set of auditory demonstration tapes was prepared at the Laboratory of Psychoacoustics, Harvard University, under the supervision of Dr. David M. Green. A number of acousticians in the USA and The Netherlands contributed to the project, which was supported by a grant from the National Science Foundation.

The Harvard tapes, as they have come to be known, were used in acoustics classrooms around the world. Unfortunately, the supply of tapes was soon depleted, and as they became increasingly difficult to acquire, teachers were obliged to depend on "copies of copies", with some deterioration in quality. Meanwhile new ideas for auditory demonstrations emerged from ongoing research in psychoacoustics.

In 1983, the Committee on Education of the Acoustical Society of America agreed that a new set of auditory demonstrations would be a valuable teaching aid, and Thomas

Rossing and Dixon Ward were asked to look into possibilities for producing them. They, in turn, sought the advice of Adrian Houtsma, William Hartmann, and others.

The project really got rolling during Rossing s research visit to IPO in 1984. A tentative list of demonstrations was prepared; Houtsma and Wil Wagenaars began production work soon afterwards. Early in 1986 a preliminary version was available for review and classroom testing.

This new set of recorded demonstrations, like the first set, is largely a Dutch-American collaboration. Nearly all the production work on the master tape had been done at the Institute for Perception Research in Eindhoven. The Compact Disc wil be produced by the Philips and DuPont Optical Company, and the text booklet will be printed at Northern Illinois University. Dr. Ira Hirsch, Central Institute for the Deaf, in St. Louis will serve as narrator for the few demonstrations, just as he for the orginal set. The project is being supported by the Acoustical Society of America, which will also assist in distributing the completed materials.

The demonstrations are grouped into following subjects:

1. Frequency Resolution and Critical Bands
2. Sound Pressure, Power and Loudness
3. Masking
4. Pitch
5. Timbre
6. Beats and Combination Tones
7. Binaural Effects
8. Echoes

AUDITORY DEMONSTRATIONS

1   Frequency Resolution and Critical Bands

1. Cancelled Harmonics

"In a complex tone consisting of 20 harmonics of a 200-Hz fundamental it is difficult, if not impossible to pick out the individual harmonics. When a harmonic is cancelled and then restored, however, it can easily be heard. In this demonstration the whole complex is presented first followed by one-second cancellations and restorations of one particular harmonic. There are three cancellations and restorations. This is done for harmonics 1 to 10."

"End of Demonstration 1."

2. Critical Bands of Masking with Noise

"First you will hear a series of 2000-Hz tones in 10 decreasing steps of 5 decibels each. Count how many steps you can hear."

"Now the signal is masked with broadband noise. Count how many steps you can hear."

"Next the noise has a bandwidth of 1000 Hz. Count how many steps you can hear now."

"Next noise with a bandwidth of 250 Hz is used."

"Finally the bandwidth is reduced to only 10 Hz."

"End of Demonstration 2."

3. Critical Bands by Loudness of Constant-Power Noise Bands

"You will hear 8 pairs of noise bands. Each pair consists of a reference noise band with a center frequency of 1000 Hz and a fixed bandwidth, followed by a comparison band with the same center frequency but a variable bandwidth. In the first pair the reference and comparison bands are identical, but in the next 7 pairs the width of the comparison band increases 15% each time, while its amplitude is decreased to keep the power always the same. Note where the comparison band begins to sound louder than the reference. The demonstration is repeated once."

"End of Demonstration 3."

2   Sound Pressure, Power and Loudness

4. The Decibel Scale Used to Express Sound Pressure Level

"Broadband noise is reduced in 10 steps of 6 decibels each."
"Broadband noise is reduced in 15 steps of 3 decibels each."
"Broadband noise is reduced in 20 steps of 1 decibel each."
"Doubling the distance from a source in a free field, while keeping its power level the same, reduces the sound pressure level 6 decibels."

"End of Demonstration 4."

5. Filtered Noise

"This is a sample of white noise."

"Now this same noise is passed through a low-pass filter with the cutoff frequency set at 10000, 4000, 2000, 1000 and 500 Hz."

"Now it is passed through a high-pass filter with the cutoff frequency set at 500, 1000, 2000, 4000 and 10000 Hz."

"Next a band-pass filter is used to give 1/3-octave noise bands centered at 500, 1000, 2000, 4000 and 8000 Hz."

"Finally you will hear samples of white noise and pink noise having the same sound power level."

"End of Demonstration 5."

6. Frequency Response of the Ear

"First a 1000-Hz calibration tone is presented. Adjust the level so that it is just barely audible."

"You will now hear tones of 7 different frequencies, each presented two times in 10 steps of 5 decibels. Count the number of steps you hear at each frequency, beginning at 125 Hz and doubling in frequency each time."

"End of Demonstration 6."

7. Loudness Scaling

"In this experiment you will be asked to rate the loudness of 20 samples of noise. Alternating with the samples will be a reference sound with constant loudness. To help establish a scale, first the reference sound is presented, followed by the strongest and weakest sounds you will hear."

"Now the twenty samples. For each one, write down a number proportional to its loudness, given that the reference sound has a loudness of 100. The reference sound precedes each sample."

"End of Demonstration 7."

8. Temporal Integration

"In this experiment the level of a broadband noise signal will decrease in 8 steps for signal durations of 1000, 300, 100, 30, 10, 3 and 1 ms. Each step is 4 decibels, except the first step, which is 16 decibels. The staircase is presented twice for each signal duration. Count the number of steps you hear in each case."

"End of Demonstration 8."

3 Masking

9. Asymmetry of Masking by Continuous Tones

"A 1200-Hz tone that increases in 10 steps of 6 decibels is used to mask a 2000-Hz tone. Then a 2000-Hz tone increasing in the same 10 steps is used to mask a 1200-Hz tone. Count the number of steps required for complete masking in each case."

"End of Demonstration 9."

10. Asymmetry of Masking by Pulsed Tones

"First a 2000-Hz test tone decreases in 8 steps while a 1200-Hz masker remains constant. Each step is 5 decibels except the first step, which is 15 decibels. Next a 1200-Hz test tone decreases while a 2000-Hz masker remains constant. The masker is heard at each step, alternating without and with the test tone. Count how many steps the test tone can be heard in each case."

"End of Demonstration 10."

11. Forward and Backward Masking

"In this demonstration the signal is a 10-ms burst of a 2000-Hz sine wave, and the masker is a 250-ms burst of noise. First the signal is heard alone, decreasing in 10 steps of 4 decibels each."

" Now the signal is followed, after a short time interval, by a burst of noise. This is termed *backward* masking, because the masker follows the signal. Three time intervals are used : 100, 20, and 0 ms. For each time interval the staircase is presented twice."

"Now the signal is preceded by the masker. This situation is called *forward* masking. Again time intervals of 100, 20, and 0 ms are used."

"End of Demonstration 11."

12. Pulsation Threshold

"If pulses of a sinusoidal signal alternate with bursts of noise, the signal is perceived as pulsating or continuous, depending upon the relative intensities of signal and noise. In this demonstration the signal is a 2000-Hz tone decreasing in 15 one-decibel steps, and the noise is 250 Hz wide, centered at 2000 Hz. For each intensity the signal is heard 4 times. Note when it begins to sound continuous."

"End of Demonstration 12."

4 Pitch

A. PITCH OF PURE TONES

13. Dependence of Pitch on Intensity

"First a 200-Hz calibration tone is presented. Adjust the level so that it is just barely audible."

"Now 6 tone pairs with identical frequencies of 200, 500, 1000, 2000, 3000 and 4000 Hz are presented; the second tone, in each case, has a level 30 decibels greater than the first."

"End of Demonstration 13."

14. Dependence of Pitch Salience on Tone Duration

"In this demonstration 1, 2, 4, 8, 16, 32, 64 and 128 periods of a 300-Hz tone are presented. The sequence is presented twice."
"This demonstration is repeated for a 1000-Hz tone."
"Finally the demonstration is repeated for a 3000-Hz tone."

"End of Demonstration 14."

15. Pitch is Influenced by Masking Noise

"A 1000-Hz tone, partially masked by a noise low-pass filtered at 900 Hz, alternates with a comparison tone. The comparison tone is adjusted until the two tones appear to have the same pitch. When the noise is turned off, the comparison tone appears to have a higher pitch than the 1000-Hz tone."

"End of Demonstration 15."

16. Octave Matching

"A 500-Hz tone alternates with a comparison tone that varies from 950 to 1050 Hz in 11 steps of 10 Hz. Which step sounds like a "good" octave? The demonstration is presented twice."

"End of Demonstration 16."

17. Stretched and Compressed Scales

"A melody played in a high register and an accompaniment in a low register are presented. In which of the three presentations do they sound in best tune?"

"End of Demonstration 17."

18. Frequency Determination (JND)

"In this demonstration 10 groups, each consisting of 4 tone pairs, are presented. In each pair there is a difference in frequency between the two tones, but the first tone may be higher or lower in frequency than the second. All pairs belonging to a group of 4 pairs have the same frequency difference. This frequency difference decreases in each successive group."

"End of Demonstration 18."

19. Linear and Logarithmic Tone Scales

"First 8-note diatonic scales, covering one octave, are presented. Alternate scales have linear and logarithmic steps."

"Next 13-note chromatic scales are presented, again alternating between scales with linear and logarithmic steps."

"End of Demonstration 19."

20. Pitch Streaming : Temporal Coherence and Fission

"In this experiment tones A and B are presented in the sequence ABA ABA. Tone A has a frequency of 2000 Hz, tone B varies from 1000 to 4000 Hz and back again to 1000 Hz. Near the crossover points, the tones appear to form a coherent pattern, characterized by a "galloping rhythm", but at large intervals tones seem isolated; this is called *fission*."

"End of Demonstration 20."

B. PITCH OF COMPLEX TONES

21. Pitch of the Missing Fundamental (Virtual Pitch)

"First a tone consisting of 10 harmonics of 200 Hz is presented, then the same tone without the fundamental, then without the first two harmonics, and so on, until only the highest 5 harmonics remain. Note that although the timbre changes, the pitch remains the same."

"End of Demonstration 21."

22. Shift of Virtual Pitch

"First a tone having partials of 800, 1000 and 1200 Hz is presented , thus giving a virtual pitch of 200 Hz. Then each partial is shifted upward in 10 steps of 20 Hz each until the frequencies are 1000, 1200 and 1400 Hz. The sequence is repeated once."
"Now a complex tone having partials of 800, 1000 and 1200 Hz is presented, followed by one having partials of 850, 1050 and 1250 Hz. As you can hear the virtual pitches are well matched by complex tones with fundamental frequencies of 200 and 210 Hz, respectively. The sequence is repeated once."

"End of Demonstration 22."

23. Masking Spectral and Virtual Pitch ("Big Ben")

"You will hear the familiar Westminster chime melody in pairs of tones. The first tone has a single frequency equal to the virtual pitch of the second complex tone."
"In the next presentation, the single frequency tone will be masked by low frequency noise, but you will hear the pitch of the complex tone."
"In the final presentation, the complex tone is masked by high-frequency noise but the low frequency tone is heard."

"End of Demonstration 23."

24. Virtual Pitch with Low and High Harmonics

"The Westminster chime melody is presented with complex tones consisting of harmonics n, n+1 and n+2 of a missing fundamental. In the first presentation, n is randomly chosen as 2, 3, or 4."

"In the next presentation, n is randomly chosen as 5, 6, or 7."

"In the final presentation, n is chosen as 8, 9, or 10."

"End of Demonstration 24."

25. Strike Note of a Chime

"An orchestral chime will be struck eight times, preceded each time by cue tones equal in frequency to the first eight partials of the chime."

"The next strike will be followed by a tone which matches the pitch of the strike note."

"End of Demonstration 25."

26. Analytic vs Synthetic Pitch

"A two-tone complex of 800 and 1000 Hz is followed by a two-tone complex of 750 and 1000 Hz. Which has the higher pitch?"

"End of Demonstration 26."

C. REPETITION PITCH

27. Scales and Melodies with Repetition Pitch

"First a 5-octave diatonic scale is presented using pairs of identical pulses. In each pair the second pulse repeats the first, with a certain time delay, which changes from 15 ms to 0.48 ms as we proceed up the scale."

"Next a 4-octave diatonic scale is presented using pulse pairs that are samples of a Poisson process. Delay times vary from 15 ms to 0.95 ms."

"Next you will hear a diatonic scale with bursts of white noise plus its repetition 15 ms to 0.95 ms later."

"End of Demonstration 27."

D. PITCH PARADOX

28. Circularity in Pitch Judgment (Shepard Pitch)

"Two examples of scales that illustrate circularity in pitch judgment are presented : first the scale used by R. N. Shepard and next a continuous version generated by J-C. Risset."

"End of Demonstration 28."

5 Timbre

29. Effect of Spectrum on Timbre

"In this demonstration a sinusoidal, triangular, square, sawtooth, and pulse waveform with the same fundamental frequency are compared."

"Next the triangular, square, sawtooth, and pulse waves are repeated, each followed by their spectra of partials with the correct amplitudes."

"Finally you will hear how sounds of a bell and a guitar can be built up by adding their partials one at a time."

"End of Demonstration 29."

30. Effect of Tone Envelope on Timbre

"You will hear a recording of a Bach chorale played on a piano."

"Now the same melody will be played backwards, from the end to the beginning."

"Now the recording as just heard will be played backwards, so that the melody is heard in the normal way, from beginning to end, but with a strange difference."

"End of Demonstration 30."

31. Change in Timbre with Transposition

"A 3-octave scale on a bassoon is presented, followed by a 3-octave scale generated by transposing the highest note in the first scale. In other words, this is how the bassoon might sound if all the notes had the same relative spectrum."

"You will now hear a melody played on a flute, followed by a bassoon, and then the two played as a duet. Then the flute is transposed 3 octaves downwards and the bassoon is transposed 3 octaves upwards. The duet now seems played by entirely different instruments."

"End of Demonstration 31."

32. Stretched Partials

"You will hear a 4-part Bach chorale played with tones having 9 harmonic partials."

"Then the same piece is played again with both melodic and harmonic scales stretched logarithmically in such a way that the octave ratio is 2.1. Melodic semitones now have the ratio 1.064 rather than 1.059."

"In the next presentation you will hear the same piece, with only the melodic scale stretched. Notice the beats and the impression of 4 mistuned instruments."

"In the final presentation only the harmonic scale is stretched. Notice again the beats and the difficulty to hear how many instruments are playing."

"End of Demonstration 32."

6 Beats and Combination Tones

33. Beats

"Two tones having frequencies of 1000 Hz and 1004 Hz are presented in the sequence 1000, 1004, both. This sequence is presented twice."

"Pairs of pure tones are presented having intervals slightly greater than an octave, a fifth and a fourth, respectively. The mistunings are such that the beat frequency always is 4 Hz when the tones of a pair are played together."

"End of Demonstration 33."

34. Distortion

"First you will hear a 440-Hz sinusoidal tone distorted by a symmetrical compressor to add mainly odd-order harmonics. It alternates with its 3rd harmonic."

"Next the 440-Hz tone is distorted asymmetrically by a half-wave rectifier. The distorted tone alternates with its 2nd harmonic."

"Now two tones of 700 and 1000 Hz distorted by a symmetrical compressor. These tones alternate with a 400-Hz pointer to the cubic difference tone."

"First you will hear a 440-Hz tone with its second harmonic added with a phase varying from $-90°$ to $90°$. This is followed by the same two tones, distorted through a square-law device. Now the phase change becomes clearly audible."

"End of Demonstration 34."

35. Combination Tones

"In this demonstration two tones of 1000 and 1200 Hz are presented. One prominent aural combination tone is the 800-Hz cubic difference tone. When an 804-Hz tone is added, it beats with this 800-Hz combination tone."

"Now the frequency of the upper tone is slowly increased from 1200 to 1600 Hz. The cubic difference tone, $2f_1 - f_2$, descends from 800 to 400 Hz. At the same time, the quadratic difference tone, $f_2 - f_1$, ascends from 200 to 600 Hz."

"End of Demonstration 35."

7 Binaural Effects

(USE HEADPHONES)

36. Binaural Beats

"A 250-Hz tone is presented to the left ear while a 251-Hz tone is presented to the right ear. In the second presentation, broadband noise is added at a level 10 decibels greater than the signal."

"End of Demonstration 36."

37. Binaural Lateralization

"Tones of 250, 500, 1000, 2000 and 4000 Hz are presented to the two ears with a constant phase difference of 45° alternating between the two ears. At low frequency the image shifts back and forth."

"Next, as the time of arrival of a click is delayed by varying amounts at one ear or the other, the apparent source of the click appears to move."

"Finally the relative intensities at the two ears of a 250-Hz and a 4000-Hz tone are varied."

"End of Demonstration 37."

38. Masking Level Differences

"First a 500-Hz staircase signal is applied to the left ear. Then it is masked with noise. In the third presentation, masking noise is applied to both ears. In the fourth presentation, both signal and noise appear in both ears. In the fifth, final presentation, again signal and noise appear in both ears, but the signal is inverted at one of the ears. For each presentation you will hear the staircase three times. Count the number of steps you can hear."

"End of Demonstration 38."

39. An Auditory Illusion

"400 and 800-Hz tone alternate between the two ears. Most listeners hear the lower tone in one ear and the higher tone in the other."

"End of Demonstration 39."

8 Echoes

40. Effect of Environment on Echoes

"First, in an anechoic room, then in a conference room, you will hear a hammer striking a brick followed by an old Scottish prayer. Playing these sounds backwards focuses our attention on the echoes that occur."

"Finally, the same sounds are recorded in a highly reverberant room with reflecting walls."

"End of Demonstration 40."

SPECIFIC PROBLEMS IN EDUCATION
AND DEVELOPMENT IN UNDERWATER ACOUSTICS

Zenon JAGODZINSKI

Technical University, Gdańsk – Poland

ABSTRACT

The geometry and propagation conditions of ultrasonic waves in the sea are such that the experiments with hydroacoustic equipment in a conventional, indoor laboratory are very limited. They may be made on a sea going ship but this is very expensive and often not quite convenient too. At the Technical University of Gdansk these experiments are made on an inland lake and a sailing or motor propelled boat is used. The procedure and scope of such experiments is described.

REQUIREMENTS OF RANGE AND ACOUSTIC
PROPERTIES OF SEA WATER

There are two reasons because of which teaching of underwater acoustics must be treated in a somewhat specific way. These are the required range and properties of the sea as a medium of propagation of acoustics signals.

The range is at least some tens of meters in echosounders or short range sonars, and up to some kilometers for deep sea echosounders and medium or long range sonars.

In comparison with other ultrasonic instruments that are used for instance in molecular acoustics, nondestructive testing, medical diagnosis etc., where the needed ranges are usually no more than some tens of centimeters, in underwater acoustics the ranges are required up to some kilometers, i.e. they are at least three orders of magnitude longer.

The required range is decisive for proper determination of all other parameters of the equipment such as the operating frequency, energy and peak pulse power, pulse length, band-width, directivity, transducers dimensions etc.

In view of this reason alone it is clear that a different approach is necessary in education and research. It is obvious that the operation of underwater acoustic equipment cannot be demonstrated or measured in a conventional, indoor laboratory even if a comparatively large anechoic tank is used.

The other reason for which a specific approach is needed is the geometry and acoustic properties of the sea.

Thinking of the sea, we usually have in mind a deep layer of water below the ship. However, for a study of propagation conditions the depth h should be compared with the horizontal distance d.

Thus it is convenient to look at this layer of water in terms of h/d ratio.

Our view may then be changed - the sea is only a very thin cover on our globe. E.g. the distance d across the Baltic Sea, from Jastrzębia Góra to Karlskrona in Sweden is about 200 km and the average depth about 50 m. So the ratio h/d is about $2.5*10-4$.

Similarly, the distance from the British Isles to Newfoundland is about 2600 km, with the average depth of about 1000 m. The ratio h/d is then about $3.8*10-4$ - similar to that across the Baltic Sea.

Of course, we do not transmit the ultrasonic signals across the Atlantic Ocean or even the Baltic Sea. We transmit them over much shorter distances but still the h/d ratio, that we have to do with, is definitely shorter then 1, with the exception of deep oceanotechnic enterprises not yet widely developed.

Fishing and geologic explorations are usually being done on shallow seas like the Baltic or the North Sea or on the continental shelves where the depth does not exceed about 200 m.

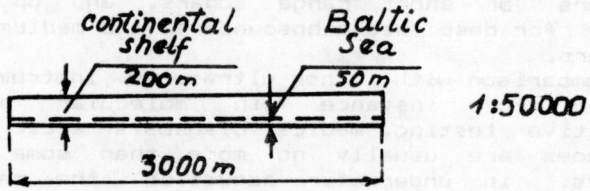

Fig.1. The continental shelf and Baltic Sea

When the ultrasonic communication equipment or sonar is used there, with the required range of about 3000 m, the ratio h/d is about 0.07, and this is a "shallow water" too, as shown in fig.1.

Such a thin layer of water forms a kind of a crude wave guide, in which specific conditions of propagation must be taken into account.

Apart from spreading and absorption that are the functions of distance only, there are multiple reflections, causing reverberations, and also, since the temperature and salinity are not uniform, there are velocity gradients.

The reverberations are the unwanted, scattered reflections that come back to the receiver and determine the signal-to-reverberation ratio, S/R.

The geometry of this effect is shown in fig.2 where, for simplicity, only first order reflections on surface and bottom are drawn.

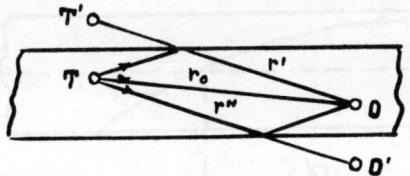

Fig.2. Surface and bottom reflections.

When the ultrasonic wave is transmitted by the transducer T, it reaches the object, e.g. the receiver O not only as the direct ray ro, but also as the rays r' and r'' reflected by the surface and bottom, respectively.

The reflected waves pass longer distances and therefore their phases are delayed. Thus the resultant wave that reaches the receiver will have a resultant amplitude and phase, depending on the reflections.

On the sea the geometric situations is practically never stable. The water surface is waving, the ship is moving and rolling. In consequence both amplitudes and phases are constantly changing.

In underwater acoustic equipment, where the centimetric wavelengths are used, even minute changes in distance result in considerable changes in amplitude and phase, and even with relatively calm waters these fluctuations may exceed 30 dB which means heavy distortion of received signals.

Such distortions are clearly visible on sonar pulses, and may be also easily observed with voice communication equipment /the "hydrotelephone"/. It is obvious, that when the signal has a more complex spectrum, the distortions will be greater.

A very good, tutorial example is the comparison of distortions in the hydrotelephone with amplitude modulation, frequency modulation and single-side-band modulation.

On electromagnetic waves, where the air may be treated as very clear medium the frequency modulation is the best.

In underwater acoustics, where different reflected waves are mixed up it is just the opposite, the simplest, single line spectrum gives best results.

The other, very important factor in the propagation of ultrasonic waves is the effect of refractions. It is the result of the velocity gradiens which depend mainly on temperature.

In late springg and summer when the upper layer of water is warmer, the ultrasonic ray, instead of reaching the distance determined by the range equation, is bent down and hits the bottom as fig.3.

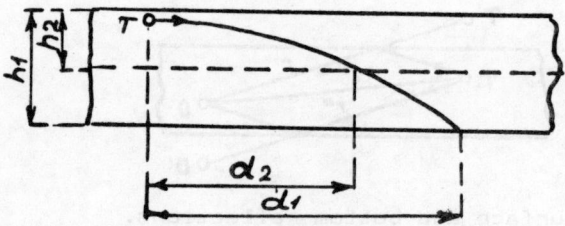

Fig.3. Refraction of ultrasonic ray.

In this way, the useful range may be drastically reduced, especially on shallow water.

The effects of reflections and refractions influence very much the operation of sonar and communication devices. In echosounders, where the beam is directed downwards this influence is of course much smaller.

THE EXPERIMENTS ON INLAND LAKE

The propagation conditions of ultrasonic waves are very important, often even decisive for efficient operation of the underwater acoustic equipment.

But they cannot be demonstrated or examined in a conventional, indoor laboratory, even a large, anechoic tank is used.

It may seem that the most straightforward solution is to take the students on board of a seagoing ship, suitably equipped with the necessary ultrasonic instruments.

It is a simple, but a very expensive proposition. Besides, it is not easy to agree the students time with the ship's research programme.

Moreover, at a closer examination it is not the best solution.

The key element, so to say the heart of any ultrasonic device is the transducer.

On a sea vessel the transducers must be built-in into the bottom, the students can not even see them and only the electric terminals are accessible.

Any change of the transducer, or even the change of its place is practically impossible because the ship should be drydocked for this operation.

Also, all the lay-out and cabling must be done in the shipyard in accordance with the marine engineering practice.

In effect all that is left for students is to operate the different knobs and observe the results without any change in the equipment.

That is why, in the author's opinion, a more educating solution is to make the experiments on an inland lake which may be treated as a "laboratory scale" model of the sea.

A boat may be used on which the transducers can be fastened with an easy to operate handle, changed and tested. No permanent cabling is needed, the connections between different parts and instruments may be designed by students as a part of their training.

The safety measures are simpler too.

Some instruments that we use in the "outdoor" laboratory of underwater acoustics are home made especially for this purpose, and a part of them was made by the students themselves as their diploma theses.

The boat we use is shown in fig.4.

It is a 7 m sailing boat with an auxiliary outboard motor. The manoevring is simpler when the motor is used, but motor makes noise, turbulence in water and sometimes causes ignition noise in amplifiers therefore, if weather permits, the sails are preferable.

Fig.4. The laboratory boat.

Sometimes, e.g. for experiments with the side-looking sonar a towed plate is used as shown in fig.5.

Fig.5. The towed plate.

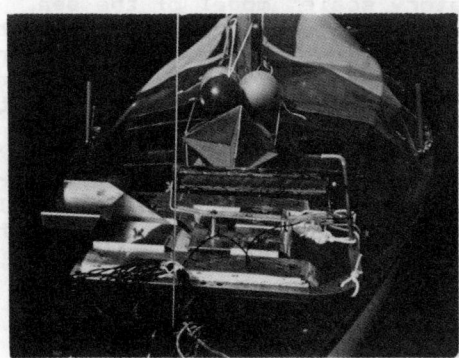

The transducers and reflectors used with the towed plate are shown in fig.6.

Fig.6. Transducers and reflectors.

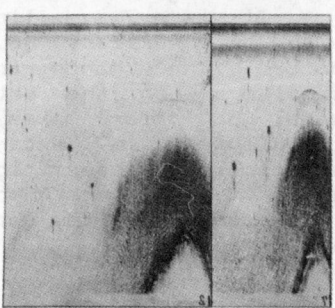

An echogram, recorded with the side looking sonar is shown in fig.7.

Fig.7. Echogram of the side-looking sonar.

On fig.8. a group of students is shown on the boat during the discussion of the results.

Fig.8. During the experiments.

The outdoor underwater acoustics laboratory is placed in the syllabus of the 8-th semestr, but it is executed during one week in June.
The programme comprises four thematic groups of experiments:
1. Propagation factors, mainly signal fluctuations, reverberations and their influence on signal distortions and range.
2. Voice communication with amplitude, frequency and single-side band modulation.
3. Echosounding and sonar.
4. Doppler effects and velocity measurements.

After twenty years of experience we are convinced that the outdoor laboratory on the inland lake is a good and efficient part of students' education.

APPARATUS FOR A PRACTICAL COURSE IN ACOUSTICS FOR UNDERGRADUATES

H. W. JONES

ENGINEERING PHYSICS DEPARTMENT
TECHNICAL UNIVERSITY OF NOVA SCOTIA, CANADA

ABSTRACT

This paper describes a set of apparatus which can be manufactured locally without great difficulty, often quite cheaply. The apparatus allows a large range of experiments to be designed which relate to basic physical acoustics, architectural acoustics and to noise control. The apparatus consists of a high frequency anechoic chamber, reverberation room, high frequency sound sources, one of which is broadband and the other which is pure tone. The sound sources can be readily manufactured in the normal departmental workshop. The apparatus was designed to work in the frequency range of 6 to 160KHz and is thus correspondingly small so that it can be easily accommodated in a laboratory. The use of oscilloscopes and relatively cheap microphones can provide the data from which measurements are made. Some of the experiments which have been used are described, but there are many more which are not and, no doubt, a little ingenuity will extend the list very considerably.

INTRODUCTION

The origins of the apparatus which is described in this paper can be traced to a project which was started some fourteen years ago[1]. At that time, it was necessary to measure the attenuation of traffic noise over the ground in order to devise strategies for environmental noise control. To meet the needs of that project, we designed a complete set of apparatus which allowed us to model the urban or rural environment at 1/80 scale. To do this we produced relatively simply and cheaply the following items:

1) a high frequency anechoic chamber which was effective in the frequency range from 6KHz to 160KHz,

2) a reverberation room which is 1/80 of the usual size,

3) a set of modulated Hartmann type ultrasonic whistles which operate over the range of frequencies indicated earlier,

4) an apparatus for measuring the acoustic impedance of surfaces at near normal and oblique angles.

The apparatus provides the basis for a range of experiments in acoustics which are not generally available and as such it may be of interest to a wider audience than those who read research reports on noise control work.

DESCRIPTION OF THE APPARATUS

Brevity demands that an adequate discussion of the principles of acoustical scaling be omitted. Those who may be interested should see references 2, 3, & 4 which discuss the matter in some detail. A detailed study of this topic is not essential to those who will use the apparatus for many of the undergraduate experiments. It may be sufficient comment to note that scaling requires a change of frequency and wavelength to match the ratio of scaling. i.e. 1/80 scaling requires frequencies to be increased by a factor of 80. Two effects do not scale:

1) absorption, either classical or molecular, and
2) the acoustical boundary layer.

The first is compensated for in the readings which are taken, if this is necessary[5]. The problems relating to the acoustical boundary layer are usually avoided by choosing frequencies which are below the threshold, i.e. at which the effect is noticeable[6].

The Hartmann Whistle was used as an omnidirectional sound source. Figure 1 show the cross-section of the whistle and figure 2 a photograph of such a whistle installed in the reverberation chamber. The sound spectrum from this source is shown in figure 3. It is to be noted that peak levels approach 130dB (re: 2×10^{-5} Pa.) which usually swamps the background noise and does necessitate hearing protection. The essential features are: a) an over-expanded nozzle supplied with compressed air (typically about 300KPa absolute), b) an adjustable depth-cavity of 0.5mm diameter bore, and c) a means for adjusting the gap between the nozzle lip and cavity lip. Nozzle and cavity are the same bore and can easily be replaced with a different sized pair. The cavity needle allows the depth to be adjusted so that an octave—change of frequency can be obtained.

The performance of these whistles has been explored to find the effect of changing, air pressure, gap size between nozzle and cavity, and cavity depth. The value of frequency (f) calculated from $\lambda = 4(\ell + 0.3\ d)$ (1) is in good agreement with experimental data. Figure 4, for a fixed whistle geometry, indicates that f varies little with pressure whereas L_t is strongly affected. Figure 5 shows that nearly spherical emission is achieved by the whistle.

It was found that three whistles are required to cover the most important frequency band (20 — 160KHz) with nozzle and cavity bore diameters of 0.46, 1.07, 1.83 mm. The supply pressure required for the 1.07 mm diameter nozzle is somewhat higher than that for the other two; 40 psi (280 KPa.) suffices.

The High Frequency Reverberation Chamber is described in the following paragraph. To obtain a diffuse sound field, a steel reverberation chamber with a volume of 0.89 m^3 and a surface area of 5.75 m^2 was constructed. To reduce standing waves, the chamber was designed with no sides parallel and no geometrical axis of symmetry. A photograph of the chamber is given in Figure 6. The prototype ultrasonic whistle was used as a sound source, mounted on the chamber wall as show in Figure 2. A solenoid valve controlling whistle air supply was used to shut off the whistle. Initially this valve did not shut fast enough to allow reverberation times down to 30 msec to be measured. After valve modifications, the whistle shut—off time was reduced to 7 msec. A 1/8 inch B & K microphone, a 2607 measurement amplifier, and a Tektronix 5L4N Spectrum Analyzer were used to measure the received sound signal. Much cheaper electronic apparatus can be used (as described later).

The basic measurement procedure consisted of setting the whistle at the required frequency and then activating the solenoid valve to shut off the whistle. The oscilloscope was triggered simultaneously and the decay of 60 dB (linear over about 50 dB) can be obtained. This allows accurate extrapolation to obtain the reverberation time (R.T.).

The reverberation time can be related to the absorption of the chamber by

$$RT \text{ (sec)} = \frac{0.161 \text{ V}}{A + 4 \text{ M V}} \tag{1}$$

where V is the chamber volume (m^3), M is the air attenuation constant (m^{-1}) and

$$A = S \, [-\ln(1 - \bar{\alpha})] \tag{2}$$

and

$$\bar{\alpha} = \frac{\Sigma \, S_i \alpha_i}{S} \tag{3}$$

S is the total surface area (m^2) of the chamber and S_i is the surface area of material in the chamber with Sabine absorption coefficient α_i. For the case of $\bar{\alpha} \ll 1$ then

$$A = S \, \bar{\alpha} = \Sigma \, S_i \alpha_i \tag{4}$$

As a check on the absorption characteristics of the empty chamber, a series of measurements were taken over the frequency range of 20 KHz to 160 KHz. It was found that the absorption of the walls (term A of Equation 1) was negligible when compared to the air absorption term (4MV). Values of A found for the empty chamber were less than 0.07 m^2 over the above frequency range; hence the sabine absorption coefficient α_i for the chamber wall is less than 0.012 over the frequency range.

As well as being used for measuring the properties of absorptive material for the anechoic chamber, the reverberation chamber has been used to measure the absorptive properties of Freon−12 (ref. 7). The whistle, enclosed in a plastic bag and air, was exhausted through the mounting pipe. The reverberation chamber was sealed and filled with Freon−12. The apparatus was flushed with freon gas and measurements made on several successive days. It was observed that the results were repeatable and this was taken as indicating that the impurity content was not varying significantly. Typical results are shown in Figure 7 along with measurements taken in air. The absorption coefficient α is related to the coefficient M by

$$M = 2\alpha \qquad (5)$$

The chamber was built of 12 gauge steel sheet, welded to angle iron at the corner and had a close fitting door. Panel resonance was not a problem with the dimensions and frequencies chosen.

<u>The Anechoic Modelling Chamber</u> is required for experiments with a continuous ultrasonic sound source. The lining of this chamber should absorb at least 99% of the acoustic energy incident on it over the range 6 KHz to 160 KHz. Several locally available fibrous materials were tested. Two inch fiberglass insulation proved to be the most suitable. Reverberation chamber tests in the frequency range 20 to 160 KHz were conducted with samples of materials placed in the closed chamber. The resulting sabine absorption coefficients for the fiberglass insulation were greater than unity. This is possible for highly absorptive materials, see reference 8. It was shown that the absorbing lining was effective to the required frequency of 6KHz. (measurements showed that the fiberglass insulation was 95% absorptive down to frequencies of 1.4KHz). To increase low frequency absorption, the fiberglass insulation was stapled to 3 inch deep wooden ribs spaced 6 inches apart in a corrugated form (as was first suggested by Rayleigh, reference 9). The fiberglass should be covered with <u>thin</u> plastic sheeting to prevent fibres floating in the air and being inhaled by the students. Figure 8 shows a typical ultrasonic anechoic wall panel.

The anechoic chamber construction consists of a 2" x 4" framework of dimensions 2.5 m high, 2.5 m in width, and 6 m long. Much smaller sizes would be more convenient as a student facility. The wall panels are made of 1/4 inch plywood with the inside surfaces covered with the corrugated fiberglass insulation. Figure 9 shows the completed chamber. It should be noted that all of the wall panels can be removed in order to have free access, and consequently, the whole of the chamber floor space can be filled by the model. If necessary the floor can be covered with fiberglass panels and planks used for access.

The Impedance Measuring Apparatus shown in figure 10 is described below. The basic method of measurement, from a St. Clair generator, consists of directing a high intensity ultrasonic beam onto the test sample and measuring the magnitude and phase of the received signal. A second measurement is taken under identical geometric conditions with the test sample replaced by a hard reflective surface. An absorptive foam block and a lead shield is used to minimize the direct sound level received by the microphone. A measurement of the phase and magnitude of this direct sound wave is also made. The three measurement situations are shown in Figure 11, p'_1, p'_2, p_3 are the complex sound pressures obtained from the magnitude and phase measurements. From these measurements, the real and imaginary parts of the reflection coefficient can be found.

A photograph of the basic physical apparatus is shown in Figure 10. High intensity plane waves are produced using a St. Clair generator (reference 10) which is essentially a solid aluminum cylinder that is driven electrodynamically in a longitudinal mode of vibration at a resonant frequency. Figure 12 shows the St. Clair generator and the various sizes of cylinders used. This ultrasonic beam is aimed at a 1/4 inch thick, 12 inch by 5 inch brass plate upon which test samples are mounted. Some typical polystyrene samples covered with cloth and paper surfaces are show in the foreground of Figure 10. The 1/8 inch B & K microphone with lead shield and the absorptive foam block are also shown. The source, receiver, and reflective plate are mounted on a mechanism which maintains equal angles of incidence and reflection ($\pm$ 0.25°) as the receiver arm is rotated to vary the grazing angle of incidence. During the measurements, the apparatus was mounted inside a high frequency anechoic chamber. This was done to achieve a free field condition and also to reduce phase changes due to turbulence and currents in the air which were considerable in the open laboratory.

A schematic of the apparatus and associated electronic equipment is shown in Figure 13. The a.c. signal from the Hewlett—Packard test oscillator after amplification is used to drive the St. Clair generator. The magnitude of the sound pressure received by the microphone is measured using a 2607 B & K measurement amplifier. The phase of the signal received by microphone was measured with respect to the output signal from the H.P. Test oscillator by comparing the two signals on a dual channel oscilloscope (a phase locked amplifier could have been used for this purpose). Due to phase instabilities caused by air currents, the relative phase of the signal was measured to within ± 0.1 radians.

DISCUSSION

The apparatus which has been described can provide relatively cheap facilities for quite a wide range of experiments. We, while we had the anechoic chamber and reverberation room used them for quite a range of experiments. For example we measured the absorption coefficient of Freon−12 ($C\ Cl_2F_2$) in the reverberation chamber using the ultrasonic whistle, as was described earlier[7]. Figure 7 shows the results obtained. It is difficult to imagine all the experiments which can be performed − one interesting one which comes to mind, for example, is the levitation of small objects in the near field of the St. Clair generator. (After Whymark Ref.11) Many experiments are obvious and hardly need discussion. Much, therefore, is left to the interest of the imagination of the instructor.

Perhaps it is worth noting that as many experiments only require the relative sound level, it is possible to make a small piezoelectric microphones for use with the apparatus. The excellent integrated circuits from manufacturers like Burr Brown Inc. or Analog Electronics Incorporated and other manufacturers make it possible to provide integrated circuit amplifiers, both linear & logarithmic, for use with the microphone. Similarly true r.m.s. integrated circuits are readily available. The use of such circuits make it possible to produce, quite cheaply, all the electronics which may be required. Nowadays it is easy to use A/D converters and create files in cheap computers for data display and processing.

CONCLUSION

This paper describes a set of apparatus which we have built and used both for research and for teaching. It is simple, it is cheap, and it is effective. It can be adapted to a large range of experiments which should be interesting and instructive to the undergraduate.

REFERENCES

1. Jones, H.W., Stredulinsky, D.C., Vermeulen, P.J. and Yu, J. "An Experimental and Theoretical study of the Modelling of Urban Noise Problems", Dept. of Transportation, Govt. of Alberta, Canada, 1976.

2. Yu, J., MSc. Thesis, University of Calgary, Calgary, Alberta, Canada, 1978.

3. Anderson, G., "Acoustic Scale Modelling of Roadway Traffic Noise", Bolt, Beranek and Newman, Cambridge Mass., Report 3630, Sept. 1978

4. Jones, H.W., "Acoustical Scale Modelling for Noise Control", to be published.

5. Jones, H.W., Stredulinsky, D.C., and Vermeulen, P.J., "Modelling of Environmental Acoustics", Proceedings of 9th International Congress on Acoustics, 1, p.1336, Madrid 1977.

6. Almgren, M., "Scale Model Simulation of Outdoor Sound Propagation Considering a Sound Speed Gradient", Proceedings, Inter−Noise '86, p.1301−1306.

7. Jones, H.W., D. Stredulinsky, and P.J. Vermeulen, "Absorption of Ultrasound in Freon−12", J. Acoust. Soc. Am., 60, No. 6, December 1976, p.1309.

8. Beranek, L.L., Noise and Vibration Control, McGraw−Hill Inc., New York, 1971, p.221.

9. Rayleigh, Baron, Textbook of Sound, A.B. Wood, 3rd Edition, 1964, G. Bell and Sons Ltd., London, England, p.545

10. Sinclair, H.W., Rev. Sci. Inst., 1941, 12, p.250

11. Whymark, R.R., "Acoustical Field Positioning for Containerless Processing Ultrasonics", 13, No. 6, Nov., 1975, pp.251−261

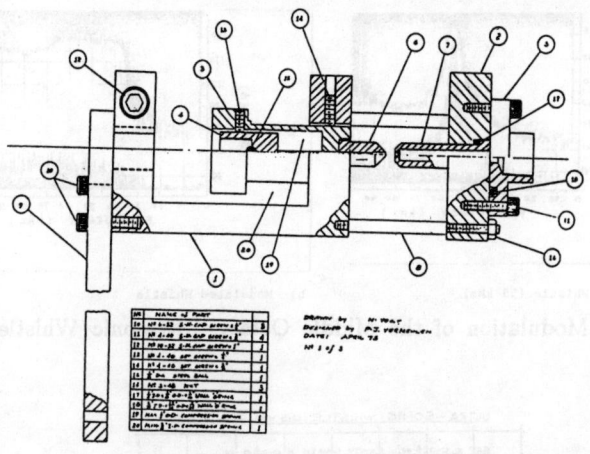

fig. 1.   Initial Ultrasonic Whistle

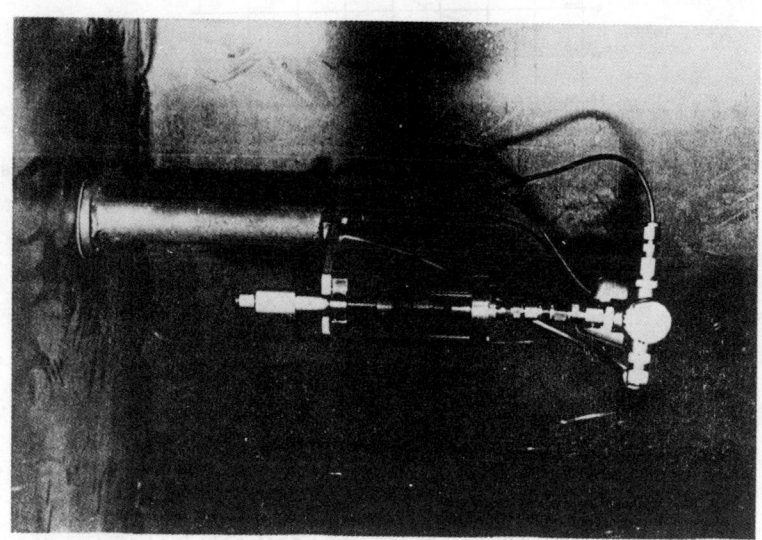

fig. 2.   Ultrasonic Whistle In Reverberation Chamber

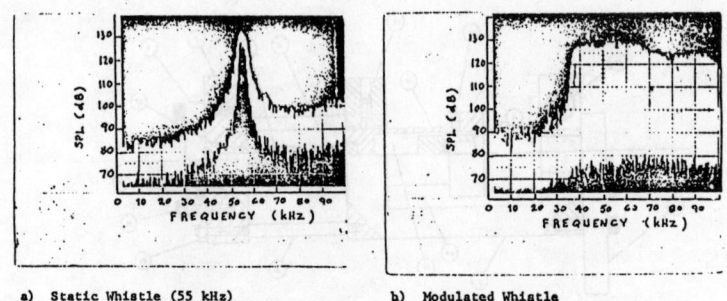

a) Static Whistle (55 kHz)   b) Modulated Whistle

fig. 3. Modulation of the Middle Octave Ultrasonic Whistle

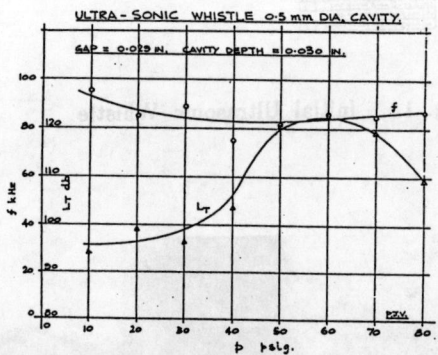

fig. 4. The Influence of the Air Supply Pressure on Ultrasonic Whistle Characteristics

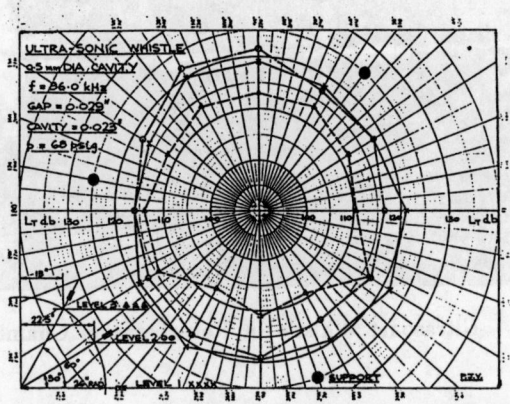

fig. 5. Directional Characteristics of the Initial Prototype Whistle

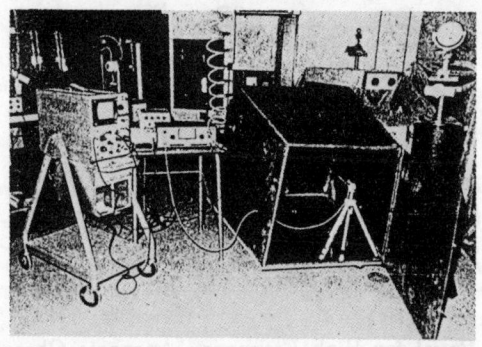

fig. 6. High Frequency Reverberation Chamber

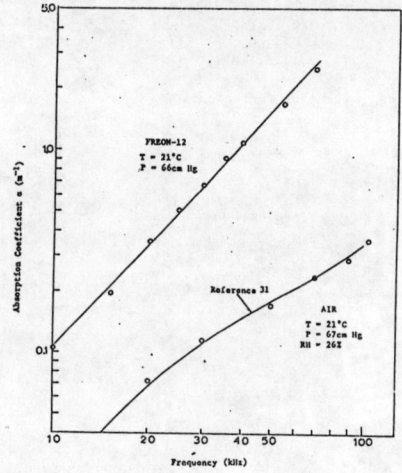

fig. 7. Absorption of Sound in Air and Freon−12
(Reference 31: J.E. Piercy, Private Communication)

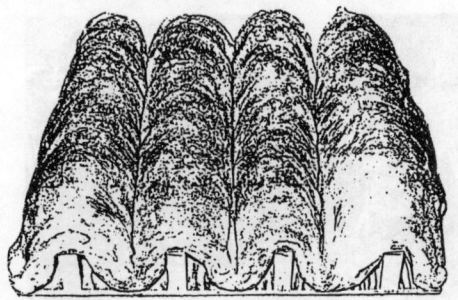

fig. 8. Panel for High Frequency Anechoic Chamber

fig. 9. Completed Anechoic Chamber

fig. 10. Impedance Measurement Apparatus

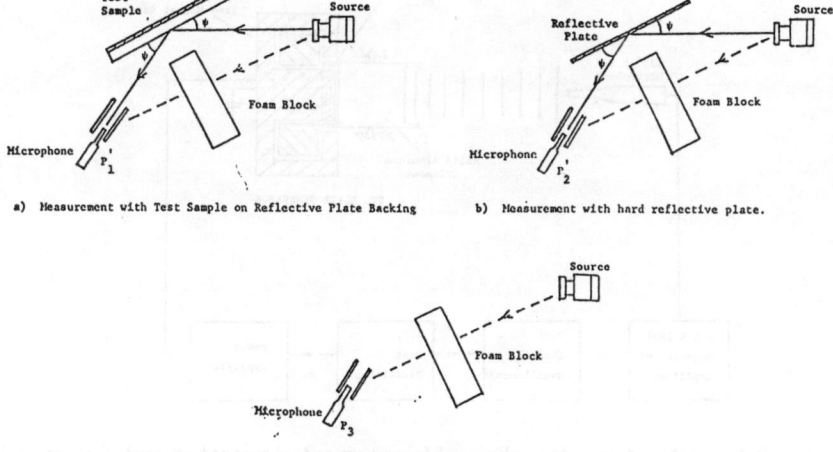

fig. 11. Impedance Apparatus Measurement Configurations

fig. 12. St. Clair Ultrasonic Generator

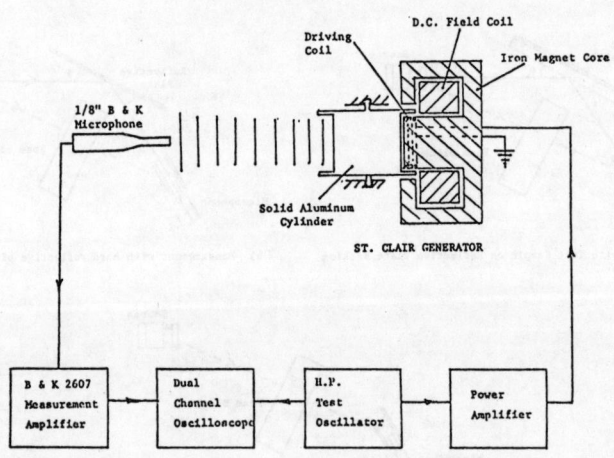

fig. 13. Schematic of the Impedence Measurement Apparatus and Associated Electronic Equipment

L'ENSEIGNEMENT SUPERIEUR DE L'ACOUSTIQUE EN FRANCE

R. LEHMANN
Faculté des Sciences du Mans
Université du Mans, France

GENERALITES

Il est bon de rappeler que l'acoustique est peut-être la plus ancienne des sciences mais, hélas, son enseignement a été, je crois, très longtemps délaissé dans la plupart des pays, et en particulier en France.

Si je prends l'exemple de l'acoustique des salles, il est utile de rappeler que son étude remonte à l'Antiquité puisque les Grecs et les Romains nous ont légué des théâtres dont l'acoustique est souvent parfaite (Epidaure en Grèce et Orange en France, pour n'en citer que deux).

Il est évident que les théâtres antiques se sont toujours présentés de manière relativement simple car ils étaient tous du même type et ne comportaient qu'un nombre réduit d'éléments distincts:
- des gradins en forme de demi-cercle qui épousaient presque toujours la cavité d'une colline,
- en bas et sur le devant une surface généralement plane, plus ou moins importante mais souvent petite, appelée orchestre,
- plus loin, une scène souvent longue et étroite qui était quelquefois fermée vers l'arrière par un mur.

Tous ces édifices, sans exception, n'avaient pas de toit et étaient réalisés en pierre.

Malheureusement, les textes anciens nous renseignent très peu sur l'art de l'architecte, sauf ceux dont s'est inspiré VITRUVE.

L'épigraphie, par contre, fournit quelques détails concernant la nature des matériaux mais ne précise jamais leur agencement. Heureusement que dans ce domaine l'archéologie supplée la littératurre.

Enfin, de nombreux travaux plus techniques ont été conduits depuis une trentaine d'années au moins; ils nous ont permis de comprendre et de mettre en évidence les connaissances de nos anciens dans le domaine de l'architecture.

Il est de rappeler, qu'à cette époque, les Grecs et les Romains ne disposaient pour leurs réalisations que de l'oreille et de la géométrie dont ils connaissaient certainement admirablement les principes les plus simples.

Ils s'étaient déjà rendus compte qu'il fallait différencier les théâtres destinés à la parole de ceux destinés à la musique pour lesquels une certaine résonance était déjà recherchée (réalisation des odéons).

On peut se poser la question de savoir si des progrès considérables ont été réalisés depuis cette époque, pratiquement et non pas théoriquement, naturellement, dans le domaine de l'acoustique des salles. L'expérience prouve que depuis plusieurs dizaines d'années, de nombreuses salles jouissant d'une excellente acoustique, ont été réalisées de par le monde mais aussi un grand nombre d'autres ont été mal conçues et c'est encore quelquefois, hélas, le cas de nos jours.

Pourtant, que de progrès techniques et scientifiques ont été réalisés depuis cette époque, plus spécialement à partir des travaux de SABINE, puis grâce à la mise en évidence du concept dit de la fonction de transfert de la modulation qui permet, en particulier, d'exécuter des approches géométriques et statistiques dans le champ réverbéré beaucoup plus précises.

De ce point de vue, il est indispensable de signaler les travaux publiés en 1985 par HOUTGAST et STEENEKEN.

L'EVOLUTION DES CONGRES INTERNATIONAUX D'ACOUSTIQUE

Il est intéressant de comparer les thèmes et les communications présentés lors des divers congrès internationaux d'acoustique organisés sous les auspices de l'Union Internationale de Physique Pure et Appliquée, de 1953 à 1986.

Le premier d'entre eux (en 1953) s'est tenu à LA HAYE (Hollande) et a réuni une cinquantaine de participants sur 3 jours consécutifs.

Lors du dernier congrès international tenu an 1986 à TORONTO (Canada) près de 1000 participants étaient rassemblés pour entendre, en plusieurs sessions simultanées, environ 700 communications présentées par plusieurs centaines d'auteurs, sur des thèmes aussi variés que différents, tels que, à titre d'exemple: la physio et la psychoacoustique, la communication parlée, l'acoustique sous-marine, l'acoustique physique, lélectroacoustique, l'ultra-acoustique, le traitement du signal sonore, etc...

On retrouve la même évolution aussi rapide en l'espace d'une trentaine d'années si on se rapporte aux thèmes principaux des conférences générales présentées lors des congrès internationaux successifs.

Seuls trois ou quatre thèmes généraux étaient envisagés lors des premiers congrès de 1953 et de 1956, alors qu'au prochain congrès prévu en 1989 en Yougoslavie, 15 thèmes généraux sont d'ores et déjà retenus par les organisateurs.

Il est bien évident que dans la plupart des pays, l'enseignement supérieur de l'acoustique a suivi cette évolution et, en particulier, en France.

L'EVOLUTION DE L'ENSEIGNEMENT SUPERIEUR DE L'ACOUSTIQUE EN FRANCE

A ma connaissance, en 1970, une seule université française (MARSEILLE) dispensait un enseignement supérieur de l'acoustique, très orienté vers le domaine de l'acoustique physique.

Aucune autre université ne dispensait d'enseignement complet dans cette discipline, sauf quelques cours orientés vers tel ou tel domaine, en fonction de la spécialisation de l'enseignement.

1. L'Université du MANS

C'est pour remédier partiellement à cet état de fait que j'eus, personnelement, l'idée de créer au MANS, dès 1972, un enseignement supérieur de l'acoustique beaucoup plus orienté vers applications pratiques. Cela n'était pas très facile, en raison de la multiplicité des domaines de l'acoustique car mon but, à l'époque, était de donner aux étudiants des notions suffisantes pour pouvoir, par la suite, s'intéresser à des domaines aussi variés que l'acoustique des salles et des bâtiments, l'électro-acoustique (réalisation de microphones et haut-parleurs), la physio-acoustique, l'ultra-acoustique, etc...

A cette époque, il n'était pas encore question de dispenser un cours de traitement du signal sonore.

De 1972 à nos jours, l'enseignement de l'acoustique à l'Université du MANS a évidemment beaucoup changé, afin de tenir compte de l'évolution des techniques et de l'utilisation massive de l'ordinateur et des analyseurs en temps réel.

Ainsi, ce troisième cycle d'acoustique tend à devenir moins appliqué et plus théorique ce qui, je le reconnais, n'est pas conforme à l'idée originale. Cet enseignement comprend des valeurs de base et des valeurs complémentaires. Les premières sont presque toutes théoriques alors que les secondes le sont totalement.

A titre indicatif, les valeurs de base sont les suivantes:
a) Acoustique physique:
   Lois fondamentales de la propagation des ondes,
   Equation d'onde dans un fluide parfait,
   Les guides d'onde et les pavillons,
   Les équations de continuité de la mécanique des milieux continus,

Les équations aux discontinuités,
Les équations générales de l'acoustique non linéaire,
La propagation en milieu dissipatif,
Les théories de la diffusion et de la diffraction,
Les théories ondulatoire, géométriques et statistiques en acoustique des salles,
L'acoustique en milieux mobiles.

b) Théorie des vibrations:
Oscillateur harmonique,
Equations du mouvement pour les systèmes à plusieurs degrés de liberté,
Oscillations libres et forcées de ces systèmes,
Etude des systèmes non linéaires,
Mécanique des milieux élastiques,
Les analogies électro-mécaniques,
Vibrations des barres,
Systèmes de protection contre les vibrations,
Transducteurs piézoélectriques,
Etude des vibrations aléatoires.

c) Acoustique de la matière condensée:
Propagation des ondes élastiques dans les solides,
Ondes élastiques de volume dans les milieux continus, illimités, limités et dans les milieux anisotropes,
Ondes élastiques de surface,
Production et détection des ultrasons,
Application des techniques ultrasonores en médicine, en électronique, en acoustique sous-marine.

d) Perception et acoustique des salles:
Anatomie et physiologie de l'oreille,
Notions fondamentales de psychoacoustique,
Ecoute binaurale, latéralisation et localisation des ondes,
Théories géométriques et statistiques de la réverbération,
Couplages dans les salles,
Evaluation subjective et critères physiques de la qualité des salles.

e) Théorie du signal:
Analyse fréquentielle (transformée de Fourier et distribution),
Convolutions,
Signaux conjugués,
Corrélations temporelles,
Echantillonnages,
Variables aléatoires,
Signaux aléatoires (processus stationnaires et stochastiques),
Applications à l'acoustique (localisation des sources,

imagerie acoustique et intesimétrie).

f) Traitement numérique du signal:
Echantillonnage des signaux continus,
Codage des signaux codés,
Filtrage,
Transformée rapide de Fourier,
Algorithmes,
Fonctions de transfert des filtres,
Filtres récursifs et non récursifs,
Synthèse et analyse de la parole.

Les valeurs complémentaires se divisent en trois parties, à savoir:

g) Electroacustique:
Analogies,
Lignes de transmissions,
Approximation des constantes localisées,
Filtres acoustiques,
Transducteurs électroacoustique (haut-parleurs, microphones et enceintes),
Les phénomènes de réciprocité,
La perception des sons musicaux,
Le fonctionnement et le rayonnement des instruments de musique.

h) Methodes mathématiques de l'acoustique:
Les fonctions spéciales (Legendre, Bessel, etc...),
Les problèmes aux limites (espaces de Hilbert, équations intégrales et aux dérivées partielles),
Les méthodes d'approximation,
La méthode des éléments finis,
Applications de toutes ces méthodes à la mécanique et à l'acoustique.

i) Micro-informatique appliquée:
Concepts de logique programmée,
Architecture d'un microprocesseur,
Eléments de programmation en assembleur,
Circuits périphériques associés aux microprocesseurs,
Structure d'un micro-ordinateur,
Contrôles de processeus micro-informatisés,
Eléments de programmation en langage évolué.

A titre indicatif, les valeurs dites complémentaires concernant les méthodes mathématiques de l'acoustique et la micro-informatique appliquée n'étaient pas enseigées lors de la création du troisième cycle d'acoustique. Il en était de même des valeurs de bases relatives au traitement du signal.

Ainsi, comme je le disais précédemment, l'évolution de l'acoustique appliquée vers l'acoustique théorique est flagrante et, peut-être, rendue nécessaire par l'évolution des techniques.

Je regrette cependant, personnelement, que les applications ne soient pas traitées de manière plus détaillée. Cela serait possible si on augmentait l'horaire de l'enseignement annuel qui est déjà très lourd puisqu'il est de l'ordre de 200 heures, auxquelles il convient d'ajouter une cinquantaine d'heures de travaux pratiques et un stage obligatoire de 4 mois en laboratoire.

A l'Université du MANS, les thèmes de recherche retenus et se rapportant au programme qui vient d'être précisé peuvent se résumer de la façon suivante:
- transducteurs électroacoustiques,
- instruments de musique à vent et ondes guidées,
- acoustique des salles,
- matériaux absorbants,
- restitution des enregistrement anciens et analyses des signaux.

D'une manière plus explicite il est possible de préciser les principaux axes de recherche des travaux en cours:
- étude d'un capteur d'impédance avec deux transducteurs réciproques,
- étude d'un giromètre acoustique,
- réseau acoustique ordonné et désordonné (conception de silencieux réactifs à réflecteurs),
- mesures acoustiques en salle et sur maquette,
- caractérisation des ondes impulsionnelles,
- diffuseur à réseau de Shroeder,
- modélisation des propriétés acoustiques de mousse plastique (application de la théorie de Biot),
- mesure de l'impédance acoustique en incidence normale et en incidence appliquée,
- modélisation du champ acoustique au voisinage d'un panneau d'impédance déterminée,
- amélioration des enregistrements anciens de parole par utilisation des techniques de traitement numérique du signal,
- mesure de la célérité du son dans les gaz en vue de déterminer certaines de leur propriétés.

2. L'Université de Lyon

Le but de l'enseignement technique dispensé dans cette université est principalement de former des spécialites dans le domaine de la génération du rayonnement et de la propagation des ondes acoustiques avec, de plus, une bonne connaissance des techniques expérimentales relatives aux sons audibles et aux ultrasons, avec application à des cas concrets relevant des divers secteurs industriels.

Plusieurs filières sont prévues, à savoir:
- rayonnement acoustique des structures,

- aéroacoustique,
- ultrasons,
- acoustique théorique.

L'ensemble des cours est réparti de la façon suivante:

a) Acoustique fondamentale:
   - éléments de base de l'acoustique linéaire (propagation en milieu infini)
   - champ acoustique en milieu fini (propagation dans les conduits, acoustique des salles),
   - notions fondamentales concernant le rayonnement et la diffraction,
   - transmission et transparence des plaques,
   - notions d'acoustique physiologique,
   - éléments de base concernant l'instrumentation.

b) Acoustique approfondie:
   - équations de propagation en milieux inhomogènes ou en mouvement,
   - approximation géométrique et parabolique,
   - exemples en acoustique aérienne et sous-marine,
   - formulation intégrale des problèmes de diffraction (application au cas des écrans),
   - rayonnement des sources étendues.

c) Compléments concernant les vibrations:
   - méthode intégrale, méthode variationnelle, matrice de transfert,
   - vibrations aléatoires et non linéaires.

d) Ultrasons:
   - propagation des ondes ultrasonores dans les solides isotropes et anisotropes finis et infinis,
   - propagation dans un solide piezoélectrique,
   - atténuation, production, détection et utilisations des ondes ultrasonores,
   - vibrations de réseau,
   - techniques nouvelles en contrôle ultrasonore.

e) Rayonnement acoustique des structures vibrantes:
   - interactions fluide-structure,
   - caractérisation du rayonnement,
   - puissance et intensimétrie acoustiques,
   - formulation intégrale et application aux plaques et coques,
   - transmission du son par les plaques et coques,
   - méthode expérimentale moderne de réduction du bruit (imagerie, antennes, intensimétrie, etc...).

f) Méthodes numériques en acoustique:
   - compléments de mathématiques (transformations intégrales - fonctions spéciales - fonction de Green),

- méthode des éléments finis,
- méthode de nature intégrale (problème du noyau et des singularités),
- méthode des éléments frontière ou de surface,

g) Aéroacoustique:
- formulation de Lighthill en milieu infini,
- approches théoretiques récentes et applications aux écoulements turbulents,
- bruits de jets infra et supersoniques (influence de la densité et de la température),
- influence des surfaces solides (théorie de Curle),
- équation de Ffowcs - Williams et Haukins,
- propagation dans les conduits,
- bruit des machines tournantes.

h) Turbulence:
- équations générales des écoulements turbulents (jets, sillages, couches limites...),
- présentation de modèles théoretiques simplifiés,
- étude de la turbulence homogène et isotrope,
- dispersion de particules à partir d'un point de mesure (problème de pollution atmosphérique),
- techniques de mesures (anémométrie à fil chaud et à laser),

i) Traitement des signaux et des images:
- Méthodes générales (convolution - échantillonnage - corrélations - détection synchrone - densité spectrale d'énergie - filtrage adapté, etc...),
- application du traitement du signal aux systèmes sonars haute fréquence,
- reconnaisance de formes,
- mise en oeuvre numérique des méthodes appliquées,
- numérisation d'une image,
- compression des informations,
- méthodes élémentaires de renforcement des contrastes,
- filtrage de Fourier et déconvolution,
- synthèse d'images tomographiques.

Les stages d'initiation à la recherche peuvent être effectués soit dans le Laboratoire de l'Université de Lyon, soit à l'Ecole Centrale de Lyon, soit à l'Institut des Sciences Appliquées de Lyon, soit à l'Institut de la Communication et de la Parole de Lyon, soit à l'Institut National de Recherche des Transports et de la Sécurité de Grenoble, soit au Centre Scientifique et Technique du Bâtiment de Grenoble.

3. Université de BORDEAUX
Il a'agit principalement, dans cette université, de former des spécialistes en acoustique ultrasonore et en

applications industrielles de caractérisation et de contrôles non-destructifs des matériaux et de structures.

L'enseignement est divisé en deux parties principales comprenant, d'une part des enseignements théoriques et, d'autre part, des enseignements technologiques.

Les premiers comprennent principalement les disciplines suivantes:
- acoustique générale relative aux fluides,
- acoustique générale relative aux solides,
- représentations intégrales des problèmes de propagation,
- propagation ultrasonore dans les milieux inhomogènes et anisotropes,
- instrumentation et traitement du signal.

Les enseignements technologiques comprennent deux parties principales, à savoir:
- technologie générale de l'acoustique,
- étude des transducteurs ultrasonores.

Les programmes de ces divers enseignements sont pratiquement identiques aux enseignement correspondants qui ont été détaillés dans les chapitres précédents.

Quant aux domaines de recherche, ils sont orientés plus principalement vers les activités suivantes:
- propagation des ondes métérialles dans les milieux homogènes et inhomogènes,
- théories d'homogénéisation et modélisation du comportement mécanique des milieux multiphasiques structurés (matériaux composites),
- instrumentation et traitement du signal ultrasonore par spectro-interférométrie assistée par ordinateur,
- théorie de l'acousto-élasticité,
- applications à la caractérisation ultrasonore des anisotropies induites par contraintes internes ou par endommagement,
- évaluation et contrôles non destructifs ultrasonores des matériaux et des structures,
- métrologie des constantes élastiques et visco-élastiques des solides isotropes et anisotropes (bois, roches, céramiques, composites, etc...),
- conception et réalisations de transducteurs et de systèmes adaptés à des problèmes spécifiques de contrôle de qualité.

4. Université de STRASBOURG

Cette université délivre un enseignement de 3ème cycle de Mécanique et d'Acoustique que l'on peut détailler de la façon suivante:

a) Mécanique approfondie des fluides:
- étude de la turbulence,
- probabilité statistique,

- équations statistiques des fluides turbulents,
- équations de l'énergie interne et de l'enthalpie.

b) Corrosion:
- bases électrochimiques de la corrosion,
- principaux mécanismes de la corrosion.

c) Biomécanique des système physiologiques:
- système de l'equilibration,
- système auditif,
- système artériel.

d) Applications en acoustique:
- interactions acousto-optiques,
- isolation et absorption acoustiques,
- acoustique générale, acoustique moléculaire.

e) Méthode de calcul numérique:
- éléments finis en mécanique,
- formalisation.

f) Propriétés arhéologiques et hydrodynamiques des matériaux fluides:
- loi fondamentale de l'hydraulique,
- écoulements internes laminaires et turbulents,
- écoulements polyphasiques,
- sédimentation et maintien en suspension de particules,
- hydrodynamique des écoulements,
- propriétées mécaniques et acoustiques des écoulements,
- initiation et analyse des méthodes modernes de mesures,
- méthodes optiques en mécanique (contrôle non destructif, interfaçage des appareils, etc...),
- rhéologie des matériaux,
- propriétés microscopiques et comportement macroscopique de suspensions et de solutions colloïdales,
- comportement mécanique des matériaux nouveaux, des matériaux composites et des cristaux liquides.

g) Modélisation mathématique et traitement du signal:
- définition et élaboration d'un modèle de diffusion,
- algorithmes de résolution et théorèmes fondamentaux,
- traitement avec réduction de données,
- traitement sans réduction de données,
- traitement non linéaire,
- étude d'un contrôleur numérique de processus,
- identification et modélisation des processus,
- asservissements,
- synthèse des régulateurs,
- application à des dispositifs de mesure et d'essais analogiques.

Les axes de recherche dépendent des laboratoires

d'accueil, mais on peut les résumer de la façon suivantes:
- écoulements turbulents,
- écoulements de liquides en milieux poreux,
- échanges fluide-fluide ou fluide-solide,
- modélisation des systèmes physiologiques normaux et pathologiques,
- rhéologie et écoulements de fluides polyphasiques,
- propriétés acoustiques et visco-élastiques de gels,
- propriétés hydronamiques et acoustiques des cristaux liquides et de certaines polymères,
- spectrométrie ultrasonore de biomolécules et de systèmes enzymatiques,
- développement des nouvelles techniques de mesures ultrasonores,
- rhéologie du béton,
- métallurgie et corrosion des matériaux,
- sédimentation fractionnée de particules solides en écoulement pulsé,
- déformation et endommagement des solides.

5. Université de MARSEILLE

Cette université délivre le plus ancien des 3èmes cycles d'acoustique dispensés en France et qui était plutôt orienté, à l'origine, vers l'acoustique physique et l'acoustique théorique.

L'orientation de l'enseignement s'est quelque peu modifiée depuis ces dernières années et tend maintenant vers un enseignement d'acoustique générale, peut-être plus particulièrement adapté aux études d'acoustique sous-marine et d'ultra-acoustique.

Le programme des cours peut être développé de la façon suivante:

a) Théorie et méthodes mathématiques de l'acoustique et des vibrations:
   Mécanique des milieux continus
   - établissement des équations de la mécanique des milieux continus fluides et solides,
   - linéarisation,
   - absorption,
   - phénomènes non linéaires.
   Acoustique théorique
   - propagation en espace libre et en espace clos,
   - propagation guidée,
   - propagation en milieu aléatoire,
   - méthodes de résolution analytique,
   - équations intégrales de frontières,
   - couplage fluide-structure,
   - approximations analytiques.
   Théorie des vibrations
   - vibrations des systèmes élastiques (poutres, membranes,

plaques, coques),
- influence de la viscosité,
- réponse d'un système vibrant à une excitation aléatoire,
- propagation des ondes dans les solides.

b) Traitement du signal et méthodes numériques:
  Processus aléatoires et traitement du signal
  - processus aléatoire et estimation,
  - modélisation et identification,
  - analyse spectrale,
  - méthodes numériques associées,
  Modélisation numérique des problèmes d'acoustique et de mécanique
  - équations intégrales de frontières,
  - méthodes d'éléments finis en élasticité,
  - méthode d'éléments finis pour les problèmes de propagation.
  Calcul numérique et informatique
  Il s'agit principalement de travaux pratiques.

c) Acoustique sous-marine et ultrasons:
  Ultrasons
  - émission -réception,
  - technologie des transducteurs piezoélectriques,
  - techniques de mesures.
  Acoustique sous-marine
  - propriétée acoustiques du milieu marin,
  - propagation en milieu marin,
  - instrumentation en acoustique sous-marine.
  Acoustique du traitement du signal à l'acoustique

d) Acoustique de l'environnement et psychoacoustique:
  Perception des sons
  - notions de hauteur,
  - notions de timbre,
  - applications à la parole et aux sons.
  Synthèse numérique des sons
  - conversion,
  - échantillonnage,
  - quantification,
  - distorsion,
  - filtrage.
  Psychoacoustique
  - système auditif,
  - notion de sonie,
  - masquage,
  - perception spatiale.
  Acoustique des salles
  - qualité d'une salle,
  - paramètres physiques d'une salle.
  Génie acoustique

- grandeur acoustique,
- unités acoustiques,
- notions de niveaux,
- transducteurs,
- acoustique urbaine,
- acoustique architecturale,
- acoustique industrielle.

Cet enseignement est complété par une dizaine de séances de travaux pratiques qui comprennent en particulier les manipulations suivantes:
- étalonnage des transducteurs,
- mesure de l'impédance d'un matériau par la méthode du tube,
- propagation acoustique au-dessus d'un sol,
- mesure de la vitesse et de l'absorption des ultrasons,
- détection ultrasonore des défauts dans les métaux,
- mesure du module d'élasticité complexe d'un matériau,
- vibrations des barreaux et des plaques,
- absorption acoustique active,
- mesures de sonie et de masquage.

Les principaux thèmes de recherche de l'équipe enseignante de ce 3ème cycle peuvent se résumer de la façon suivante:
- perception de la parole dans le bruit,
- localisation auditive,
- détection audiophonique sous-marine,
- informatique musicale,
- étude des matériaux composites à base de fibre de verre ou de carbonne,
- modélisation mécanique,
- propagation et diffraction des ondes acoustiques,
- couplage entre vibrations des solides élastiques et champ acoustique,
- intensimétrie acoustique,
- propagation en milieu non homogène,
- propagation sous-marine,
- imagerie acoustique,
- diffusion acoustique,
- absorption acoustique active (application aux casques anti-bruits et aux véhicules).

6. Conservatoire Nationale des Arts et Métiers de PARIS

Il s'agit principalement d'un enseignement en formation continue qui conduit à un diplôme d'études supérieures techniques d'acoustique. Par la suite, les étudiants peuvent préparer un diplôme d'ingénieur du C.N.A.M.

Il s'agit essentiellement d'un enseignement d'acoustique générale, plus particulièrement orienté vers l'acoustique physique et vers les applications que l'on peut définir de la façon suivante:

a) Dynamique des vibrations:
   Systèmes à un degré de liberté
   - analogies électriques, mécaniques et acoustiques,
   - oscillations libres, oscillations amorties, oscillations forcées,
   - régimes transitoires et régimes permanents,
   - adaptation et transfert d'énergie entre excitateurs et résonateurs.

   Systèmes à plusieurs degrés de liberté
   - systèmes linéaires à couplages homogènes,
   - couplages électromécaniques et mécano-acoustiques,
   - circuits à plusieurs degrés de liberté: propagation dans les milieux discontinus.

   Vibrations élémentaires des corps solides
   - masses quasi-ponctuelles, cordes, membranes, barres et plaques,
   - détermination des caractéristiques vibratoires: fonctions de transfert, analyse modale,
   - étude des couplages et de l'amortissement.

b) Propagation:
   Propagation des ondes planes
   - équations fondamentales: équation d'état, équation d'EULER, conservation de la masse. L'équation d'onde à une dimension et l'équation d'HELMHOLTZ,
   - généralités sur les phénomènes de propagation: onde plane progressive et onde plane stationnaire,
   - les équations pratiques de l'acoustique.

   Propagation des ondes sphériques
   - l'équation d'onde en coordonnées sphériques,
   - potentiel des vitesses, onde de pression et onde de vitesse.

   Rayonnement et directivité des sources
   - les impédances en acoustique: impédance spécifique, impédance de rayonnement, impédance ramenée,
   - énergie rayonnée par une source d'onde plane, par une sphère pulsante,
   - monopôles et dipôles: modélisation des sources complexes,
   - rayonnement et directivité d'un piston plan encastré,
   - directivité des sources.

c) Electroacoustique:
   Généralités
   - les transducteurs: systèmes électrodynamiques, électrostatiques, piézoélectriques. Principes et technologies,
   - les microphones utilisés pour la mesure et la prise de son,
   - applications physiques, biologiques et médicales.

d) Mesures acoustiques:
   Mesures vibratoires des sources
   - mesures des vibrations: accéléromètres,
   - analyse et traitement du signal vibratoire,
   - sources étalons: générateurs, machines à frapper, etc...
   Mesures des champs acoustiques:
   - choix et utilisation du matériel,
   - microphones de mesure: qualités, défauts et limites d'utilisation, étalonage,
   - filtres, adaptateurs, atténuateurs et amplificateurs,
   - appareils de mesure: sonomètres, fréquencemètres, distorsiomètres, compteurs, indicateurs, enregistreurs...,
   - les analyseurs en temps réel,
   - le traitement du signal acoustique.

e) Acoustique des salles non couplées:
   - Généralités: approche géométrique et ondulatoire de l'acoustique des salles,
   - étude du champ acoustique dans un local,
   - durée de réverbération,
   - détermination des caractéristiques acoustiques d'une salle.

f) Acoustique des salles couplées et environnement:
   - étude des champs stationnaires couplés,
   - incidence sur l'environnement de l'énergie sonore rayonnée par un local,
   - transferts d'énergie réverbérée entre plusieurs locaux,
   - champs réverbérés transitoires dans les salles couplées,
   - couplages électroacoustiques.

g) Critères d'appréciation subjective des salles:
   - intelligibilité,
   - perception musicale,
   - caractérisation acoustique d'une salle: les critères objectifs.

h) Acoustique industrielle:
   - réflexion et absorption sur les obstacles,
   - impédance de parois sous incidence normale et incidence oblique,
   - transparence acoustique. Fréquence critique et loi de masse,
   - les matériaux utilisés pour l'isolement acoustique: amortissement et résonance,
   - rayonnement des structures,
   - les méthodes de diagnostic et traitement des bruits en milieu industriel.

i) Perception des sons et des bruits:

- anatomie et physiologie,
- notions fondamentales de psychoacoustique,
- écoute critique,
- effets du bruit sur l'organisme: effets auditifs et non auditifs.

j) Prise de son et sonorisation des salles:
- matériel d'enregistrement et de reproduction des sons,
- techniques de la prise de son: choix et utilisation des metériels,
- techniques stéréophoniques.

7. Université de PARIS VI

La physique moderne utilise les ondes acoustiques dans un domaine spectral très étendu: la longueur d'onde peut varier du micron (microscope acoustique) à quelques centaines de mètres (géophysique) en passant par le domaine millimétrique des applications industrielles ou médicales, et métrique de l'acoustique sous-marine.

Cependant, les phénomènes concernés ont des parentés conceptuelles; une modélisation commune est donc possible qui constitue l'objet de l'acoustique physique. Cette discipline s'intéresse également aux relations entre les ondes acoustiques et la matière qui en est le support.

Le programme de l'enseignement se divise en deux parties que l'on peut résumer de la façon suivante:

a) Acoustique fondamentale:
- génération et détection des ultrasons,
- propagation, diffraction, diffusion, absorption,
- thermo-dynamique des solides avec variables électromécaniques,
- acoustique de FOURIER,
- imagerie et holographie acoustiques,
- interaction des ondes élastiques avec d'autres ondes (ondes lumineuses-ondes élastiques), acoustique non linéaire, cavitation,
- utilisation des ultrasons pour le diagnostic des propriétés de la matière,
- ondes guidées (ondes de surface).

b) Applications des ultrasons:
<u>Techniques échographiques appliquées:</u>
- à la détection sous-marine (sonar),
- au contrôle non destructif des matériaux,
- à la prospection géologique,
- au diagnostic médical.
<u>Techniques de traitement du signal par ondes élastiques:</u>
- dispositif opto-acoustique,
- spectroscopie ultrasonore dans les pièces métalliques

et dans les tissus mous,
- application des ultrasons en chimie.

V. CONCLUSIONS

Je me suis efforcé, dans cet exposé quelque peu aride, d'indiqueur les grandes lignes des enseignements dispensés dans ce que nous appelons en FRANCE, les "trosièmes cycles" qui se situent à un niveau Bac + 5.

Ces enseignement sont ouverts aux étudiants titulaires d'une maîtrise de sciences physiques, de sciences mathématiques, d'un diplôme d'ingénieur ou de tout diplôme jugé équivalent.

Ils ont pour but principal de copmléter les formations théoriques de base déjà acquises par les étudiants au cours de leur premier cycle d'enseignement supérieur et de leur donner une formation aussi étendue que possible aux techniques expérimentales récentes dans le domaine de l'acoustique et de la mécanique vibratoire.

La plupart de ces enseignements cherchent également à donner aux auditeurs une formation aux techniques numériques modernes, telles que les éléments finis et le traitement du signal.

Le premier diplôme obtenu est le diplôme d'Etudes Approfondies d'Acoustique qui comprend une partie théorique relative aux programmes exposés dans les chapitres précédents et une partie pratique consistant en un travail de laboratoire, d'une durée relativement courte (environ 3 mois à temps plein).

Ensuite, les étudiants qui le souhaitent, peuvent poursuivre des recherches dans un domaine spécialisé afin d'obtenir, en 3 ans environ, une thèse de Doctorat.

Il y a une quinzaine d'années, l'enseignement donné à l'Université de MARSEILLE était très orienté, comme déjà indiqué précédemment, vers l'acoustique physique et l'acoustique théorique, tandis que celui donné à l'Université du MANS était beaucoup plus orienté vers l'acoustique appliquée.

Depuis, des enseignements de plus similaires ont été créés à LYON, BORDEAUX, TOULOUSE, MARSEILLE et PARIS et bien que des différences non négligeables existent encore dans les programmes, ils sont maintenant assez homogènes. (Les informations concernant TOULOUSE ne me sont pas parvenues en temps utile).

A titre d'exemple, les bases théoriques de l'acoustique, le traitement du signal, l'acoustique des salles et la physio-acoustique sont enseignés de manière quasiment identique dans toutes les universités.

Les différences qui apparaissent concernant plus spécialement l'électroacoustique, la métrologie acoustique, l'acoustique musicale, l'acoustique des bâtiments et l'ultra-acoustique.

Ainsi, chaque année, sur l'ensemble du territoire français, une centaine d'étudiants obtiennent leurs diplômes d'études approfondies d'acoustique ce qui, à mon sens, est suffisant pour les besoins actuels.

D'autre part, on peut estimer à une trentaine par an, le nombre de thèses de doctorat qui sont soutenues dans le domaine de l'acoustique.

A mon avis, les horaires représentent la difficulté majeure car, d'une manière générale, ils sont limités à 200h environ pour l'année d'étude, ce qui est insuffisant si on désire dispenser un enseignement d'acoustique très général pour permettre aux étudiants de se spécialiser, éventuellement, dans le domaine de leur choix.

ACOUSTICS AND PHYSICS

Wilhelm Løchstøer,
University of Oslo
Norway

This conference is dedicated to "Prospects of Modern Acoustics Education and Development", and so directed towards the future. Then I have to apologize for addressing you more or less as a "voice of the past".
Firstly I am retired, and thus without any responsibility for any form of education.
Secondly the Acoustic Group at the Institute of Physics, University of Oslo where I have been working for the last 40 years is now deleted. Owing to the well known fact of limited resources, the people now responsible for the Institute have decided to give priority to other fields of physics than acoustics. Sorry, but other times, other interests and meanings.
Therefore, what I have to say you is more a summary of experiences from the past, than a program for the future. After all it is my hope that knowledge of the past may be worthwile for the work with the future, even it is not my intention to try to "freeze" the situation to what it has been.
In his opening address to the conference INTERNOISE, I think it was in Edinburgh, professor Ingerslev, the president of INCE (Institution of Noise Control Engineering) said that acoustics from being a part of physics had developed to an independent part of science. I can not give my conditionless agreement to that, if "independent" means to separate acoustics from physics. In my opinion, based on my experience, acoustics must never loose the contact with physics. Even if acoustics has grown immensely in practical importance, they are still the basic laws of mechanical vibrations and waves that determine it all. It is physics, if you like it or not.
What bothers me in the situation today is just that it is a trend within acoustics to loose the connection with physics. In many educational institutions acoustics is only taught as a course in, for instance, electronics and so on. And for environmental people, concerned about noise, noise is only a pollution in the air and not waves generated and

propagated in the air. In this way, I am afraid, acoustics will be split in several specialities, which apparently have nothing to do each with other, with no common base. After all it all obeys the laws of generation and propagation of mechanical vibration and waves in all types of media.

For many people acoustics only means what we can hear, that means a subjective detection. What then about infrasound or ultrasound Ultrasound, for instance, has in recent times become very important in medicine, both for diagnosis and treatment. Is it medicine? Is it acoustics? Is it physics? I think it must be a collaboration, where physics has to play an important part.

Even in ISO (International Standardization Organization) TC43: Acoustics, the question has been raised: What is acoustics? And what is physics? No answer yet. Liaison.

May be I am influenced by heritage when I so strongly recommend that acoustics never looses contact with physics. Both as a student and teacher I have my experience from the Institute of Physics at the University of Oslo. This Institute has, until recently, been a "center of gravity" in the education system of acousticians in Norway. And there acoustics has been taught as a part of physics, or on a base of physics.

The system in the institute is, that students who wish to specialize in physics (including acoustics) first have to take a basic education in science. This is normalized to 3 1/2 years full time work. This basic course in science then must include a general course in physics, normalized to 1 year full time work. In practice this basic course is divided in 4 half-time courses over 2 years. This is partly to give the students a longer "time-contact" with physics, partly to give the students the possibility to take other courses, for instance in mathematics, paralell with their study of physics. The general course in physics include mechanics, heat, electricity, optics, nuclear physics, statistical physics and so on.

The students might choose other courses in physics as well, and they often do. But there are certain restrictions in the choise of courses during the first 3 1/2 years, to secure a certain broadness, to prevent the education to be too specialized.

For students intending to study acoustics it was very common to have the necessary general course in physics and a special course in electronics, and a general course in mathematics combined with a special course in differential equations. And they very often had some courses in chemistry too. So they were well acquainted with physical thinking, physical ideas and methods when they started their studies in acoustics.

After finishing this general study in science, the undergraduate students might continue as advanced students in special parts of physics, among them acoustics (normalized 2 years full time work).

I have to mention that our institute is organized in "scientific groups". Until recently one of these groups was the "acoustic group". Now it is deleted, as mentioned before.

As advanced student a student has to do a specified scientific work under the supervision of a scientist. So there is a certain restriction in the choice for the students, as the possibilities are limited to the existing scientific groups in the institute. It is possible to do this scientific work outside the institute under the responsibility of the institute. In acoustics, for example we have had a very close connection to the institute of audiology at the State Hospital.

The scientific work of each student must necessarily be concentrated on very narrow fields of acoustics. The supervisor has the responsibility to give a "course" in that narrow field. In addition all students in acoustics had to take a course in general acoustics (half time during one year approximately). This course in general acoustics has been my responsibility during last 30 years. In addition the advanced students had to choose some courses in advanced physics, mostly theoretical physics.

So, you see, the students from our institute were not very specialized acousticians, but rather physicists with acoustics. May be they could not tell you "do it in this or this way" immediately after they had finished their studies. But in the long run I think they will be useful. I am very sorry, and a bit worried, that the possibility of "that sort of acousticians" now is finished, and so is the Norvegian Acoustical Society in a letter send to the University. In Norway the education possibilities in acoustics now are rather limited, I think.

At the University in Bergen they have a course in acoustics at the institute of physics. But the group there is rather limited, and even if they are very skillfull, they are rather specialized on non-linear acoustics. At the technical university in Trondheim the education in acoustics are only special courses given in electrical engineering and in building engineering. Even if many people are engaged in acoustics there, the formal education is rather limited and specialized. At the institute of audiology at the State Hospital they now are very much concerned with the problem of education as the possibility in collaboration with the Institute of Physics is now disappeared. They are very interested in the connection with physics.

Symptomatic is the telephone I had a few days before I left for this conference. It was the Chairman of the Union for technical personnel at the "hearing central" (we call it). He asked me to give a lecture at their annual meeting of sound waves. For, he said "these people know everything about electronics in audiometers and hearing aids, but they know nothing about sound waves and waves in enclosures".

As mentioned earlier it is not my intention to "freeze" the past, and to say "do it as we did, and everything will be all right". Far from that. But I think that it is important that at least some acousticians keep close contact with physics, and that this ought to be a possibility in the educational system. There are many reasons for that, I think. Here I will only mention a few.

A student will probably be active in 30-40 years after he has finished his studies. What is a speciality or interesting topic today, may be routine or "out" in a few years. During this time of 30-40 years, acoustical problems and technics will probably change. But the fundamental physical phenomena will remain the same. A person with knowledge of the physical phenomena will therefore easier and better adapt himself to the new situation, and be able to go to the "bottom" of the problems. I will also mention that the development of new methods and new instruments to a great extend depends upon insight in the fundamental physical phenomena.

At last I will mention that I think it very important that the people engaged in education work, not only are specialists telling the students do it in this way or this way, but have a knowledge of the fundamental phenomena, and they are physicists.

So my experience and my advise for the future education: keep close contact between acoustics and physics. How this will be done, nearest time will show.

PLACE OF ACOUSTICS IN THE TECHNICAL UNIVERSITIES

I.MALECKI

Polish Academy of Sciences

Warsaw

Poland

ABSTRACT

Teaching of acoustics since XIX century. Problem of ultrasonics after II World War. New problems in mid 70ties like the ultrasonics diagnostic and the speech control of computers. Kind of lectures: a) general lectures, b) basis specialistic lectures, c) application specialistic lectures, d) lectures for non-specialistic, e) facultative lectures.
The curriculam of lectures for faculties: a) building engineering and architecture, b) mechanical, c) electronics and telecommunication. Items of diploma and doctor's theses.
Teaching of acoustics in high schools of developing countries.

1. A Short History

In universities of XIX and in the begining of XX century acoustics was represented in a very modest way during general physics lectures. Obviously it was acoustics of audible sounds only. Specifically some information was presented on the structure and action of ear, propagation of sound waves in air and structure of musical instruments. Usually acoustics was not considered to be the separate branch of physics but rather part of mechanics. In the course of lectures on the architecture problems of room acoustics were treated in purely descriptive way as related to specific halls.

In the mid 20-ties of XX century in connection with the progress of radio broadcasting special interest evolved in the acoustics of broadcasting studies, while improvements in construction of microphones and loudspeakers required proper development of electroacoustics. It was the reason why during lectures on radio engineering in some universities problems of acoustic were treated, paying interest mainly on the electroacoustic transducers.

Problems of ultrasonic waves only after II World War found relation with practice in echo ranging and nondestructive material testing. Till this time those topics were only marginally treated in lectures on physics, usually apart from section on acoustics. From 1945 on technical applications of ultrasonics waves are mentioned in specialistic lectures, e.g. on metallurgy (non-destructive testing of rolled sheet) and sea navigation (sea bottom depth finding). At the same time follows the development of audioacoustic and a number of its branches acquired high social and economic importance. In 60-ties grows noise hazard and social consciousness of this phenomenon. It resulted in introduction of noise control problems to the lectures presented on mechanical engineering, architecture and building engineering faculties.

In mid 70-ties new problems emerged. Ultrasonic engineering is still more and more used in industry, while ultrasonic apparatus for medical diagnostic are finding wider application in the public health service. In this connection ultrasonic apparatus are treated in more detailed manner in lectures.

The very beginning of 80-ties is marked with substantial increase of interest in the acoustic of the speech on account of progress being made in the speech control of computer systems. In this field studies on phonetics, acoustics and computer science have to be combined requiring common approach not only in research work but as well on the part of university lectures.

As anyone can see the place of acoustics and its part on the higher technical schools has changed recently to the great extent and this process was not completed yet. For instance one can assume, that optoacoustics and quantacoustics will find broad applications and will be introduced into the subject of lecture

2. Aims of Lectures on Acoustics

Depending on faculty, particular subjects and the year of studies specific aims of lectures on acoustics differ.
One can distinguish the following kinds of lectures:
a) lectures supplementing general knowledge of student in the field of physics and not related directly to future work of the school graduate,
b) specialistic basic lectures, giving the student basis

for masering specialistic subjects from the field of acoustics applications,
c) lectures giving information required for independent design of acoustic objects and electroacoustic apparatus,
d) lectures giving information from acoustics for students from other departments in order to enable them taking into account acoustic requirements during solving the problems from their field,
e) lectures on selected branches of acoustics for students with special personal interests in those subjects, apart from normal course of their studies (e.g. musical instruments).

Each of those groups of lectures requires individual treatment and frequently even lecturers of different knowledge and approach to the subject. At the same time depending of those lectures into general curriculum could be widely diversified

3. General Lectures

As a general rule those lectures are given within the framework of lectures on general physics at I or II year of studies. The most extensive are lectures in physics being held on chemistry, electronics and telecommunication faculties while the shortest are given on architecture faculty. Those lectures are compulsory for all students of given faculty.

Depending on general approach to the course of physics problems of acoustics are:
1) separated into individual section, or
2) presented in several parts of the course of lectures.

In the first variant usually apart from information of strictly phisical nature, e.g. covering propagation of sound waves, there are discussed in a short manner such problems as sound perception, harmfulness of noise, room acoustics and the most important applications of ultrasounds (e.g. Bruxelles, Paris VI, Delft, Budapest, Beograd, Aachen and Goettingen universities). In some higher schools (e.g. Wrocław, Praha) in general course of physics in the section on fundamentals of electronics some electroacoustic transducers are discussed.

In the second variant sound waves are discussed within the framework of general wave theory (e.g. Genf, Leuven and Bucarest universities) and vibration of acoustic systems in the section on dynamics of mechanical systems.

It seems that the first variant praiseworthy as it enables complex approach to acoustics including problems going beyond classical course of physics.

4. Basic Specialistic Lectures

Those are complete lectures or parts of lectures describing theoretical and experimental foundations of acoustics. Such lectures are aimed mainly at giving information necessary for work of future engineers as the designers of acoustic objects (rooms, sound amplification systems, radio and TV broadcasting studios, soundcontroled buildings), designers of the sound and ultrasonic acoustic apparatus and equipment (loudspeakers, microphones, musical instruments, measuring and diagnostic apparatus, sonar) or responsible for operation of those devices.

First part of this course of lectures is common for all those branches and could be separated in the program as an individual lecture. The second part being the basis for some branches of applications of acoustics and ultrasonic engineering is ususlly combined in one unit with discussion of those applications.

The main subject of the first part of lectures are usual:
a) propagation of sound waves in gases, liquids and solids,
b) mechanical vibratory systems,
c) electroacoustics transducers,
d) electro-mechano-acoustic analogies,
e) processing of acoustic signals,
f) acoustic measurements,

Such lectures are given usually on III year of studies, mainly on electronics, telecommunication and technical physics departments. On other departments those subjects are included in specialistic lectures of application nature. Such lecture courses last for 1 or 2 semesters (e.g. Liège, Leuven, Strasburg, Praha, Brno, Lyngby, Budapest, Gdańsk and Zagreb universities).

The second part of lectures closely related to applications, rather seldom carried out as a separate lecture covers only those theoretical problems, which are required for solution of specific group of practical problems related to the line of studies on respective department.

One can distinguish the following groups of subjects:
on architecture and building engineering departments
a) room acoustics,
b) propagation of sounds through partition and constructions,
c) free space acoustics,
d) sound perception,
e) noise effect on organism,

on mechanical departments
a) impact and flow sound sources,
b) noise hazard in industry,

c) attenuation of sounds in partitions and constructions,
d) ultrasonic non-destructive material testing methods,

on electronics and telecommunication departments, including computer science departments
a) acoustic structure of the speach,
b) processing of speech signals,
c) ultrasonic transducers,
d) measurments of acoustic signals.

5. Application Specialistic Lectures

Lectures from this group are together with laboratories and design practice held mainly at III or IV years of studies for small teams of students trained to be in the future design engineers, constructors or operation managers.

One can distinguish the following groups of specialy lists.

Building engineering and architecture

1) Designers of lecture rooms and microphone studies.
2) Planner of acoustics of towns and communication networks.
3) Designer of sound amplification systems to be used outdoors.
4) Designers of acoustic sound insulation in buildings.

Each one from those groups should be familiarized with the knowledge presented in above mentioned lectures for architecture and building engineering departments. Apart from this each group should be trained in specific topics, such as:

for group 1
a) history of architecture of concert halls and auditoriums,
b) technique of microphone reception,
c) properies of sound absorptive materials;

for group 2
a) traffic noise,
b) regulations on premissible noise levels,
c) noise measurments;

for group 3
a) construction of microphones and loudspeakers,
b) method of calculation of acoustic fields,
c) subjective assessment of acoustic quality;

for group 4
a) calculation of insulation of building structures,

b) economic calculation of utilization of insulation.

Mechanical departments

1) Designer of noiseless machinery and equipment.
2) Supervisors of working conditions.
3) Supervisors of production control.

For each one from those groups is essential the knowledge presented in above mentioned lectures for mechanical departments. Apart from this additional specialistic lectures e.g. on specific topics as follows:

for group 1
a) theory of vibration of complex mechanical systems,
b) action of conjugate electro-mechano-magnetic fields,
c) properties of vibration damping materials;

for group 2
a) mechanism of interaction of sounds and vibration on human organism,
b) acoustics of factory halls,
c) regulations on noise control;

for group 3
a) ultrasonic signals in non-homogeneous media,
b) standard values of permissible effects,
c) apparatus for automatic ultrasonic inspection.

Electronics and telecommunication departments

1) Supervisors of electroacoustic apparatus.
2) Designers of microphones and loudspeakers.
3) Designers of tape recording apparatus.
4) Audio operators working in radio, film, TV and phonographic industry.
5) Designers and operators of the apparatus for medical and industrial inspection and diagnostics.
6) Designers and operators of ultrasonic scientific apparatus.
7) Designers and operators of voice-operated computers.
8) Designers and operators of echo ranging apparatus.

For each one from those groups is essential the knowledge presented in above mentioned lectures. Apart from this additional specialistic, e.g. on specific topics as follows:

for group 1
a) structure of electroacoustic channel,
b) telephonometric measurments,
c) subjective assessment of sound quality;

for group 2
a) construction of microphones and loudspeakers,

b) assessment of quality of electroacoustic apparatus,
c) mechanical vibrating system;

for group 3
a) construction of sound-recording and reproducing apparatus,
b) physical chemistry of magnetic materials,
c) permissible distortions during recording and reproduction;

for group 4
a) acoustics of musical instruments,
b) sound control apparatus,
c) subjective acoustic effects;

for group 5
a) construction of electronic control and diagnostic apparatus,
b) propagation of ultrasounds in non-homogeneous media,
c) standards for ultrasonic control,
d) foundations of metallurgy or medicine;

for group 6
a) information on molecular acoustics, quanta acoustics and sonochemistry,
b) ultrasonic mesaurments of non-acoustic values;

for group 7
a) analysis and synthesis of the speech,
b) principles of control of voice-operated digital equipment,
c) comparative acoustics of various languages;

for group 8
a) principles of sea navigation,
b) underwater acoustics,
c) apparatus for underwater signalling.

Presented division of subject matter is rather detailed and hard to implementation in practice, as it would be necessary to give lectures to very small teams (2-4 students). Therefore such teams are joined together in greater groups for giving them lectures and special subject of study is treated at the stage of preparation of diploma.

6. Lectures for non-specialists

Those lectures are aimed at conveying data sufficient for future engineer for getting general orientation about the role played by acoustics in technical objects being constructed or supervised by him and about the range of specialistic works that should be carried out by

acousticians or specific electroacoustic apparatus that should be bought.
Such information could be presented in special lectures or contained within the lectures of more generalnature.

Architecture and building engineering faculties

General information are required on:
a) room acoustics,
b) sound insulation,
c) noise control.

If not comprised in individual lecture the best place for discussing those subjects are lectures on building technical insulation, general building engineering and house planning.

Mechanical faculties

General information are required on:

a) noise and vibration hazards and their control,
b) ultrasonic control of production.

The first point is generally included in general course of lectures on machine building and in lectures on industrial safety. The second point sometimes is treated during specialistic lectures, e.g. on organization of production or on monitoring of metallurgical processes.

Electronics and telecommunications faculties

General information are required on:
a) acoustics of speech,
b) room acoustics, especially of microphone studies,
c) electroacoustic transducers,
d) acoustic measurments,
e) applications of ultrasounds.

Because of vast subject matter frequently there are given separate lectures on acoustics and electroacoustics.

7. Facultative lectures

Within the program of individual studies there are sometimes held topics suited to personal interests of students, not closely related to the course of studies. As it is known from experience the main interests are:

a) subjects related to computer techniques - control of voice operated computers,
b) musical acoustics

c) the latest applications of acoustics - hypersounds.

However, such interested groups of students are very small, too small to arrange special lectures and new initiatives are limited to laboratory works carried out at the margin of main course of studies.

8. Theses of diploma

Education of acousticians through lectures, classes and laboratories is completed with theses of diploma. Those theses have widely diversified character:
a) theoretical theses generally cover determination of complex acoustic fields and vibrating systems,
b) laboratory theses are orientated mainly on investigations of propagation of waves in solids, liquids and gases, and testing of materials and systems being used in ultrasonic transducers,
c) constructional theses cover generally construction of measuring acoustic apparatus, special apparatus for medicine and industry and construction of models of the microphones and loudspeakers,
d) design theses for the most part cover acoustic designs of rooms, sound amplification systems and insulation systems of building structures.

Theses are prepared during last semester of studies, completed with final examination.

9. Doctor's Theses

The chances for finding the subjects for doctor's theses are different in various branches of acoustics. In some economically and technically important branches, where relatively many experts are required (e.g. manufacture of sound amplification systems) chances for getting doctor's thesis are minimal. On the other hand, in branches of rather very narrow range of current applications (e.g. quanta acoustics) there exists a number of problems waiting for explanation and in this connection eligible for doctor's theses. In this situation requirement of getting done the doctor's thesis in other to get right for giving lectures on acoustics and electroacoustics at university should be treated in a flexible way. The additional difficulty is the fact, that for doctor's thesis in physical or psychological acoustics it is hard to find at candidate's own department the professor conferring a degree.

10. Teaching of Acoustics in High Shools of Developing Countries

The question arises if lectures on acoustics should have the same contents and volume all round the world or there exists the need for introduction some features of lectures, specific for a given region. Undoubtedly information on fundations of acoustics, such as propagation of waves, physiology of hearing and electroacoustic transducers are of universal nature and should be lectured irrespective of latitude. On the other hand problems related to applications of acoustics should be presented with proper attention paid to local conditions.

The most essential factor is forecasted place of work of future graduate. In highly developed countries he will be employed mainly in industrial research and development laboratories and sophisticated production. It does not mean however that in developing countries it is useless to teach some limited number of industrial engineers. It is needed both for technical and political reasons. It is essential to have at disposal native specialistics capable of:
a) authoritative assessment of imported technical equipment,
b) in the future creation of domestic electroacoustic industry,
c) initiating of introduction of ultrasonic methods in the industry, medicine and fishery.

However majority of acousticians - engineers in developing countries will be working operating electroacoustic apparatus (radio, TV, echo ranging equipment), sound aplification systems and construction of acoustic objects (rooms, stadiums).

When giving in developing countries lectures on applications of acoustics one should take into account specific features of those countries, especially with respect to building engineering and noise control. For instance standard values of permissible noise levels stimulated in European countries are not suitable in southern countries because of different way of life and weather conditions. The same refers to insulation of buildings.

The lectures of high schools of developing countries apart from didactic duties are under an obligation to carry out reserch. Because of spareness of personnal and apparatus in majority of developing countries research can be carried out in a limited field. However even such limited work could be of importance for world progress in acoustics, especially when aimed at specific problems of the region. At the first place one should put here investigations on acoustic structure of languages of a given region, as nobody else would in position to do it. Another interesting subject would be investigations of specific conditions of propagation of sound waves, e.g. in tropical water regions. Of course volume of lectures on acoustics must be adapted to general curriculum and personal possibilities of lecturers. In majority of universities of developing countries specialists in acoustics are missing. In this connection obligation of teaching foundations of acoustics is put usually on physicists. It is also much-desired that lectures on electronics should have minimum information on acoustics, to be included in their lectures, and the same applies to lecturers on general building.

ACADEMIC PROGRAMS IN ACOUSTICS
AT THE TECHNICAL UNIVERSITY OF WROCŁAW

WOJCIECH MAJEWSKI AND JANUSZ ZALEWSKI

Institute of Telecommunication and Acoustics
Technical University of Wrocław
Wrocław, Poland

Abstract - The problems connected with academic programs in acoustics, especially those concerning the education of acoustical engineers, are discussed. A graduate program in acoustics leading to the masters degree in electronics with emphasis on telecommunication acoustics and industrial acoustics is presented.

1. INTRODUCTION

A major objective of the academic program in acoustics is to educate specialists fully qualified to various kinds of activities in the area of acoustics and its applications. It sounds very simple, but to work out the academic program in acoustics is a difficult task due to an interdisciplinary character of acoustics. This interdisciplinary character requires to include in the program the knowledge of many diverse scientific disciplines. This is especially true as far as academic programs for acoustical engineers are concerned. The students have to acquire a large range of basic acoustical knowledge concerning foundations of acoustics, electroacoustics, speech and hearing, musical acoustics, ultrasounds, infrasounds, underwateracoustics, acoustic signal processing, etc., and a large range of applied technical knowledge based on these foundations, concerning electroacoustic systems, sound recording and reproduction, noise abatement, building acoustics, man-machine communication by voice and many others. It is not possible to include such range of education in any existing system of education at the technical universities in Poland. The optimal solution would be to create a Faculty of Acoustics at one of the technical universities, what however is not possible at present. Thus, the only way of educating acoustical engineers is to do it within the section of acoustics at selected faculties of technical universities.

In the present paper we shall present a model of education of acoustical engineers developed at the Technical University of Wrocław during several years of trials, changes and improvements. The education of specialists in technical acoustics is being performed at the Faculty of Electronics for over thirty years. The Faculty of Electronics provides a proper basic knowledge of mathematics, physics, circuit theory, electronic networks, electronic measurements, digital technique and computers, etc. The selection of Electronics Faculty to place within it the section of acoustics results from two basic reasons:
- the program of mathematics is the largest, the level of mathematics the highest and satisfactory for the purposes of teaching acoustics,
- the electrical engineering courses provide sufficient background for the courses in electroacoustic transducers, electroacoustic devices and systems, ultrasound technique, sound recording and reproduction, man-machine communication and many others.

While working out and implementing subsequent versions of academic programs in acoustics at the Technical University of Wrocław, the needs of national economy were taken into account what influenced the profiles of graduates. The present needs require the graduates of the acoustics section of our university to be specialized in telecommunication acoustics and industrial acoustics. The graduate should be prepared to a professional activity and development of his knowledge in the range of conversion, transmission and reproduction of acoustic signals and evaluation of quality of these processes, as well as in the range of designing electroacoustic transducers, devices, systems and measuring equipment. He should also be fully qualified to work in the area of noise and vibration control, ultrasound technique and speech communication.

The two already described basic necessities, i.e. the necessity to place the acoustics section in one of the traditional faculties at technical universities and the necessity to form the profiles of the graduates according to the diverse needs of national economy, forced our faculty members to look for many compromises while working out the academic programs in acoustics. On the other hand, the continuous improvements of the programs based on the acquired experiences permitted our faculty members to work out a satisfactory model of education in acoustical engineering, the the details of which are presented below.

2. CHARACTERISTIC OF ACADEMIC PROGRAM IN ACOUSTICS

The Institute of Telecommunication and Acoustics at the Technical University of Wrocław offers graduate programs in acoustics at the Faculty of Electronics leading to the masters degree in electronics with emphasis on telecommunication acoustics and industrial acoustics.

The studies at the Faculty of Electronics last five years /10 semesters/ and occupy 4890 hours of classes, 2145 of which concerns the specialization in the acoustics section. Since each semester contains 15 weeks of classes, the number of 2145 hours assigned to specialization corresponds to 143 hours of classes on a weekly basis /143 x 15 = 2145/. The distribution of these 143 hours in particular semesters is presented in Fig.1 in a form of weekly diagram of courses. In this figure each course is described by its title and a

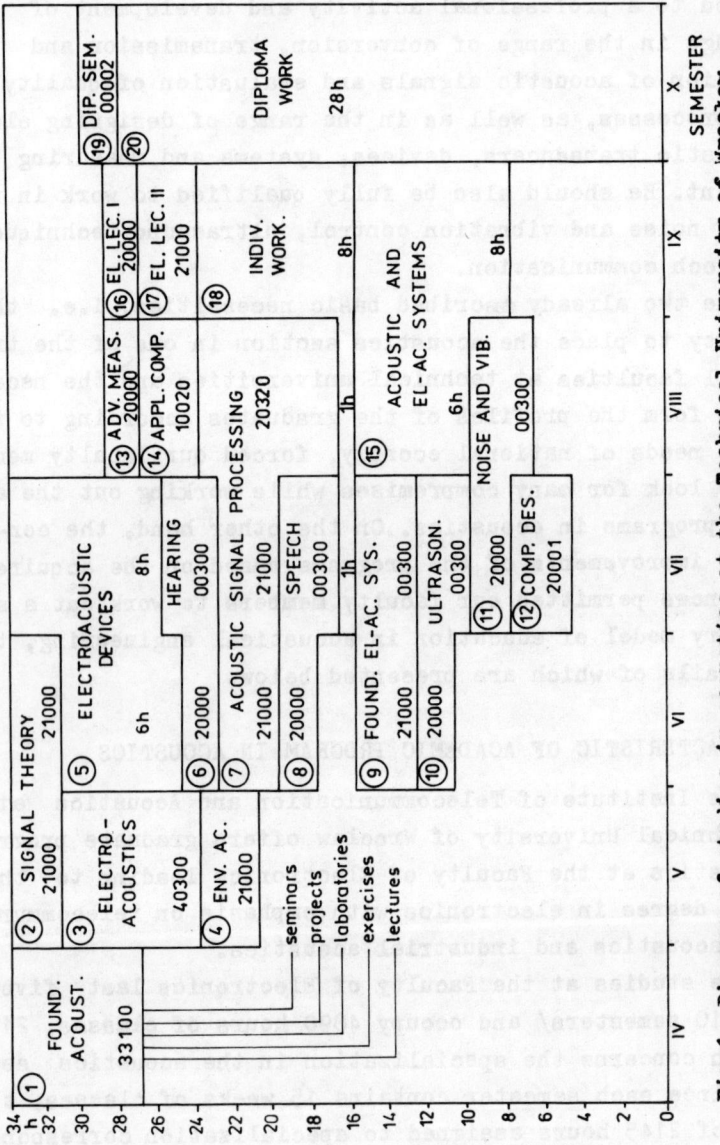

Fig.1. Diagram of studies in acoustics at the Technical University of Wrocław.

number consisting of five digits. The first digit presents the amount of lecture hours per week in given semester, the second - the amount of exercises, the third - laboratories, the fourth - projects and the last - seminars. The courses that are run on a basis of individual work are described by means of a total number of hours per week. The empty space below the described courses presents the number of hours occupied by courses common for all sections of electronics, i.e. the courses in mathematics, physics, circuit theory, as well as humanities, foreign languages, physical education, etc. These type of courses occupy also the first three semesters that are not shown in the graph.

Basic functions in the educational program for the section of acoustics at the Faculty of Electronics are performed by the following subjects:

- **Foundations of acoustics**

  It is a one-semester course /33100 - sem.IV/ initiating the education in the section of acoustics. The purpose of this course is to get the students acquainted with vibration theory, sound waves and quantities characterizing the wave and the medium, acoustic radiation, acoustic structures, etc.

- **Electroacoustics**

  It is also a one-semester course /40300 - sem.V/. The purpose of this course is to acquaint students with the bases of converting electrical energy into acoustic one and vice versa, with electroacoustic transducers /microphones, earphones and loudspeakers/ and measurements of their properties.

- **Acoustic signal processing**

  It is a three-semesters course /21000 - sem.VI, 21000 - sem.VII, 20320 - sem.VIII/ encompassing the theoretical bases of digital analysis and processing of acoustic signals in digital systems. A prerequisitive knowledge is provided by Signal theory course /21000 - sem.V and VI/.

A detailed knowledge of particular areas of acoustics necessary to form the required profile of graduates is provided by the following courses:
- Environmental acoustics /21000 - sem.V/,
- Speech /20000 - sem.VI, 00300 - sem.VII/,
- Hearing /20000 - sem.VI, 00300 - sem.VII/,
- Ultrasounds /20000 - sem.VI, 00300 - sem.VII/,
- Noise and vibration /20000 - sem.VII, 00300 - sem.VIII/,
- Foundations of electroacoustic systems /21000 - sem.VI, 00300 - sem.VII/.

The methodological knowledge connected with the application of computers and processors in acoustics is provided by the courses:
- Computer aided designing in acoustics /20001 - sem.VII/,
- Advanced measurement methods /20000 - sem.VIII/,
- Applications of computers in acoustics /10020 - sem.VIII/.

An important function in profiling the acoustical engineers is performed by an individualized problematically oriented education within the range of two following subjects:
- **Electroacoustic devices**

Semester	Selected lectures	Problems	Open lab.	Seminar
VI	4h	2h	-	-
VII	2h	4h	1h	1h

This subject embraces the issues of analog and digital devices of sound channels, devices for speech processing and transmission, ultrasound devices and equipment for acoustic and electroacoustic measurements. Each student receives from his superviser a problem to be solved during two semesters, making use of specialized lectures, literature and experiments in open laboratory. Advised by his superviser the student selects in each semester two specialized lectures closely connected with the problem to be solved. The open laboratory permits the student to carry out ex-

periments concerning the problem, while the seminar permits to exchange information between the students and to cooperate in solving the problems.
- Acoustic and electroacoustic systems

Semester	Selected lectures	Problems	Open lab.	Seminar
VIII	4h	2h	-	-
IX	2h	4h	1h	1h

This subject embraces the issues of analog and digital systems of sound signals, ultrasound systems, systems for vibration and noise abatement, systems for man-man and man-machine communication by voice. The rules of realization are identical as in case of Electroacoustic devices.

Two elective monographic courses /21000 and 20000 - sem.IX/ permit the students to enlarge and deepen their knowledge according to their interests, actual trends in science and technology or the needs of national economy. The elective lectures may have different character with emphasis on theoretical aspects of acoustics, methodological issues or applications.

The diploma thesis is worked basically in sem. X /28h/, but a preliminary work toward it is performed in preceding semesters /sem.VII - 1h, sem.VIII - 1h, sem.IX - 8h/. The bases and presumptions of diploma thesis and the final results of it are presented during the Diploma seminar /00002 - sem.X/.

The area of knowledge and the field of professional abilities of the graduate is determined by the range of obligatory and elective courses, by the topics of his individualized problematically oriented education and by his individual work and research training in the laboratories.

While working out the presented academic program in acoustics at the Technical University of Wrocław our faculty members had two following objectives in mind:
- to educate a versatile graduate who could be employed

in a variety of places: industrial plants, research and development institutes, design bureaus, environmental protection units and technical units connected with telecommunication, radio and television;
- to educate a self-dependent and active graduate, specializing him in an intellectual and investigative climate where interdisciplinary considerations tend to develop as a natural part of the academic and research training that he receives.

We hope that both objectives have been achieved.

ACKNOWLEDGEMENT

The authors of the present paper acknowledge a substantial contribution of Professor Janusz Renowski and his co-workers from the Acoustics Unit of our Institute in working out the presented academic program.

Acoustical foundations of training medical students
and postgraduates

Millner, R.; H. J. Hein
Institute of Applied Biophysics, Faculty of Medicine
Martin-Luther-University Halle, GDR

In the education of medical students we teach Medical
Physics and Biophysics. In research work we deal with ultrasound diagnostics and its physical fundamentals.
In this topic the acoustics and ultrasound physics are
included. In the past, the acoustics in medicine was concentrated on the investigations of the voice and hearing
organ to understand the underlying physiological processes.
That was the starting point where the field of physiological
acoustics is developed from. The phenomenological description produced limited possibilities to understand the natural
procedures by diagnostics and therapy. To day modern medicine
has undertaken efforts to explore the interactions between
mechano-acoustical vibrations and the human body for diagnostics and therapy. Hence two completely different starting
points in physics emerged.

In classical physics knowledge was obtained essentially by
phenomenological descriptions linked to morphological observations. However, the introduction of vibrations into the
human body caused interactions. This application brought
about completely new realizations.

The acoustically evoked potentials are striking examples.
The intention of our remarks is to outline that it is not
sufficiently enough to teach the medical students phenomenological acoustics, only. The student must certainly be able
to describe the phenomena of acoustical procedures, but the
students have to be able to learn interfering in unknown
acoustical procedures by experimental methods, too. This
approach is obviousely evident when explaining structures of
biological tissues by ultrasound effects.
Investigations like this - are found in medicine and technology e. g. non-destructive testing and opto-acoustics.

Subsequently, we are to outline that acoustics exceeds physics, even when considering the applications in medicine and biology. In this field, acoustics frequently has a mediating role between diagnostics and therapy as well as between knowledge in theoretical and experimental medicine.
This item is considered when educating students of medicine. Acoustics is teached in the basic course. It is integrated into medical physics.
The lessons are to give an overview of producing stimulated effects, i. e. the explanation of various inter-actions between mechanical vibrations (till to very high frequencies) and biological tissues.
Knowledge obtained by the students within the basic course enables them to understand clinical applications. On the other hand, it is intended to motivate the students by demonstrating these clinical applications.
Additionally, the students have the chance to become familiar with technical elements of acoustics in a course of practical exercises. We also deliver information on a high level for physicians and engineers working in clinical practice or research. This includes the students' work. Their elaborates are related to one topic only and belong to some selected problems of research.
Table 1 depicts the different steps in educating medical students in acoustics and the obtained results. On the left side, below, the final result is a specialized otolaryngologist and on the right side, below, a physician well trained in ultrasound diagnostics.
They frequently work in groups and contribute to solve a more complex problem (e. g. ultrasound diagnostics).
For the highly specialized application in clinical medicine it is required to continue the training in postgraduate courses. This postgraduate education is included in the well defined specialist's training for physicians. Within this period, the trainee has to elaborate a thesis for graduation.

Table 2 gives an overview on the postgraduate training for physicists and engineers (left) and for physicians (right). Due to our experiences, none of the older physicians who did

CLASSIFICATION OF ACOUSTICS

FACULTY OF MEDICINE UNIV. HALLE - GDR / 250 STUD.

1. STAGE

BASIS COURSE MED. PHYS.
LECTURE AND SEMINAR
1./2. TERM FOR ALL STUD.

2. STAGE

PRACTICAL EXERCISES IN ACOUSTICS
3. TERM FOR ALL STUD.

3. STAGE

SPEC. TRAINING US - METHODS		
4. TERM FOR 10 - 20 STUD.		
SEMINAR THEORY	DEMONSTRATION	EXERCISES PRACT.

4. STAGE

PHYSIOLOGICAL ACOUSTICS		CLINICAL - US APPLICATION
	RESEARCH WITH GRADUATION	
CLIN. ACOUST.	BIOPHYS. ULTRASOUND	- INSTRUCTION
- EXERCISES	WITH CLINICAL APPLICATION	- DEMONSTRATION
- INSTRUCTION	DIPLOMA	
RESEARCH GRADUATION	DISSERTATION A + B	
OTOLARYNGOLOGIST		SPEC. PHYSICIAN WITH SPEC. KNOWLEDGE IN ULTRASOUND

Table 1: Overview about several steps for teaching medical students in acoustics and ultrasound physics

POSTGRADUATE TRAINING IN ULTRASOUND PHYSICS AND DIAGNOSTICS

PHYSICIST A. BIOMED. ENGN.	PHYSICIAN
INTENTION: SPECIALIZED IN MED. PHYS. OR BIOMED. ENGNG.	DOCTOR FOR OBSTETRICS, CARDIOLOGY (SPECIALIZED IN APPLICATION OF SONOGRAPHY)
DURATION: 4 YEARS	DURATION: 3 - 4 YEARS

TRAINING COURSES:
 15 - 20 % FOR ULTRASOUND IN
 - METHOD. ULTRASOUND
 - BIOPHYSICS "
 - PHYSIOLOGY
 - BIOMED. ENGN.
	SPEC. COURSES
	 - BIOPHYSICAL PHENOMENA
	 - EVALUATION OF
	   . A - B - SCAN PICTURES
	   . MATH. TREATMENT
	   . DOPPLER
	 - PRACTICAL EXERCISES

SPEC. CERTIFICATION	IN PREPARATION: CERTIFICATION FOR US-OBSTETRICS ECHOCARDIOLOGY

Table 2: Overview on postgraduate education

not specialize in ultrasound diagnostics is able to interprete a B-scan picture or a functional Doppler signal due to his knowledge of morphology and anatomy.
Conventional anatomy was not related to ultrasound diagnostics (not even to computerized tomography) since the M. D. obtains the information in ultrasound diagnostics from a body section and not from a volume element.
We tried to inspire our teaching staff in anatomy to consider this fact. The next lines will point out some examples evidencing the relevance of acoustics for the basic training of the students and for the postgraduate training.

There are certainly only a limited number of students who are interested in an education for "ultrasound specialist". Yet, ultrasound diagnostics is the first step in diagnozing a disease in several clinical disciplines.
In some cases, the scanning picture or the Doppler signal represent well-defined information:

1. In surgery of
   cystic processes
   processes of the thoracal wall
   vessel anomalies
   heart disease

2. To monitor diagnostic interventions (e. g. visualizing the position of a biopsy needle in the humans's body)

3. Ultrasound is able to provide a reliable basis for further diagnostics. i. e. CT or NMR.

That is the reason why a physician should have fundamental knowledge in ultrasound.
From our point of view this situation is equally the same all over the world. At present, the medical students are not so strongly interested in physics.
Therefore it is achieved to give examples for the superiority of this method compared to others. When stimulating the student's interest in this way, we can continuously proceed to derive the laws of physics by modelling. We think this to be

the best chance to get an insight into the character of the effects and of the underlying information processes.

One example is the processing of B-scan pictures. They are at least interpreted as a primary physical information, transferred to clinical morphology, structure and function.

When comparing the methodical approach in radiology it can be said that an interpretation in this discipline is much more easier for one physical effect only, i. e. x-ray absorption, has to be interpreted.

In ultrasound it is the opposite. One has to consider reflection, absorption, back-scattering, sound velocity, diffraction, and the echo's structure.

In detail it means that the B-scan technique has to be explained as a sectioning technique. The B-scan picture, e. g., is generated by the transfer from an echogram to a distribution of pixels (gray-scale pixels). The matter is to be interpreted in time and in frequency. Additionally, picture cosmetics, scanning processes, and the loss of information have to be illustrated.

Furthermore, the performance of the technical device, the propagation of the sound waves as well as the interaction with organs and structures of the human body should become plausible.

These lessons are very helpful for understanding the complexity of information used by the physician to filter his diagnostic essentials.

The lessons start with the basic training course and they are supported by clinical demonstrations.

What is the most efficacious way of teaching medical students in ultrasound as a biophysical entity?

Probably there are various similarities in education when comparing our approach with foreign countries. One striking difference is the fact that the medical students in the GDR have to deliver a diploma thesis before they can obtain their M. D. graduation which certainly also requires (another) thesis. This diploma thesis should not exceed 20 - 40 pages dealing with one topic of experimental, theoretical or practical medicine. For preparation they join any institute or clinic to

COOPERATION BETWEEN MEDICAL AND PHYSICAL STUDENTS IN EDUCATION AND RESEARCH

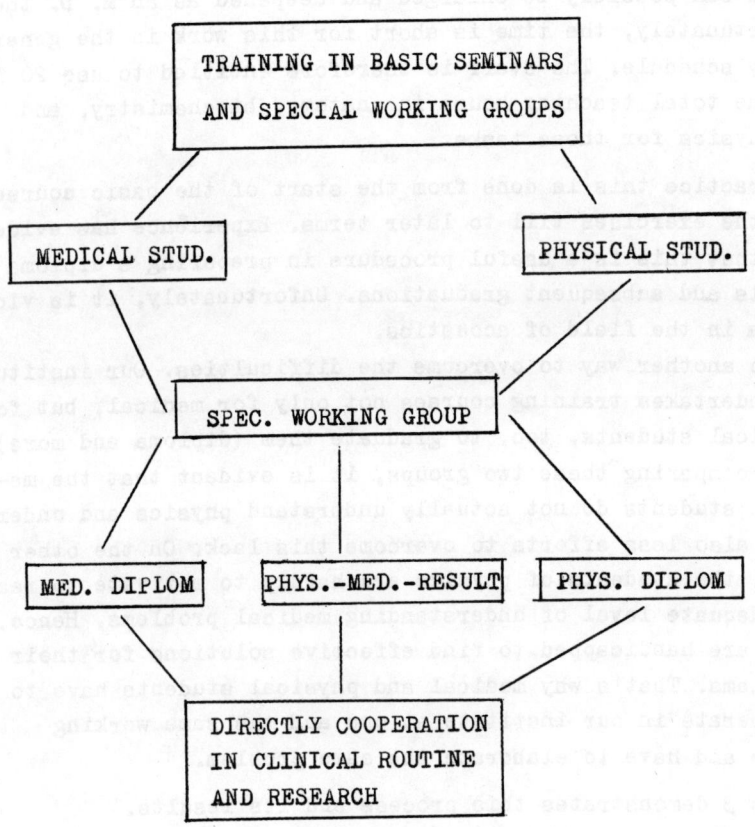

Table 3: Principe of cooperation between medical and physical students and the results

become familiar with their special research interests. One prerequisite for working successfully is that each student has to make the experiences of his own in the field of his interest. That is a great support for the diploma thesis which can possibly be enlarged and deepened as an M. D. thesis. Unfortunately, the time is short for this work in the general study schedule. The staff is therefore entitled to use 20 % of the total teaching hours in anatomy, biochemistry, and biophysics for those tasks.

In practice this is done from the start of the basic course and the exercises till to later terms. Experience has evidenced that this is a useful procedure in preparing a diploma thesis and subsequent graduations. Unfortunately, it is vice versa in the field of acoustics.

We go another way to overcome the difficulties. Our institute undertakes training courses not only for medical, but for physical students, too, to graduate them (diploma and more). When comparing these two groups, it is evident that the medical students do not actually understand physics and undertake also less efforts to overcome this lack. On the other side, the students of physics are hardly to motivate to reach an adequate level of understanding medical problems. Hence, they are handicapped to find effective solutions for their problems. That's why medical and physical students have to co-operate in our institute at one and the same working place and have to elaborate the same problem.

Table 3 demonstrates this process and its results.
In this way we can be more successful in a shorter period of time. There is also a competiting behaviour observable within the groups. The proof of the effectivity of this co-operation are several publications and reports on biophysical effects and their applications, respectively. The following topics are included:

- Ultrasound spectroscopy
  . Measurements of damping, back-scattering in pathological scrotal tissue.
  . The same approach, but applied to liver tissue.

- Investigations of B-scan picture analyses for the characte-

rization of tissues
  - Application in placental tissue
- Improvement of B-scan methods
- Correlation analysis of reflected ultrasound signals
  - The assessment of Doppler signals to characterize blood flow and the vessel wall
  - Evaluation of ultrasound effects at low intensities.

This approach results in a remarkable effect:
The student feels responsible, is therefore not only interested in an academic gain, but wants to see the outcome of his activities in practice.
This is related in all cases to a stronger engagement of the students because it is obviously attractive for them to report on their results at a scientific meeting. Every year a scientific competition takes place at our university for students of different branches.
Some of them were also awarded prizes for their work on sonography. For there is a growing demand of skilled personnel in the application of ultrasound, an education is rather attractive. When they were educated at a biophysical institute, it is easier for the students to find an attractive position when they leave.
Sometime it becomes necessary from our point of view to involve engineers or mathematicians in the work of a larger group - depending on the problem under investigation.
In this respect we have to consider that the role of computer science was steadily increased in acoustics.

In essence, why is it useful to educate students of medicine in acoustics and ultrasound physics in particular?

1. The share of biomedical techniques which are ultrasound-based is increasing.
2. Biological investigations of the inter-actions between mechanical energy and human tissue opens a completely new realm in clinical basic research.

How to teach most successful?

1. We should try to give the students an adequate motivation in the basic courses.

2. It proved to be fruitful to involve various disciplines in one working group (e. g. physicists, engineers, and mathematicians).

INTEREST IN MUSIC AND PROGRESS IN ACOUSTICS

Andrzej Rakowski
Academy of Music in Warsaw
Department of Music Acoustics
Okólnik 2, 00-368 Warszawa, Poland

ABSTRACT

The historical links between music and acoustics have been shown to be an important factor both in the development of modern theory of music and in the progress in the science of hearing. The influence of musical experience was particularly strong in investigating pitch and timbre of sound, in studying auditory memory, and in evaluating the effect of a sensory dissonance.

1. INTRODUCTION

Connections between music and acoustics have a very important tradition and long history. They begin in antiquity and continue through Middle Ages until very recent times. For centuries music has been giving inspiration for scientific research in the area of sound and hearing. The Greek word "akoustikós" - "referred to hearing" was at the origin of the name for the whole discipline, and "to hear" meant very often "to hear music".

Some 40 - 30 years ago it looked like those links with hearing were getting more and more loose. Acoustics reached a new stage where its interest was centered at the realms outside the possible perception through human auditory system; ultrasonics, physical acoustics, infrasounds and others attracted more attention. Musical acoustics was considered something awfully old-fashioned, a nineteenth-century antiquity which had been totally explored by Helmholtz.

The above situation changed when a group of tallented physicists interested in music started investigating the physics of musical instruments. Bacus, Benade, Cremer, Hutchins, Lottermoser, Meyer, Rossing and many others significantly pushed forward the science of complex vibrating systems employed to produce musical sounds.

Another impact for progress in acoustics through interest in musical phenomena was obtained in the field of auditory perception. There was also a continuous demand to improve acoustic technology for electroacoustic transmission of music.

These facts, as well as many other phenomena, show that the links between music and acoustics continue to be an important feature for both domains of human activity. In the present paper we shall confine ourselves to the links between music and the part of acoustics that derives directly from the original term "akoustikós": The science of hearing.

2. MUSIC, ACOUSTICS AND MATHEMATICS

The Greek philosopher of 6th century B.C., Pythagoras, is often quoted as the first to establish mathematical relations between musical tones. Using a simple device with one string, called monochord, he found that musical intervals of an octave and of a pure fifth correspond to simple ratios of the string length 1 : 2 and 2 : 3. This observation laid foundations of very complex and numerous mathematical theories of musical scales. The most important of them, the Pythagorean scale, derives all relations between tone frequencies from the above mentioned intervals of the pure fifth and the octave.

Another important scale was based upon the discovery of Didymus /1 c. B.C./ that the relation of string lengths 4 : 5 corresponds to the particularly well-sounding major third. All these discoveries and theoretical speculations concerning musical scales were based on very limited knowledge of underlying acoustic phenomena. General tendency derived from Pythagorean philosophy was to minimize the empirical, psychoacoustic factors in music and to overestimate the magic of numbers. The followers of Pythagoras considered natural numbers as a criterion of truth and a key for understanding the phenomena of nature. The same idea was also expressed by Plato /4 c. B.C./, for whom "irrational work could not be called art". Plato derided the efforts of those musicians who tried to establish musical scales through listening to the chords rather than through theoretical calculations, and who "believed in their ears instead of believing in their mind".

The Pythagorean philosophy of numbers strongly affected the beliefs of late antiquity and Middle Ages. St.Augustin wrote that "the science of music is the science of numbers". Later, music was considered a part of mathematics and it was not until the times of Helmholtz that the problem of musical scales was treated in a purely acoustic way.

Helmholtz /1863/ was the first to draw attention to the consequences of the harmonic structure of sound. This was particularly important in explaining the phenomenon of musical consonance. It has been long known that simultaneously sounding tones in intervalic relations of an octave, a pure fifth or a pure fourth create the sensation of smoothness called consonance. Opposite sensation, a dissonance, appeared for such intervals as a minor of major second and a minor of major seventh. The so-far-accepted explanation was taken directly from Pythagorean philosophy of numbers. In the Pythagorean scale frequency ratios for octaves and fifths could be expressed as simple relations of small natural numbers 2 : 1 and 3 : 2, while similar relations for seconds and sevenths were much more complicated. Minor and major seconds were expressed as ratios 256 : 243 and 9 : 8; minor and major seventh as ratios 16 : 9 and 243 : 128. In terms of Pythagorean philosophy more complicated meant "less perfect", so musical intervals of the second and the seventh were counted dissonances.

The explanation offered by Helmholtz was deprived of any mythological context. Dissonance, in his view, occoured when notes differed from the simple ratios, because the harmonic spectral components then produced fluctuations in intensity /or beats/ through their interaction. The resulting sensation of roughness was perceived as dissonance. Interestingly, Plomp and Levelt /1965/ describing some earlier investigations and making their own experiments found that traditional division between consonances and dissonances does not apply for musical intervals composed of simultaneously sounding pure tones. Pairs of such tones were described as discordant when they stimulated overlapping regions of the basilar membrane /i.e. when they were separated by less than a "critical band"/. Pairs of pure tones stimulating nonoverlapping loci were judged to be consonant regardless of whether or not their frequencies were in the ratios of small whole numbers. The interval of a major seventh elicited exactly the same sensation of smoothness and consonance as that of an octave. What a bitter disappointment for the still-active defenders of the Pythagorean philosophy in music!

Another interesting example of the interplay between music, acoustics and mathematics is offered by the case of electronic music in its early stage of development. An aesthetic philosophy and musical technology developed in the early fifties by the group of young German composers and physicists /W.Meyer-Appler, H.Eimert, E.Stockhausen and soem others grouped around the Electronic Music Studio, Köln/ followed the trends iniciated by Schönberg in his 12-tone music system. Schönberg replaced traditional harmony and tonal relations in music by introducing to each composition an autonomic organisation based on arbitrarily chosen "series" of twelve intervals. Eimert and others, fascinated by the possibility to "compose" electronically spectra of individual tones, introduced mathematically designed series of relations between partial tone frequencies of complex tones as unifying factor of a given composition. The same series of numbers could have been used to organize time relations and intervalic structure of melody. By those means electronic music composers strived to achieve a "total serialism" in music.

This time again, a conflict between music and mathematics broke out in the field of sensory perception. Mathematically designed series of partial tone frequencies or sequences of time intervals were neither distinguished from each other, nor remembered or recognized. Most of the carefully designed musical constructions could not be perceived and so remained unnoticed by the listeners. The break-down of the idea of total serializm in music was a disappointment to some avanguard composers and philosophers but at the same time it gave an important impact on the acoustic and psychological research concerning the limitations of auditory perception.

3. MUSIC AND PSYCHOACOUSTICS

The abilities for the perception of sound have been frequently investigated in relation to music. There is a striking analogy between the elements of a musical composition and the physical or psychological attributes of sound. In a musical composition following elements can be separately analysed: musical dynamics, melody, harmony and instrumentation, rhythm and tempo. Corresponding psychological attributes of a single tone are: loudness, pitch, timbre and perceived duration; their physical correlates /though remaining in complex relations with the former ones/ are: intensity, frequency, spectrum and physical duration.

The objective of psychoacoustics is to find out the existing relations between psychological and physical parameters of sound; these relations appeared to be particularly complicated in the domain of pitch. The so-called "theory of hearing", mostly concerning explanation of the way in which pitch is perceived, developed under strong influence of the experimental evidence taken from music.

The first scientific theory of hearing was formulated by Helmholtz /1863/. He offered the resonance theory, in its early, premature stage called "Theory of the harp in the ear". Musical analogy of the observed phenomena was obvious, because the transverse fibers of the basilar membrane in the cochlea were compared to resonating strings of a musical instrument.

The Helmholtz's resonance theory was later essentially changed and improved, particularly by Békésy /1942/ and Zwislocki /1950/, however its main assumption saying that the perception of pitch was associated with the place of maximum stimulation of nerve cells in the cochlea has remained unchanged until today. Its main supplement and to some degree contradiction was presented by Schouten /1940/. Again it was induced by observations of a musical nature. Schouten noticed that the formation of pitch in the sound of bells is not necessarily connected with the existence of a physical component of corresponding frequency. He called that phenomenon "pitch of the residue" and iniciated a long·series of investigations by many authors who showed that in some conditions pitch is dependent on frequency of oscillations /time between succesive beats/

and not on the place of stimulation of Corti organ cells.

The development of psychoacoustics and, particularly, the growth of music-induced research on pitch perception were particularly strong within the last two decades. One of the main contributors in this field was Reinier Plomp with a series of works started with the paper based on his doctoral dissertation /1967/ and summed up with the book "Aspects of tone sensation" /1976/. Modern view on the theory of hearing was also presented in three different ways by Houtsma and Goldstein /1972/, Wightman /1973/ and Terhardt /1974/. The first and the latter work particularly strongly referred to musical phenomena.

Apart of its influence on the theory of pitch perception musical approach was especially valuable in reference to the auditory assessment of timbre. The works which should be mentioned here are those concerning auditory analysis /Plomp 1964/, multidimensional scaling of musical timbre /Grey 1977/, the use of Long Term Average Spectra in musical research /Jansson and Sundberg 1975/76/ and the specification of timbre in singing /Sundberg 1977/.

One of the important conclusions based upon investigating the perception of music was that perceiving both music and natural languages might be taken as parts of the same general process of human auditory communication. Similar tendency has been observed in both domains concerning generative approach /Lehrdal and Jackendoff 1983/ categorical perception /Burns and Ward 1978/ and phoneme-formation phenomena /Rakowski and Miśkiewicz 1985/. The above mentioned similarity appeared to be particularly useful in investigating the role of short-term and long-term memory in the formation of auditory codes. Such investigations were relatively simpler in the domain of unidimentionally changing sensations, such as musical pitch /Rakowski and Morawska 1987/; still they brought some more general conclusions.

Of particular value in studying the ultimate power of auditory functions through psychoacoustic experiments appears using selected, highly trained musical subjects. Their life-long musical training may be taken for an extention of the routine latoratory practice, which leads to particularly accurate results in measuring differential thresholds of hearing /Nordmark 1968, Rakowski 1971/. It should be noted that applying most accurate methods of measurement and using most sensitive, highly trained subjects is a fully justified way of conducting psychoacoustic experiments in which maximum reduction of noise is desirable /Stevens 1961/.

Another important feature of musically-trained subjects is the existance in their long-term memory of highly overlearned standards for timbres of musical instruments, musical intervals and, in some cases, also the standards for absolute pitch of tones of the musical scale. This fact creates an unique possibility of investigating the properties of auditory memory which, though measured using specific musical material, nevertheless may reveal some general properties of hearing /Rakowski and Hirsh 1980, Rakowski 1983/.

4. CONCLUSIONS

In conclusion it may be said that the links between music and acoustics appeared stumulating in the development of both domains. Acoustically-based explanation of some phenomena of musical hearing was important both for the theory of musical scales and for setting the rules of musical composition. Acoustic design for new musical instruments, improvement of the acoustics of concert halls and, above all, the powerful influence of newly developed electronic and digital acoustic technology created an entirely new situation for the whole area of musical activity.

On the other hand, interest in music was an important factor inducing acoustic research in many fields. In the present paper only some problems concerning the research on hearing were mentioned, though physical investigations of musical instruments were certainly not less important. Furthermore musical requirements were of great importance in the development of room acoustics, sound recording, and sound-synthesis technology. The interest in music appeared to be a powerful factor inducing progress in acoustics.

REFERENCES

Békésy G. von /1942/. Über die Schwingungen den Schneckentrennwand beim Präparat und Ohrenmodell. Akust.Zeits. 7, 173-186.

Burns,E.M., Ward,W.D. /1978/. Categorical perception - phenomenon or epiphenomenon: Evidence from experiments in the perception of melodic musical intervals. J.Acoust.Soc.Am. 63, 456-468.

Grey,J.M. /1977/. Multidimensional perceptual scaling of musical timbres. J.Acoust.Soc.Am. 61, 1270-1277.

Helmholtz,H. von /1863/. Die Lehre von den Tonempfindungen als physiologische Grundlage für die Theorie der Musik. Verlag F.Vieweg-Sohn, Braunschweig.

Houtsma,A.J.M., Goldstein,J.L. /1972/. The central origin of the pitch of complex tones: Evidence from musical interval recognition. J.Acoust.Soc.Am. 51, 520-529.

Jansson,E.V. and Sundberg,J. /1975/1976/. Long-Time-Average spectra applied to analysis of music. Acustica 34, 15-19 and 269-274.

Lerdahl,F., Jackendoff,R. /1983/. A Generative Theory of Tonal Music. The MIT Press, Boston.

Nordmark,J.O. /1968/. Mechanisms of frequency discrimination. J.Acoust. Soc.Am. 44, 1533-1540.

Plomp,R. /1964/. The ear as frequency analyser. J.Acoust.Soc.Am. 36, 1628-1636.

Plomp,R., Levelt,W.J.M. /1965/. Tonal consonance and critical bandwidth. J.Acoust.Soc.Am. 38, 548-560.

Plomp,R. /1967/. Pitch of complex tones. J.Acoust.Soc.Am. 41, 1926-1933.

Plomp,R. /1976/. Aspects of Tone Sensation. Academic Press, New York.

Rakowski,A. /1971/. Pitch discrimination at the threshold of hearing. Rep. 7th Internat.Congr.Acoustics, Budapest, 20-H-6.

Rakowski,A. /1983/. Pitch discrimination and musical interval recognition in backward masking. In: R.Klinke and R.Hartman /Eds./, Hearing - Psychological Bases and Psychophysics, 315-320, Springer, Berlin/Heidelberg/New York/Tokyo.

Rakowski,A., Hirsh,I.J. /1980/. Poststimulatory pitch shifts for pure tones. J.Acoust.Soc.Am. 68, 467-474.

Rakowski,A., Miśkiewicz,A. /1985/. Deviations from equal temperament in tuning isolated musical intervals. Archives of Acoustics, 10, 95--104.

Rakowski,A., Morawska-Büngeler,M. /1987/. In search for the criteria of absolute pitch. Archives of Acoustics 12, /2/, in print.

Schouten,J.F. /1938/. The perception of pitch. Philips Techn. Rev. 5, 286-294.

Stevens,S.S. /1961/. Is there a quantal threshold? In: W.A.Rosenblith /Ed./, Sensory Communication, MIT Press, Boston.

Sundberg,J. /1977/. The Acoustics of the singing voice. Sci.Am. 236, 82-91.

Terhardt E. /1974/. A contribution to the theory of pitch, consonance and harmony. J.Acoust.Soc.Am. 55, 1061-1069.

Wightman,F.L. /1973/. The pattern-transformation model of pitch. J.Acoust. Soc.Am. 54, 407-416.

Zwislocki,J.J. /1950/. Theory of the acoustical action of the cochlea. J.Acoust.Soc.Am. 22, 778-784.

PHYSICAL ACOUSTICS AT THE FERMI SUMMER SCHOOL IN VARENNA

D.SETTE
Università "La Sapienza", Roma, ITALY

1. Interdisciplinarity is one of the main characteristics of the part of Science that is indicated as Acoustics, and which has its core in the physical processes of generation and wave propagation of mechanical disturbances. The figure, taken by a R.B.Lindsay classical book, indicates clearly such a characteristic of a wide field of Acoustics.

For the reason the physics of mechanical waves is great interest to large number of scientistis and technologists whose main interest however is directed in other directions. This fact may explane the apparent contrast between the rather exiguous number of students which takes at colleges Acoustics as the major line of their formation. and which intend to become acoustical physicists and the much larger number of specialists which need to be concerned with subjects in advanced acoustic research.

The need exists, therefore, of devising methods by which the interaction between reserchers, especially young reserchers, in various fields, from one side, and scientists involved in advanced acoustic research, from the other, can be ehanced. A possible method is that of specialized summer schools.

The Italian Physical Society has been running the Fermi Summer School since 1953 in a nice place, the Villa Monastero (fig. 2) in Varenna on Como's lake. It is a School which at turn considers the fields of Physics where recent reserch has produced important advancements and innovations.

The main characteristics of these Schools are those:
1) of charging as lecturers those physicists which, at the time, are more responsabile of the latest developments in the subjects examined at the particular school; 2) of gathering as students young researchers active with various interests in the field; 3) of giving space to the students willing to present the results of their significative research; 4) of creating a clymate of continous fruitful interaction among partecipants: lecturers and students. The creation of a relaxed and friedly clymate is very important in this kind of school, and I will return later on this subject by relating an episode occurred in Varenna.

A similar action is also carried on in Italy at the Maiorana Center in Erice in Sicily. It differs from the Varenna school for its more direct orientation towards applications, of the Schools in Acoustics.

. I wish here to concentrate the attention on the courses in Acoustics held in Varenna, for which I have a direct experience. Going through the subjects of the lectures and the seminars, you will easily recognize the partecipation of top level lecturers and the quality of the students which, at times, have offered seminars.

The first School in Acoustics was held in 1962 and devoted to: Dispersion and absorption of sound by molecular processes. The list of subjects and the lecturers are indicated (table I). The development of ultrasonic spectroscopy of fluids has been thoroughly presented by the authors of the main researches in the field and some seminars have illustrated related subject in solids.

The second School was held in 1974 under the title "New Direction in Physical Acoustics" (table II). Various subjects were considered from molecular acoustic, to sound in superfluid helium, in liquid crystals, in solids, to echos of phonons, to the propagation in random and fluctuating media, to the effects of finite amplitude waves in fluids, to surface waves and to acoustic in space.

In the third School,"Frontiers in Physical Acoustics", held in 1984, a number of subjects have been considered (table III) which indicate new directions of future and fundamental activity. Nonlinearity, the physical origin of noise, chaotic and turbulent behaviour of simple non linear systems, transition to chaos, cavitation, are of this kind as well as the near-field holography, the scanning acoustic microscope, the acoustic engines. The propagation in porous media, the applications of acoustic in space were also considered. I wish to underline the fact that the results of recent researches made by students were discussed in some seminars, as the one concerning the discover of a non propagating hydrodynamic soliton.

3. I would like to terminate this short presentation with a real story which testify the plesant and productive atmosfere among partecipants at the School, who where living togheter in the small and attractive town of Varenna.

The first and the second Schools were held in periods (respectively 6th-18th and 5th-17th August) which included August 15th, a rather peculiar holiday in Italy whose origin goes back to Roman time: <u>feriae Augusti</u>, now <u>Ferragosto</u>,
The time is favourable for amusements in a little town and various fanny entertainiment and contests are organized by a town Committee for tourism. One of these contexts has attracted the interest of a group of partecipants to the School. A rope was tended between two

poles which hold electric lamps: a ham (prosciutto) was placed to hang in a point along the rope and the partecipants to the game where requested to guess the distance from ground of the lowest part of the prosciutto; the winner is the person who has made the closer guess to the successive experimental determination. It was interesting and funny to see a bunch of physicists making guesses, trying to use a method of measurements based on an aligment of the three points:the lowest point of the prosciutto, the top of the head of a physicist standing still, and the foot of another physicists laying on the ground with the head in contact with the legs of the collegue (fig. 3): from the knowledge of the heights of the two physicists and the determination by steps of a linear dimension on the ground, the unknown height was evaluated.

At the first school the success was there and we enjoyed the delicious prosciutto in a successive gathering at the School.

The difficulty came at the second school when other seven competitors arrived,with luck, to guesses having the same small deviation from the experiment. The prosciutto had to be devided and this operation could not be performed at that late hour in the night. The prosciutto was left to the local tourism office.

Mantaining the cheerful spirit,we asked for an official statement of deposit. This was easily obtained and here it is (fig.4).
The document may lead to various considerations, but it mainly furnishes a nice glimpse on the Varenna atmosphere.
The officer has written:
"We state that Mr.Enrico Fermi has won the prosciutto contest togheter with other seven competitors".

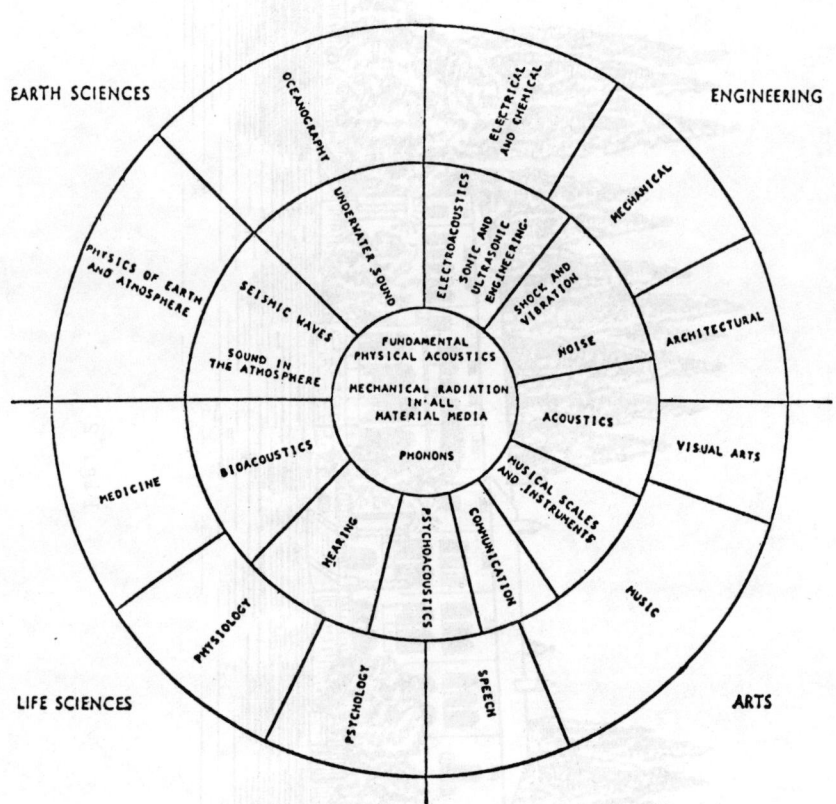

Fig. 1

Fig. 2

XXVII COURSE
Director: D. SETTE

Dispersion and absorption of sound by molecular processes.

SUBJECTS

K. F. HERZFELD - Theory of fluids.
H. O. KNESER - Relaxation thermique dans les gaz.
M. GREENSPAN - Translational dispersion in gases.
J. LAMB - Dispersion and absorption of sound by molecular processes: thermal relaxation in liquids.
T. A. LITOVITZ - Ultrasonic relaxation in liquids.
K. TAMM - Acoustic relaxation in electrolyte solutions.
W. P. MASON - Absorption and dispersion of sound in polymer solutions, polymer liquids, rubbers and solid polymers.
K. F. HERZFELD - Theories of relaxation times.
H.-J. BAUER - Phenomenological theory of multiple relaxation processes.
C. G. SLUIJTER - Rotational relaxation in the sound absorption of hydrogen isotopes.
R. T. BEYER - Measurements of ultrasonic velocity and absorption in liquids at high frequencies, at high temperatures and under high hydrostatic pressure.
R. T. BEYER - Measurements of ultrasonic velocity and absorption in liquids under high sound intensities.
L. LIEBERMANN - Determination of the kinetics of chemical reactions by sound absorption and dispersion.
J. M. STEVELS - The dielectric properties of quartz crystals and fused silica in relation to their imperfections.
R. W. B. STEPHENS - Ultrasonic attenuation in liquid metals.
A. J. KOVACS - Recouvrance de volume des verres; comparaison entre les expériences et une théorie phénoménologique.
A. J. KOVACS, J. D. FERRY - Variations des paramètres viscoélastiques du polyacétate de vinyle dans le domaine de sa transition vitreuse.
P. G. BORDONI - Relaxation of lattice imperfections in solids.

Table II

1974
5-17 August

LXIII COURSE
Director: D. SETTE

New directions in physical acoustics.

SUBJECTS

R. B. LINDSAY - Historical development of physical acoustics and future perspectives.
C. J. MONTROSE - Correlation functions in molecular acoustics.
S. YIP - High-frequency short-wavelength fluctuations in fluids.
C. J. MONTROSE - Light scattering and molecular acoustics.
I. RUDNICK - Physical acoustics at UCLA in the study of superfluid helium.
S. CANDAU, P. MARTINOTY - Absorption des ondes ultrasonores longitudinales et transversales dans les cristaux liquides.
W. P. MASON - Acoustical properties of solids.
E. F. CAROME - Superconducting transducers for use in the 50 to 1000 GHz range.
K. DRANSFELD - Production and detection of very-high-frequency sound waves.
J. JOFFRIN, A. LEVELUT - Les échos de phonons.
J. J. McCOY - Wave propagation in random media.
R. R. GOODMAN - Propagation in fluctuating media.
H. O. BERKTAY - Finite-amplitude effects in acoustic propagation in fluids.
R. W. B. STEPHENS - Finite-amplitude propagation in solids.
J. DE KLERK - A physical approach to elastic surface waves.
C. ATZENI, L. MASOTTI - Surface acoustic-wave devices.
T. G. WANG - Acoustics in space.

Table III

1984

XCIII COURSE

Director : D. SETTE

FRONTIERS IN PHYSICAL ACOUSTICS

SUBJECTS

D.G. CRIGHTON - Basic theoretical nonlinear acoustics.

J.E. FFOWCS WILLIAMS - Physical origins of acoustical noise.

R.H.G. HELLEMAN - Chaotic and turbulent behaviour of simple nonlinear systems.

M. GIGLIO, S. MUSAZZI AND U. PERINI - Transition to chaotic convection.

W. LAUTERBORN - Acoustic turbulence.

A. PROSPERETTI - Physics of acoustic cavitation.

R. KEOLIAN AND I. RUDNICK - The role of phase locking in quasi-periodic surface waves on liquid helium and water.

J. WU, R. KEOLIAN AND I. RUDNICK - Discovery of a non-propagating hydrodynamic soliton.

I. TOLSTOY - Long-wavelength scatter from rough surfaces.

D.L. JOHNSON - Recent developments in the acoustic properties of porous media.

H.N.V. TEMPERLEY - Reflection of pressure pulses at free surfaces of water.

T.G. WANG - Applications of acoustics in space.

J.D. MAYNARD - Near-field acoustic holography.

J.E. HEISERMAN AND C.F. QUATE - The scanning acoustic microscope.

J.C. WHEATLEY - Intrinsically irreversible or natural engines.

E.F. CAROME - Acousto-optic transduction in optical fibers and in fiber optic acoustic devices.

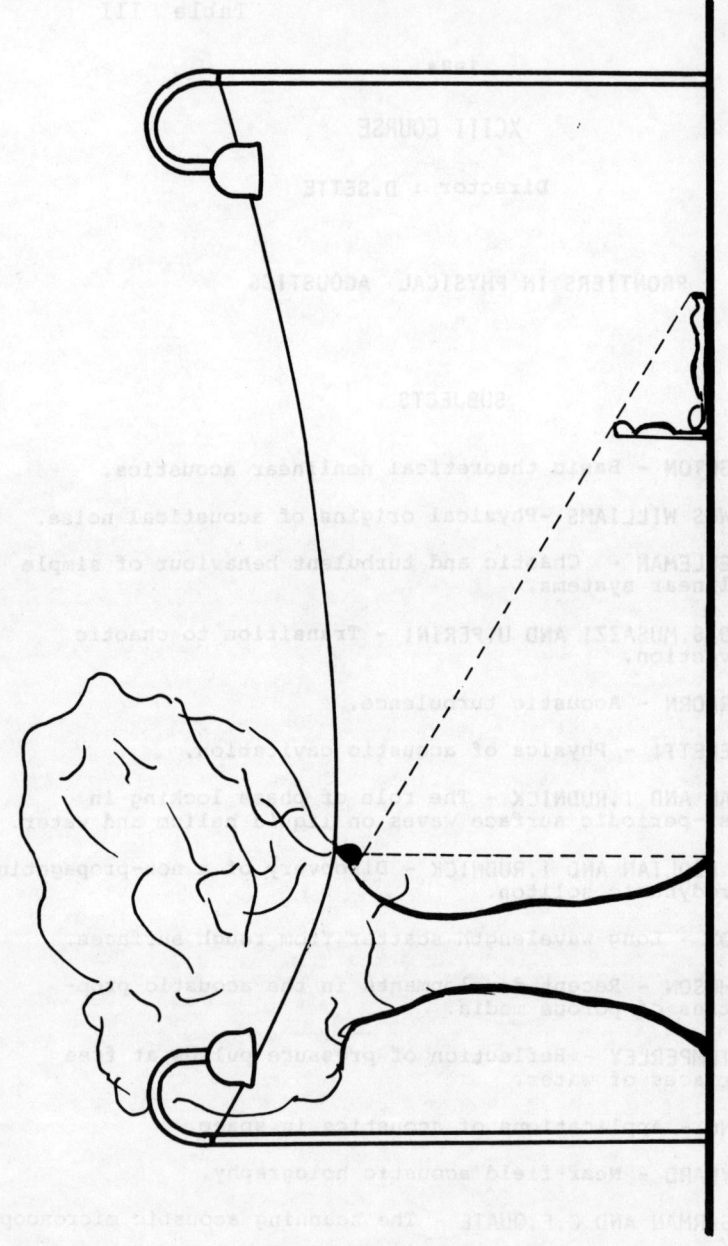

Fig. 3

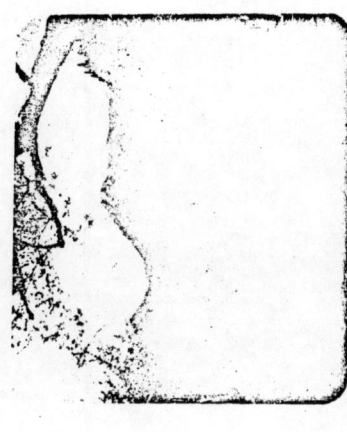

Varenna 1 5 agosto 1874.-

SI DICHIARA

che il Signor Enrico Fermi, ha vinto la gara del prosciutto unitamente ad altri sette concorrenti.

TE-LAB A SUCCESSFUL PHILOSOPHY OF UNIFIED LAB'S AT THE DEPARTMENT OF TECHNICAL ELECTRONICS

G. Schommartz

Wilhelm Pieck Universität, Rostock
Sektion Technische Elektronik,
GDR

Text of the paper not available in time for printing

Computers in Modern Acoustics Education and Research[*]

WILLIAM J. STRONG

Department of Physics and Astronomy

Brigham Young University

Provo, Utah 84604, U.S.A.

ABSTRACT

The paper will describe computer simulations for (1) a simple mass and spring vibrator, (2) a compound mass and spring vibrator, and (3) a player-clarinet system. The simple vibrator simulation illustrates analytic examples found in textbooks. The compound vibrator simulation extends typical textbook examples to include losses and a wide variety of parameterizations. The player-clarinet simulation deals with a nonlinear system in which analytic approaches are of limited value. All three cases are illustrated with computer generated graphs.

INTRODUCTION

In a recent article on graduate education in physics Beasley and Jones (1986) state that "computational physics, which addresses problems by a mixture of theory and computer simulation that were simply not accessible to physicists in the past, may arguably be a new subfield of physics". Whether or not computational physics is a new subfield of physics, computer applications in physics education and research are widespread. Computers play a very significant role in today's experimental physics. Computers play an increasing role in bridging between theory and experiment in the form of computer simulations or "computer experiments."

---

[*]Talk presented at "The Conference on Prospects in Modern Acoustics Education and Development," Gdańsk, Poland, May 19-21, 1987.

Statements similar to the above can justifiably be made about the role of computers in acoustics. One could consider a wide range of computer applications in acoustics education and research. The present paper, however, is addressed to three possible applications of computer simulation in acoustics—two aimed primarily at education and the third at research.

The first education oriented application is the computer simulation of simple mass and spring vibrators. Simple vibrators are commonly presented in textbooks with analytic solutions and graphical outputs for typical vibrator parameters and initial conditions. One aim of the simple vibrator simulation is to permit the student to compare simulation results with results obtained analytically for a case in which the analytical results are easily obtained. A second aim of the simple vibrator simulation is to permit the student to explore the vibrator behavior under a wide range of parameterizations, including both driven and freely vibrating conditions.

The second education oriented application is the computer simulation of a compound mass and spring vibrator. Compound vibrators involving coupled masses appear less often in textbooks than simple vibrators. When considered they are most often treated as lossless systems whether driven or freely vibrating. The aim of the compound vibrator simulation is to permit the student to explore the behavior of compound vibrators under realistic conditions which include losses and a wide range of parameterizations.

The research oriented application is the computer simulation of a player-clarinet system. In such a system analytic solutions are not possible because of complexities in the system, including nonlinearities. Ultimately, of course, the simulation results must be compared with experimental results to determine their validity.

VIBRATOR SIMULATION PROGRAM

The vibrator to be simulated, illustrated in Fig. 1, consists of masses M1 and M2 attached to supports by springs of stiffnesses K1 and K2, respectively. The masses are coupled to each other by a spring of stiffness K12, and have resistances (not shown) of R1 and R2. The program simulates a simple vibrator when K12 = 0 and a compound vibrator when K12 ≠ 0.

The simulation program requests the user to input the values of M1, R1, K1, M2, R2, K2, and K12 (in cgs units). The program then requests the user to specify the initial values of Y1, U1, Y2, and U2 which are the displacements and velocities of the masses. The driver frequency and amplitude are then requested. The driver is assumed to be sinusoidal with a starting phase of zero.

The simulation program then determines three frequencies—driver frequency, frequency of M1 with K1 and K12, frequency of M2 with K2 and K12—and specifies to the user the largest of the three. The user is then requested to specify the sampling frequency to be used in the simulation program based on a knowledge of the highest frequency to be simulated. The choice of the sampling frequency is rather crucial. If it is chosen too low the numerical simulation of the system is unstable and the simulation results are not reliable. If it is chosen too high excessive computations are made and the simulation is inefficient. (It should be noted also that the sampling frequency required is dependent on the algorithm used to numerically integrate the differential equations.) It is better to set the sampling frequency too high than to set it too low. As a rule of thumb the sampling frequency was chosen to be about 100 times the highest frequency to be simulated in the examples that follow. This seemed to be a reasonable choice except in cases where R1 and R2 were set so low that the system resonances became very narrow.

The waveform display algorithm was written so that four waveforms (of seven possible) could be displayed on the CRT. Examples of waveform displays of driven or freely vibrating systems are shown in the figures of the following sections. The figure captions specify the beginning and ending times of the waveforms, which waveforms are displayed, and various parameters of the simulation.

The determination of admittances in the simulations involved a rather inefficient computational approach. A frequency range over which admittance was to be determined was specified along with the frequency increment within this range. For each admittance value the system was driven with a sinusoidal force of unity amplitude and at the specified frequency. The velocity maximum was determined over a two cycle interval in the simulation response. This was done successively until the difference between two successive velocity maxima divided by their sum was less than 0.001. The velocity maximum so determined was taken as the admittance of the system at the specified frequency. (Admittance is defined as the ratio of velocity to force and the force was of unity amplitude). Admittances so determined were written to a disk file from which they could be displayed by an admittance display algorithm. Admittance values were calculated at each of 900 frequencies for the admittance displays shown in the following sections.

The simulations to be described were carried out on a Digital Equipment Corporation VAXstation II which provides significant compute capabilities and either high resolution CRT or laser printer graphics output. However, personal computers with very good compute capabilities and reasonable CRT and printer resolution are becoming readily available so that the simulations to be described should be accessible to a large number of users.

SIMPLE VIBRATOR SIMULATION

In Fig. 2 we see an example of the display for a simulation of the vibrator system as specified in the figure caption. Since the driver amplitude is set to zero the drive function is a constant horizontal line as shown in the upper curve. Since K12 is set to zero, M2 is not coupled to M1 and the parameters M2, R2, and K2 have no meaning. Mass M1 is 1 gm, resistance R1 is 2 acoustial ohms, and spring constant K1 is 4000 dynes/cm. Mass M1 is started with an initial positive displacement and zero velocity and is free to execute damped oscillatory motion. The displacement Y1, velocity U1, and acceleration A1, for mass M1 are shown in the lower three curves.

The input admittance of the simple vibrator of Fig. 2 is shown in Fig. 3. The single admittance peak is typical for a simple vibrator. From the figure we see that the simple vibrator's natural frequency is near 10 Hz.

Results of a driven simple vibrator simulation are shown in Fig. 4. With the driver frequency set at 5 Hz which is lower than the natural frequency of 10 Hz, the displacement tends to be in phase with the driver. The oscillatory motion initially shows the presence of both the driver and natural frequencies, but shows the driver frequency predominating at later times. When the system is driven near its natural frequency (Fig. 5) the oscillatory motion shows the presence of only one frequency and a monotonic increase in amplitude. The velocity is in phase with the driver in this condition. With the driver frequency set at 15 Hz which is higher than the natural frequency of 10 Hz (Fig. 6) the acceleration tends to be in phase with the driver. The oscillatory motion initially shows the presence of both the driver and natural frequencies but at later times, as the vibrator approaches steady state condition, the driver frequency predominates.

When the resistance of the system is increased the oscillatory motion approaches steady state much more rapidly (Fig. 7) than is the case with less damping shown previously in Fig. 5.

COMPOUND VIBRATOR SIMULATION

The input admittance (ratio of driven mass velocity to driver force) for a compound vibrator is shown in Fig. 8. The driven mass M1 is 1 gm; its associated resistance R1 and spring constant K1 are 2 ohms and 4000 dynes/cm, respectively. The coupled mass M2 is 1.2 gm; its associated resistance R2 and spring constant K2 are 2 ohms and 3500 dynes/cm, respectively. The coupling spring constant K12 is 400 dynes/cm. The double admittance peak is typical for a compound vibrator. The much smaller peak at the lower natural frequency is consistent with analytical results for the lossless case in which the smaller mass is being driven.

The transfer admittance (ratio of coupled mass velocity to driver force) for the compound vibrator of Fig. 8 is shown in Fig. 9. The two admittance peaks occur at the same frequencies as the input admittance peaks of Fig. 8. However, the two peaks are of essentially equal amplitude which is to be expected on the basis of analytical results for a lossless case.

The input admittance shown in Fig. 10 is for the case in which the driven mass M1 is 1.2 gm with K1 = 3500 dynes/cm and the coupled mass M2 is 1 gm with K2 = 4000 dynes/cm. The much smaller peak at the higher natural frequency is consistent with analytical results for the lossless case. Note how the admittance peaks shift for the different conditions between Fig. 8 and Fig. 10.

It is possible to start a compound vibrator with appropriate initial conditions so that it will vibrate at one of its natural frequencies. The response of the compound vibrator (M1 = 1 gm, M2 = 1.2 gm, K1 = 4000 dynes/cm, K2 = 3500 dynes/cm, and K12 = 400 dynes/cm), started with the special initial values of Y1 = 1.0 cm and Y2 = 3.14 cm, is shown in Fig. 11. The masses move in phase in this lower frequency mode. The displacements Y1 and Y2 of masses M1 and M2, respectively, are shown in the figure.

The response of the compound vibrator just described, but started with the special initial values of Y = 1.0 cm and Y2 = -0.27 cm is shown in Fig. 12. The masses move out of phase in this higher frequency mode.

The response of the compound vibrator described above, but started with the arbitrary initial conditions of Y1 = 1 cm and Y2 = 1 cm, is shown in Fig. 13. In this case the masses move with no simple phase relationship. In addition, we see the presence of two frequencies.

Another case of interest is that of a coupled oscillator with identical masses and springs. The response of such an oscillator (M1 = M2 = 1 gm, K1 = K2 = 4000 dynes/cm, and K12 = 400 dynes/cm), started with the initial conditions Y1 = 1 cm and Y2 = 0 cm, is shown in Fig. 14. We see the characteristic "beating" response as the energy is coupled from one mass to the other and back again.

Fig. 15 illustrates simulation results for a compound vibrator (M1 = 1, M2 = 1.2, K1 = 4000, K2 = 3500, K12 = 400) driven at 8.92 Hz which is near its lower natural frequency. The two masses move in phase at this frequency. Fig. 16 illustrates simulation results for the compound vibrator just described, but driven at 10.68 Hz which is near it upper natural frequency. The two masses move with opposing phases at this frequency.

PLAYER-CLARINET SIMULATION

In analytical, experimental and simulation studies of the clarinet preceding the 1980's the player's windway was ignored. However, in the last few years considerable interest has been centered on the role of the player's windway in clarinet tone production. Hoejke (1986) has provided a very nice analytical and experimental treatment of the role of the player's windway in musical instrument tone production. Sommerfeldt (1986) carried out a simulation of clarinet tone production which incorporated a model of the player's windway. The simulation reported here is that of Sommerfeldt.

Fig. 17 illustrates how the player's windway was modelled. At the left of the figure P_ℓ is the lung pressure and R_ℓ the lung resistance. The bronchi, trachea, and vocal tract were represented with lossy cylindrical tubes having yielding walls. The tubes are shown in the figure in terms of lumped element resistances, inertances, compliances, and conductances. At the right of the figure L_R and R_R are the inertance and resistance of the reed aperture, respectively. PT is the impulse response of the instrument.

The air flow through the reed aperture is a nonlinear function of the aperture area and the pressure drop across the aperture. The reed motion exhibits nonlinear behavior when it impacts on the mouthpiece. Nonlinearities become especially pronounced for large amplitude vibrations. Simulation models may be particularly useful for studying large amplitude vibrations in nonlinear systems such as the player–clarinet.

The actual instrument simulated consisted of a cylindrical tube with seven tone holes and a clarinet mouthpiece. The lung pressure was taken to be 40,000 dynes/cm^2 in all examples shown below. This corresponds to loud blowing of the instrument. The simulation results are for the written note G4 with four different vocal tract configurations. Fig. 18 shows reed opening, air flow through the reed aperture, pressure in the player's mouth, and pressure in the mouthpiece of the instrument for a case in which the player's windway is assumed to have an impedance of zero. In this case the mouth pressure is constant and equal to the lung pressure. This is typical of simulations carried out previous to the work of Sommerfeldt. The reed aperture is seen to close for about half of the cycle for this condition of loud blowing. The reed motion follows the tube pressure very closely when the reed aperture is open. The smaller and more rapid fluctuations in the tube pressure are due to reflections from the place where the tapered mouthpiece joins the cylindrical tube. They are considerably stronger in the simulation than any observed experimentally on the clarinet system. (This deficiency in the model has not yet been resolved.) These unrealistic fluctuations are seen to carry over into the reed motion and air flow.

Fig. 19 illustrates simulation results for a lung pressure of 40,000 dynes/cm^2 and an /a/-shaped vocal tract. The player's windway has finite impedance in this case which results in small fluctuations in the player's mouth pressure. The major impedance peak of the player's windway in this case is small in amplitude by comparison with impedance peaks of the instrument. The influence of the player's windway on the reed motion is insignificant in this case as evidenced by comparing the reed motion to that of Fig. 18.

Fig. 20 illustrates simulation results for an /i/-shaped vocal tract with a resonance near 800 Hz. This tract shape produces significantly larger pressure fluctuations in the mouth because its largest amplitude peak is some 20 dB higher than that of the /a/-shaped tract. Since the 800 Hz resonance is not close to either the second harmonic ($\simeq$700Hz) or third harmonic ($\simeq$1050Hz) of the mouth pressure it plays a rather insignificant role in controlling the reed vibration. This can be observed by comparing the reed motion, air flow, and tube pressures of Figs. 19 and 20.

Simulation results for an /i/-shaped vocal tract with a resonance near 1050 Hz are illustrated in Fig. 21. Pressure fluctuations in the mouth are very large. The alignment of the player's windway resonance with the third harmonic of the tube is seen to produce significant effects on the reed opening and air flow waveforms. In fact, in one case reported by Sommerfeldt the effect of the player's windway was so strong as to cause the vibration to occur at three times the frequency of the fundamental mode of the tube. This was also demonstrated on an actual instrument.

DISCUSSION

Computer simulations have been described for extending analytical formulations of simple acoustical systems and for conducting research on nonlinear systems. Computer simulation can be a powerful method for bridging between experiment and theory. The tools for implementing computer simulation of acoustical systems are becoming readily available. Questions then arise as to what is the proper role of computer simulation in the acoustics curriculum. Is it a method whose time has come? Does it tend to divorce the student too much from physical reality? Does it encourage the student to avoid hands-on experience in the laboratory? These and related issues should be discussed thoroughly with the aim of better defining the role of computer simulation in the acoustics curriculum.

ACKNOWLEDGEMENTS

Scott Sommerfeldt provided the figures relating to the clarinet simulation. Irvin Bassett assisted in the preparation of the other figures.

REFERENCES

Beasley, M.R. and L.W. Jones, 1986, "Education for Research," Physics Today **39** (June), 36-44.

Hoekje, P.L. 1986. "Intercomponent Energy Exchange and Upstream/Downstream Symmetry in Nonlinear Self-Sustained Oscillations of Reed Instruments" (Ph.D. Thesis, Case Western Reserve University).

Sommerfeldt, S.D. 1986. "Simulation of a Player-Clarinet System (M.S. Thesis, Brigham Young University).

Fig. 1. Compound vibrator system to be simulated. M1 and M2 are masses, K1, K2, and K12 are springs. R1 and R2 are resistances not shown.

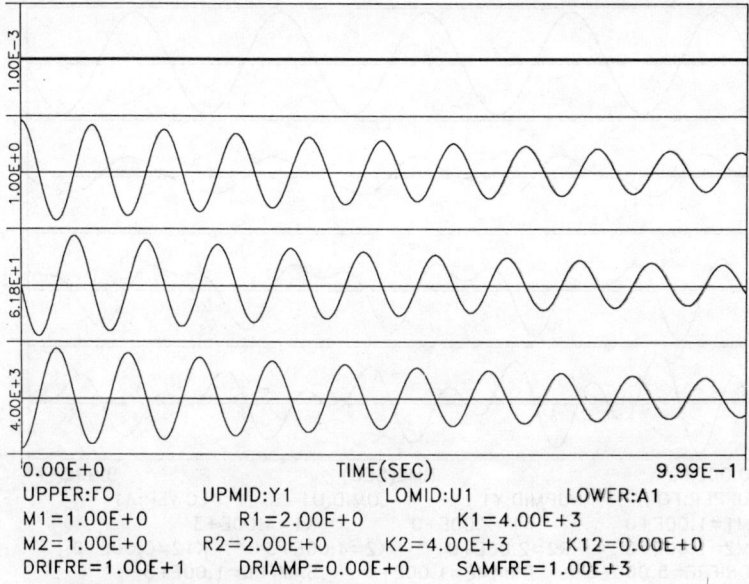

Fig. 2. Free, damped oscillation of a simple vibrator with parameters described in the text and shown in the figure.

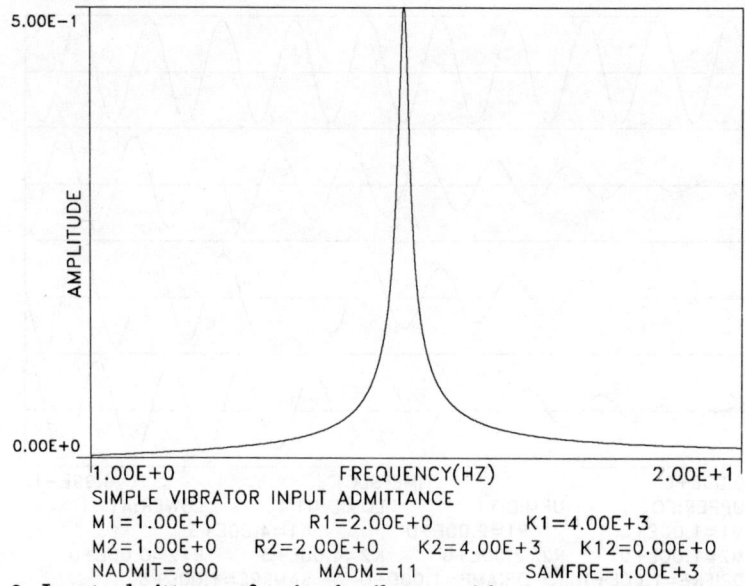

Fig. 3. Input admittance of a simple vibrator with parameters described in the text and shown in the figure.

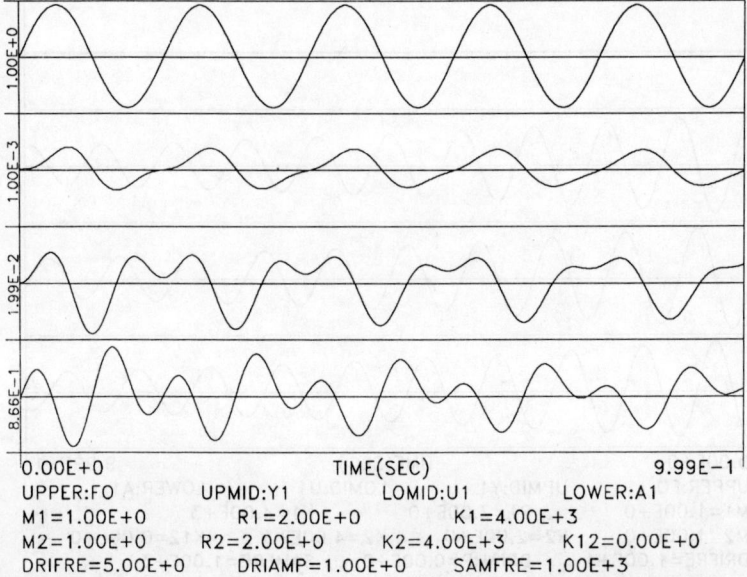

Fig. 4. Driven simple vibrator: normal mode frequency is approximately 10 Hz and driver frequency is 5 Hz.

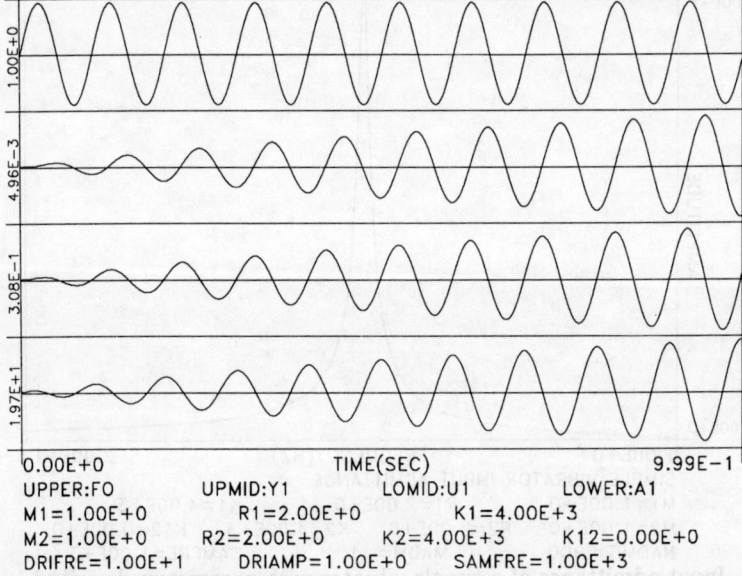

Fig. 5. Driven simple vibrator: normal mode frequency is approximately 10 Hz and driver frequency is 10 Hz.

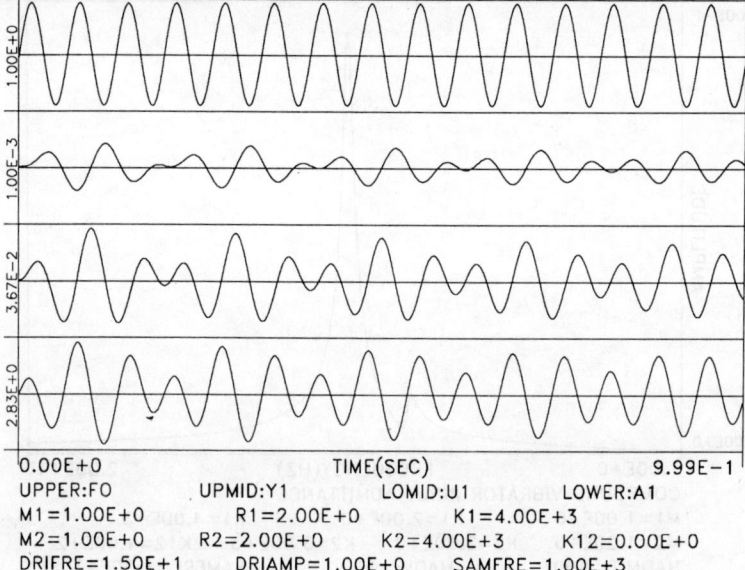

Fig. 6. Driven simple vibrator: normal mode frequency is approximately 10 Hz and driver frequency is 15 Hz.

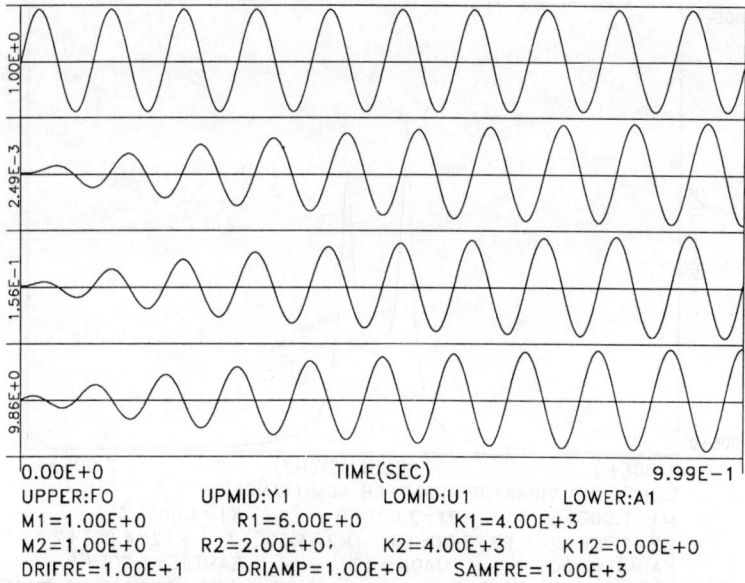

Fig. 7. Driven simple vibrator: normal mode frequency is approximately 10 Hz and driver frequency is 10 Hz. Resistance is greater than in Fig. 5.

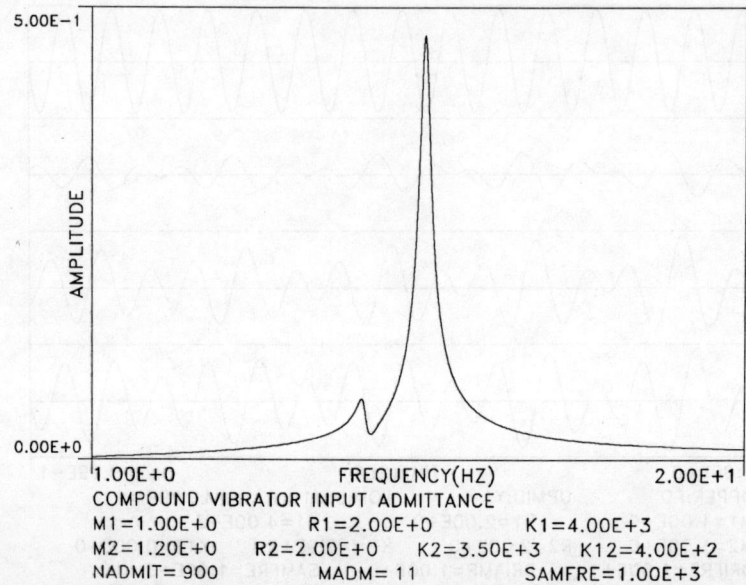

Fig. 8. Input admittance of a compound vibrator with parameters described in the text and shown in the figure.

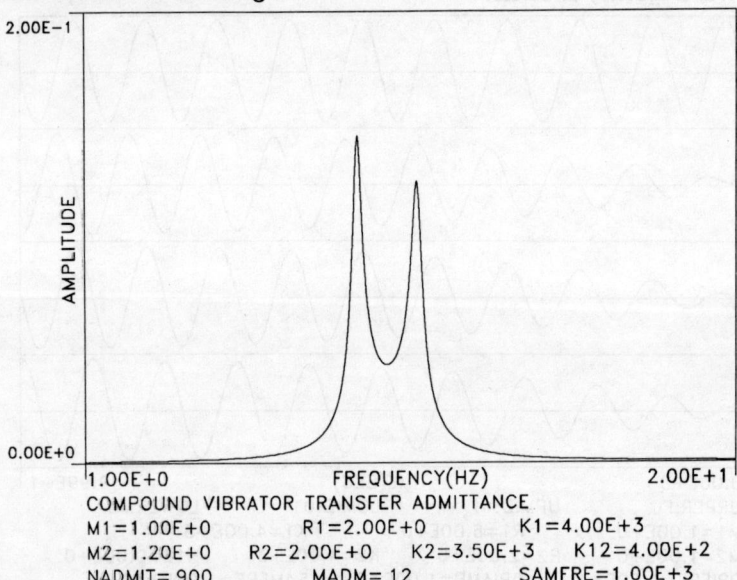

Fig. 9. Transfer admittance of a compound vibrator with parameters described in the text and shown in the figure.

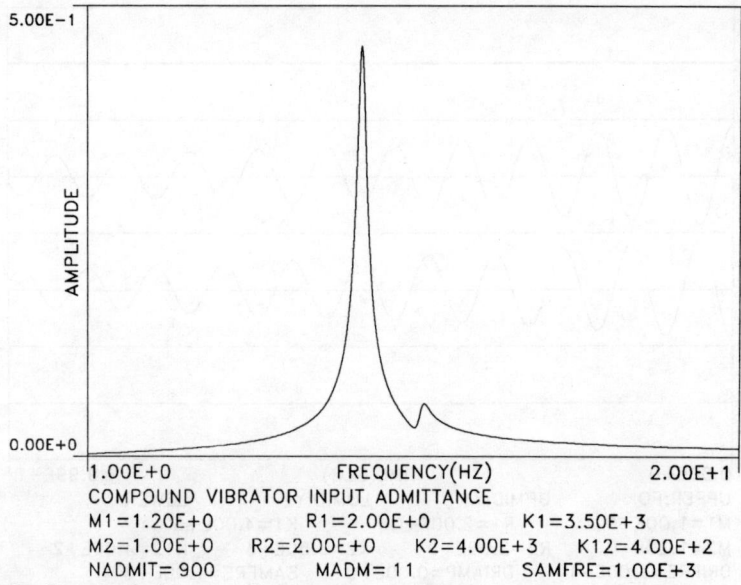

Fig. 10. Input admittance of a compound vibrator with parameters described in the text and shown in the figure.

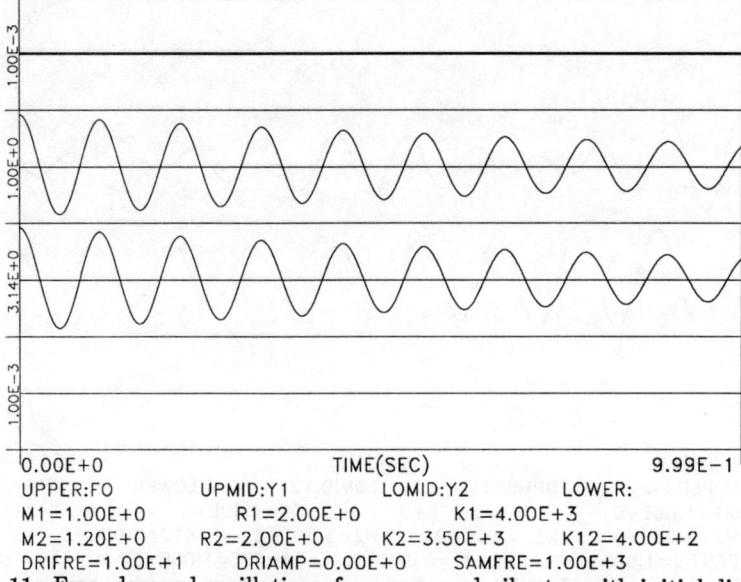

Fig. 11. Free, damped oscillation of a compound vibrator with initial displacements of Y1 = 1 cm and Y2 = 3.14 cm.

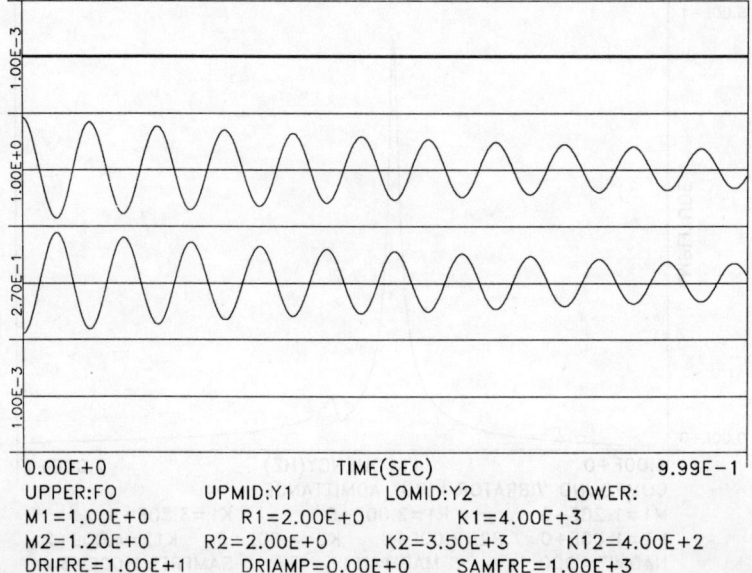

Fig. 12. Free, damped oscillation of a compound vibrator with initial displacements of Y1 = 1 cm and Y2 = -0.27 cm.

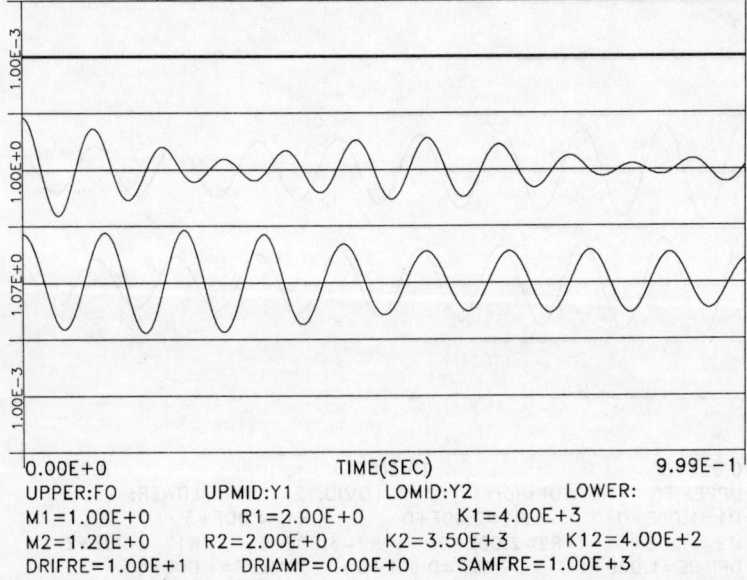

Fig. 13. Free, damped oscillation of a compound vibrator with initial displacements of Y1 = 1 cm and Y2 = -1 cm.

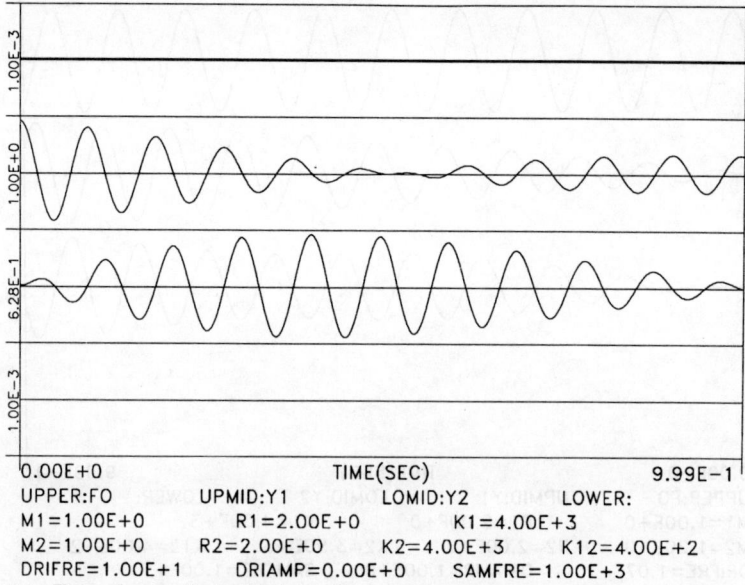

Fig. 14. Free, damped oscillation of a compound vibrator identical with masses and springs; initial displacements are Y1 = 1 and Y2 = 0.

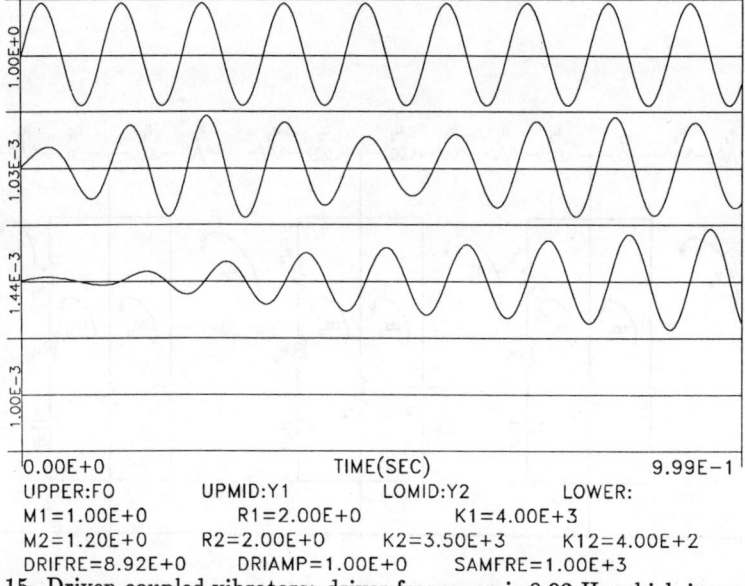

Fig. 15. Driven coupled vibrators: driver frequency is 8.92 Hz which is near the first normal mode frequency.

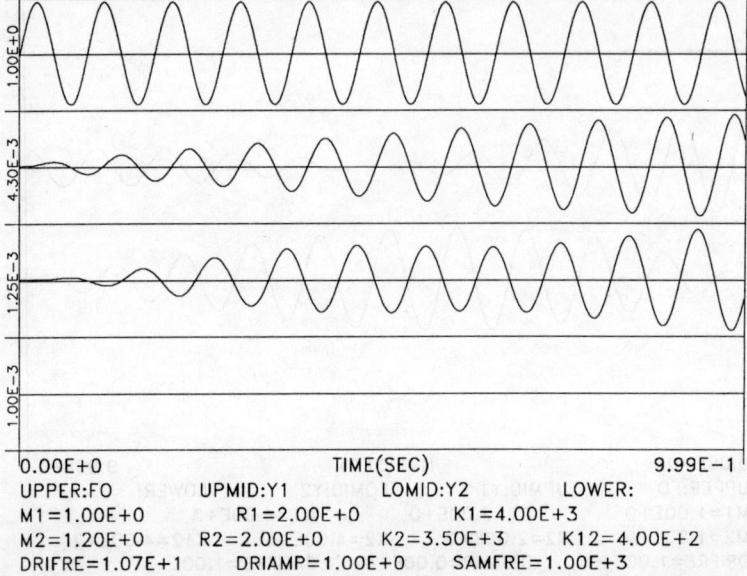

Fig. 16. Driven coupled vibrators: driver frequency is 10.68 Hz which is near the second normal mode frequency.

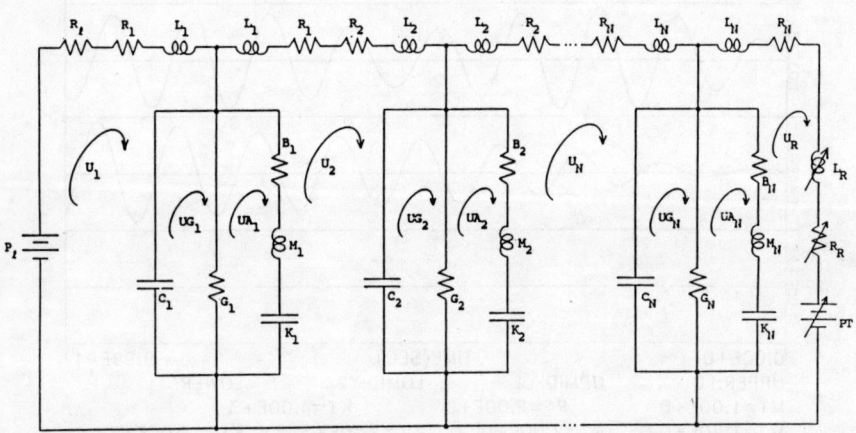

Fig. 17. Electrical circuit analog of the player – clarinet simulation model. (From Sommerfeldt, 1986, Fig. 1.)

245

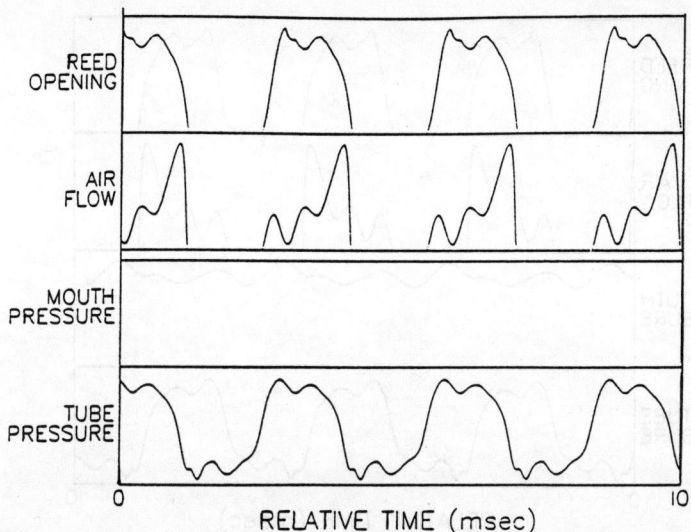

Fig. 18. Calculated waveforms for the written note G4 with a constant mouth pressure of 40,000 dynes/cm^2. (From Sommerfeldt, 1986, Fig. H1.)

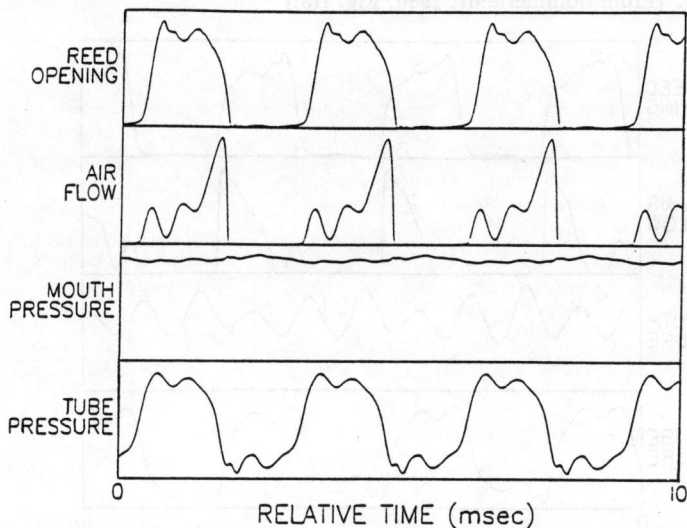

Fig. 19. Calculated waveforms for the written note G4 with a constant lung pressure of 40,000 dynes/cm^2 and an /a/-shaped vocal tract. (From Sommerfeldt, 1986, Fig. H2.)

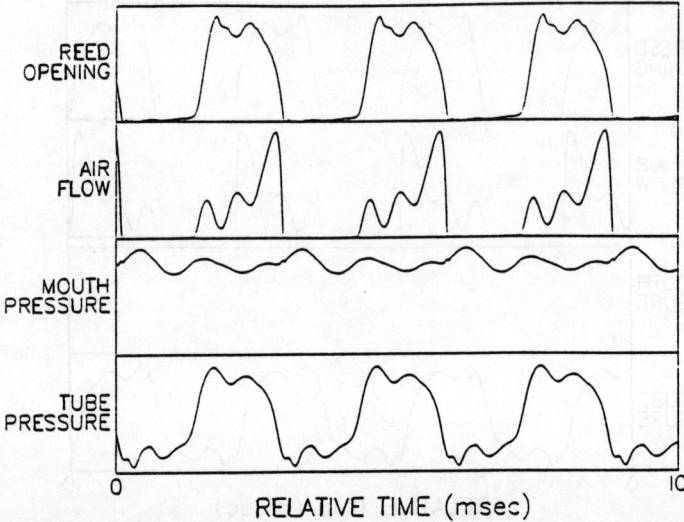

Fig. 20. Calculated waveforms for the written note G4 with a constant lung pressure of 40,000 dynes/cm^2 and an /i/-shaped vocal tract having a resonance near 800 Hz. (From Sommerfeldt, 1986, Fig. H3.)

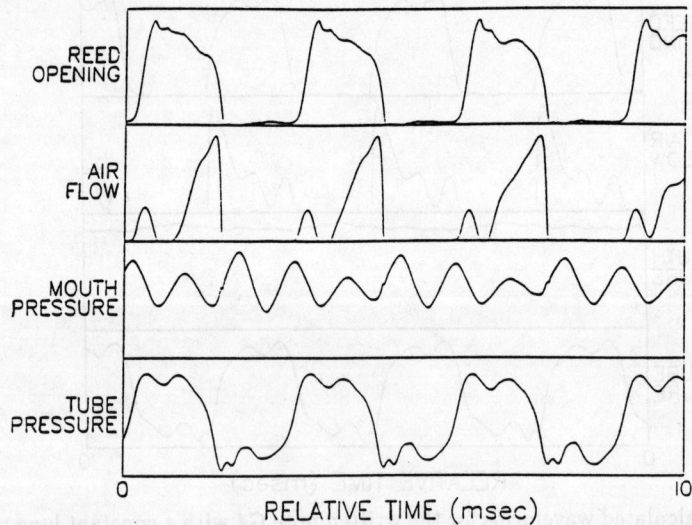

Fig. 21. Calculated waveforms for the written note G4 with a constant lung pressure of 40,000 dynes/cm^2 and an /i/-shaped vocal tract having a resonance near 1050 Hz. (From Sommerfeldt, 1986, Fig. H4.)

ACOUSTICAL EDUCATION AND INDUSTRIAL APPLICATION

A.S. Śliwiński

Institute of Experimental Physics
University of Gdańsk
Wita Stwosza 57, 80-952 Gdańsk, Poland

ABSTRACT

In acoustics one can observe better coupling between science and technology than in other branches of physics, however a gap between university education and industrial needs may be noticed, too. The problem how to decrease the gap is discussed on the background of teaching acoustics within educational program of physics at the University of Gdańsk.

1. INTRODUCTION

What is a way to find optimal coupling between education and practical application of knowledge? - it is an immemorial question and always actual problem.

The aim of this lecture is to show how the education in acoustics is useful for practice, particularly for industry. What one could expect and how that looks like in our case at the University of Gdańsk.

It is known more or less commonly that between achievements of physics and applications, particularly in industry, there exists an evident gap greater or smaller in various countries, more or less distinct in different branches of physics |1,2|, visible in acoustics, too.

Acoustics is one of those branches of physics where applications are relatively wide and where coupling between science and technology is not so bad. It reflects itself in technical acoustics activity widely developed.

An evidence of a good coupling between pure physical acoustics research and technical applications are topics of ICA Congresses |3| taking place every three years (last 12-th ICA Congress in Toronto in 1986) where more than 60% of papers concern technical applications. Also in Poland during annual Open Seminars on Acoustics (already 34 were held) and many professional meetings exchange between scientifical and technical ideas brings good effects. One could give here many interesting examples from other countries too.

In the organizing field such an integrating role is played by national and international acoustical committees, commissions, scientifical societes and other bodies initiating organization of schools, courses, workshops, meetings of professional groups, standard organizations etc. Our meeting here gives an opportunity to see how the acoustical coherence looks like on the level of education and to discuss the problem wider and improve the situation.

Often, objections are formulated that education of students in science, mainly at universities does not keep up with social needs. It is evident in a few fields like for instance an imperfection of education of teachers of physics in schools, a weak matching of graduates to their activity in industry, inappropriate education of research workers for further self-dependent work keeping the coupling with industry, too slow and insufficient penetration of new achievements to the society in general and to industry in particular, and others. The matter looks differently in various countries, however there are many common problems. A good education should meet and help to solve those difficulties.

At the end of my talk I would like to show these elements of the educational organization at our University and the modest (insufficient in our feeling) technical base we use in the educational process for graduating in physics with the specialization in acoustics which have to provide for the physics - industry coupling. It is difficult, however, to evaluate efficiency of such a programme because the results of activities depends only partly by itself. Important is the realization of the programme and there may be various ways to succeed with that in practice.

2. REASONS OF THE ACADEMIC - INDUSTRIAL GAP AND HOW ONE CAN DECREASE IT IN THE PROCESS OF EDUCATION AS EXEMPLIFIED BY ACOUSTICS

The existence of such a gap in physics was stated many times |1,2|; for example D. Lundquist (1980) said: "Industry and university are seen to exist side by side with little reference to each other. This is perhaps less serious for large industries, which have big, well-equipped and well funded laboratories, than for small industries which are unable to support R (research) and D (development) on their own. Overall, it is found that universities tend to discourage the younger physicists from being interested in the application of their research, and industry tends to remain in ignorance of developments that could mould their future production.

Factors impeding the transfer of new knowledge through to a marketable product are:
- Inertia in the acceptance of new systems;
- Suspicion expressed by oppinion formers and politicians;
- Financial problems of supporting R and D with its inherent risk;
- Long development times between research and final application.

Among the problems discouraging a better interaction are, on the university side
- Lack of experience in conducting projects on to application;
- Lack of experience in taking cost-efficiency into account;
- Reluctance to accept time constraints or long term committments
- Unclear rules governing external cooperation.

On the industry side:
- Inadequate understanding of university conditions;
- Financial demands with no certain return;
- Low priority given to research in management evaluations.

Amongst the general recommendations that are made, are a plea for universities to modify their student training programmes to focus more on applied physics and to set up information secretariate on their research. Industry is urged to recognize the importance of keeping closely in touch with the academic sphere. Personal contacts are vital, and arrangements need to be made for both university teachers as well as trainees to spend time in an industrial environment. Conversely R and D people in industry should be encouraged to give courses in universities and work there on appropriate subjects.

Governments can do much by channelling funds into joint projects, covering the risks inherent in projects which could be of national importance and establishing collective R and D institutes in special fields - of especial value to the smaller industries. Certain countries have made a success of "Science Parks" which facilitate the creation of spin-off companies from universities, but all governments can make a real contribution by eliminating needless bureaucracy" (the end of quotation).

Those are very general factors and can be related to any domain of science, so they are adequate to acoustics, too and they influence its development. The progress of science depends on education and importance of those factors must be seen already on this level. One should explain to teachers and students that educational programs should be formed and realized to decrease the academic - industry gap.

In acoustics as was mentioned we have some positive examples which improve relations science - industry. Among others it results from the interdisciplinary character of acoustics. It seems that the specificity of applications for example in noise domain favors the coupling between science and industry, like for instance in the case of noise abatement in factories arises necessity of solving many scientific problems. Relations between fundamental research and application are in many cases rather good. As further examples one can mention such domains like acoustical emission - the method used for material examination, diagnostics of machineries, building structures, and rock masses (all kinds of acoustical wave-lengths are used from infra - to ultrasound) etc. Processes of radiation consider in relation: source of vibrations - sound field, widely examined as a fundamental problem but also as a real technical one in noise abatement. Ultrasonic technology, ultrasonic diagnostics (in industry and in medicine), underwater applications, electroacoustical equipment, they are wide domains of applications. And many others.

Existence of specialized acoustical laboratories in industry, laboratories affiliated at universities cultivating applied acoustics or constructing specialized equipement (like Technical University in Gdańsk), existence of special firms producing acoustical equipment (Bruel-Kjaer and many others) existence of many consulting firms, also, service for acoustical measurements by scientifical institutions, they are all essential elements keeping strong coupling between acoustics and industry.

Very helpful for that coupling is the international cooperation in norms and standard organizations |5|. The participation and role of physicists and technologists working at universities as well as in industry in the domain of methodology of acoustical research, in preparation of measuring methods and in application adaptations based on commonly elaborated norms is also a good example, here. Table 1 - (see Fig. 5 in |5|) - shows an example of the composition of the plenary meeting of ISO/TC 43/SC1 in 1974.

Table 1

	25% :	University Acoustical Labs (Prof. Dr)
65% mostly highly specialized	30% :	High graduate Acousticians belonging to big independent (national or not) research or testing Labs
	10% :	Acousticians belonging heavy Industry (Ge, Co, IBM,...)
	20% :	People in charge of acoustical problems in national institutions for standardization
35% not specialized	10% :	Manufacturers
	5% :	Governmental Officers

Composition of the plenary meeting of ISO/TC 43/SC1 in 1974
(96 delegates) - after P. François |5|

The education program of acoustics at universities or at technical high schools should form a flexible compromise between the fundamental knowledge and application by the possibly strong contact of a student with practice, with industry. The more students will be convinced about the need of keeping bonds between pure science and application (through exchange programs with industry, practices, diploma projects etc.) the better it is. The education program should be opened as for the inflow of new elements of fundamental science as well as for new information from industry. So, a reasonable feedback is required there.

3. TECHNICAL POSSIBILITIES OF UNIVERSITIES AND THE COUPLING OF ACOUSTICS WITH INDUSTRY

It is known of old that an optimal solution for a good education would be a case when students during their studies meet an equipment in the labs at least on the same level as possesed by industry where they would work after graduated (if they hit there). Whereas, often the situation is quite opposite and in many universities the technical base does not follow the technological progress. Economical and financial reasons do not promise any good for a better state.

Recently in England near London the new type of a kind of university has been errected or founded by world leading firms in computer industry. The foundators have obliged themselves to equip the new academic centre (teaching students) not only with their newest products but straight to build up their laboratories at the area of the centre. The task of those labs is double: There are places for elaborating new conceptions of devices construction for the production in straight cooperation with scientists and students and to educate students who are attending in these new ideas and solutions "in statu nascendi". The essential is this, that in the frame of such coupling one has not only connection of research with education (as it has been traditionally at universities) but with application of science as well. It is also essential, that the idea is a global one in the sense that this academic centre has not be narrowly specialized but to form a modern multi directional university.

In the already existing universities similar initiatives can be also undertaken to increase better coupling with industry. Generally however, it is easier to find foundators to support concrete projects adapted to the actual potentials of the university. It is good when there exists a specialized laboratory at the university which can take up the project and solve the task together with the people from industry. Then there is an opportunity to equip the labs by the industry. When such lab. exists at the university students can easier be trained for industrial needs.

As I already mentioned the erection of the so called environmental laboratories for measurement service and applied departments undertaking production of unique equipment form such conditions. A kind of such solution is the situation at our University of Gdańsk which will be described as follows.

4. TEACHING ACOUSTICIANS AT THE UNIVERSITY OF GDANSK

I think it is useful to present the general organization scheme of the university to see how the acoustics is situated among other units (table 2). Education of acousticians is performed in direction of physics at the Faculty of Mathematics, Physics and Chemistry in the frame of applied physics specialization. The specialization in acoustics takes place after the 3-rd year of study (Tabl. 3). Then students have specialistic and monographic lectures, seminars and exercises (auditorial as well as laboratorial) (Table 4,5) in acoustics. Students are graduated with M.Sc. degree after final (magister) examination and positive judgement of their Master Thesis. The projects for Thesis are undertaken after the 7 or 8 semester of study (during 4-th year od study) in acoustics very often in relation to industrial needs. The coupling of acoustics with industry in our program of study is mainly realized with the cooperation between the Department of Applied Physics and Environmental Acoustics and Spectroscopy Laboratory. The Lab serves acoustical measurements and unique constructions for industry and also it participates in, as we call, central projects controled by the Office of central research planning (Tabl. 6).

Tables 6 - 8 exemplify the coupling on the level of education, too. In the table 7 the titles of M.Sc. Thesis in acoustics performed in 1981-86 are given. The numbers of graduated during this term are presented in the

Table 2

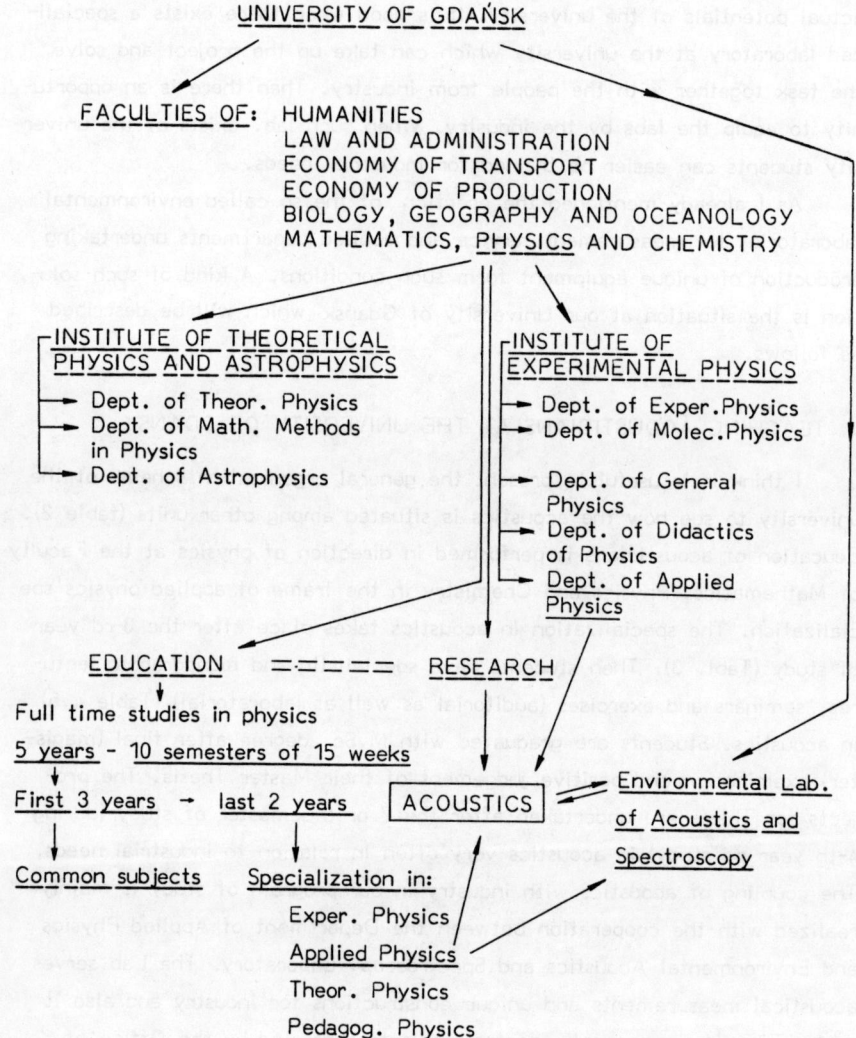

Table 3

Acoustics in the syllabus for students of physics

First 3 years common subjects

Elements of acoustics in:
- lecture "Fundamentals of Physics"
- 1-st level physical lab. (5 exercises - each 2 1/4 h)
- 2-nd level physical lab. (7 exercises - each 18 h)
- Industrial practice (1 month)

last 2 years: Specialization subjects:

4-th year:
- "Fundamentals of acoustics" (lecture 60 h, exerc. and it 30 h)
- "Applied acoustics" (lecture 30 h, exerc. 15 h)
- Seminars (60 h)
- 3-rd level lab in physics (90 h for acoustics)
- Spec. Lab in physics (in acoustics - 90 h)

5-th year:
- Spec. lectures:
  - e.g. "Molecular acoustics" (30 h)
  - "Acoustooptics" (30 h)
  - or "Physics of the sea" (underwater acoustics)
  - or others
- Diploma lab (unlimited in hours)
- Diploma seminar (60 h)

Table 4

3-rd Level Lab Topics

1. Sound field measurements
2. FFT analysis of noise
3. Statistical analysis of noise
4. Optical holography in acoustical measurements
5. Visco-elastic properties and acoustical insulation

Table 5

Special Acoustics Lab Topics

1. Noise measurements at a working place in industrial conditions (different methods of analysis, interpretation for different time and frequency characteristics, application of norms and criteria).
2. Measurements of acoustical power at different conditions and for different sources.
3. Acoustooptical interaction in isotropic media.

Table 6

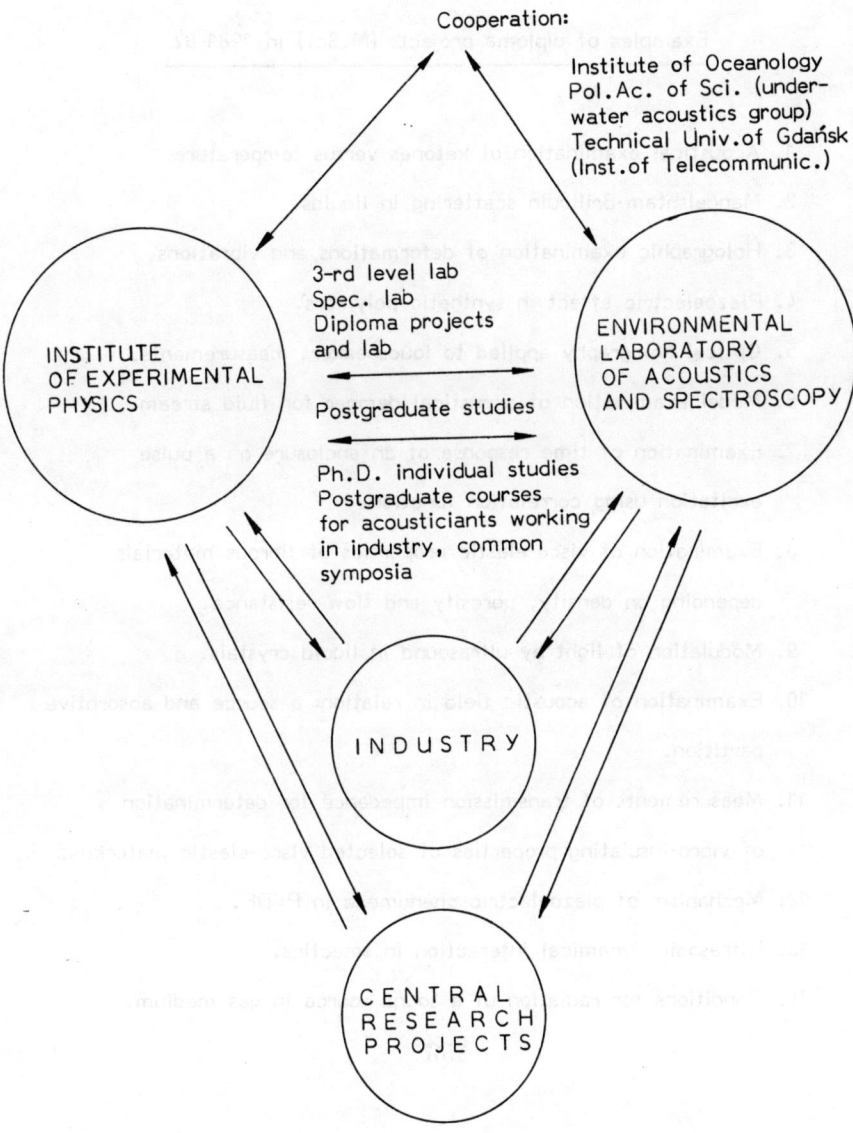

Table 7

Examples of diploma projects (M.Sc.) in 1981-87

1. Acoustical examination of ketones versus temperature.
2. Mandelshtam-Brillouin scattering in liquids.
3. Holographic examination of deformations and vibrations.
4. Piezoelectric effect in synthetic polymers.
5. Optical holography applied to loudspeakers measurements.
6. Model examination of acoustical damper for fluid stream.
7. Examination of time response of an enclosure on a pulse excitation using correlation function.
8. Examination of visco-elastic properties of fibrous materials depending on density, porosity and flow resistance.
9. Modulation of light by ultrasound in liquid crystals.
10. Examination of acoustic field in relation: a source and absorptive partition.
11. Measurements of transmission impedance for determination of vibro-insulating properties of selected visco-elastic materials.
12. Mechanism of piezoelectric phenomena in PVDF.
13. Ultrasonic dynamical interaction in smectics.
14. Conditions for radiation of a sound source in gas medium.

Table 8

Number of diploma projects in acoustics in last 5 years

	Dept.of Appl.Phys.	Environ.Lab.of Acoust.	Spectr.
1982	2	2	
1983	1	2	
1984	2	1	
1985	1	1	
1986	2	1	

Number of students in acoustics in 1986/87

4-th year	7
5-th year	3

Table 8 where the first column is referred to the Department of Applied Physics and the second one to the Environmental Acoustics and Spectroscopy Lab. At the bottom of the Table 8 the actual numbers of students of the 4-th and 5-th year of study who are preparing their M.Sc. Thesis in acoustics are given.

In general for a few years the decrease in the number of students studying physics is observed. It can also be noticed in acoustics.

5. EDUCATION AND DEVELOPMENT OF SCIENTIFIC AND TEACHING STAFF

Education and development of scientific and teaching staff belongs to one of the fundamental functions of universities. Programs of study should provide training of students for future scientific career, too. That is possible when students activities at higher semesters like seminars, special laboratories and diploma projects are in good coupling with research carried on at the university. Therefore, it is important to construct the teaching scheme in a way to match it to the staff and apparatus base of the university.

Problems of postgraduate education are important, too. The education of scientific staff takes place in a way of receiving Ph.D. and D.Sc. degrees |6|. A successful career depends very much on the proper relations between teaching and research at the university and their coupling with industry, too. We organize systematically 3 semester postgraduate training studies in physics for secondary school teachers and from time to time for industrial people in acoustics depending on actual needs. Postgraduate studies give a good opportunity to teach professional people not only about fundamental knowledge they need but about the newest achievements in acoustics, too (table 9).

Also as an example of possibilities of the postgraduate education in acoustics at our University the list of titles of Ph.D. Thesis defended in last 15 years is presented in the Table 10.

6. CONCLUSIONS

The existing gap between physics education and industry is evident in acoustics, too. However, acoustics is the branch of physics where the coupling with industry is relatively good. There exist many factors integrating education, research and technology resulting from an interdisciplinary character of acoustics.

Educational programs of acoustics at schools of the university level should form a flexible compromise between a fundamental knowledge and application by a possibly strong contact of students with the practice.

Table 9

Postgraduate course for acousticians working in industry (example)

Lectures: 1. Acoustical field in isotropic media.
2. Acoustooptical field in inhomogeneous media.
3. Electro-acoustical analogies.
4. Vibrations of plates and membranes.
5. Acoustical sources in different media.
6. Scattering and diffraction phenomena.
7. Standardization in acoustics.
8. Vibro- and acousto-insulating materials.
9. Acoustical measurements in real conditions.
10. Noise abatement problems.

Lab. exercises: 1. Acoustic pressure and intensity measurements.
2. Sound analysis.
3. Measurements of noise at the place of work.
4. Acoustical impedance measurements and material characterization.
5. Reverberation time measurements.

Table 10

Topics of Ph.D. Thesis in acoustics at the University of Gdańsk

in 1970 - 1986

1. M. Łabowski — Examination of amplitude and phase fluctuation of an ultrasonic wave in inhomogeneous medium using laser light. (1970)

2. J. Józefowska — Approximative theory of laser light diffraction containing a 3-rd harmonic by an ultrasonic wave in liquid. (1973)

3. W. Boch — Examination of solutions of zinc oxide in methyl alcohole using molecular acoustics methods. (1975)

4. A. Muszyński — Attenuation of a shock wave by the air partition formed in liquid medium. (1975)

5. S. Olszański — Influence of ultrasonics on physical properties of hypo-eutectic alloys Pb/Sb. (1977)

6. P. Tymański — Acoustical underwater channel in South Baltic Sea. (1977)

7. M. Brzozowska — The influence of an undulation of the sea surface and the bottom on the sound propagation in the shallow sea. (1977)

8. Z. Klusek — The characteristic elements of the sea noise in South Baltic Sea. (1977)

9. M. Kosmol — Nonlinear effects in ultrasonic - laser light diffraction. (1978)

10. M. Jasiński — Association of imidazole (1,3 diazole) and pyrosole in p-xylene examined with molecular acoustic methods. (1979)

11. B. Linde — Acoustical relaxation in organic liquids of cyclic structure. (1979)

12. I. Wojciechowska — Holography interferometric examination of vibrations of ultrasonic transducers radiating into liquids. (1979)

13. P. Kwiek — Optical holographic method used in investigation of ultrasonic fields. (1979)

14. P. Miecznik — Examination of the structure stabilization of six--methyl-phosphor-threeamid solution in water with ultrasonic spectroscopy method. (1981)

15. M. Witkowska-Borysewicz

   Examination of acoustooptical effects in nematic liquid crystals oriented in magnetic field. (1981)

16. Z. Dukiewicz	Determination of independent participation of structural sound sources in dispersive medium on the model of a ship. (1982)	
17. W. Bandera	Using of mechanical impedance method for determination of the complex Young modulus of viscoelastic materials samples dynamically loaded by infinite mass. (1982)	
18. S. Weyna	Criteria of evaluation and application of determination methods of vibroacoustical characteristics for partitions of ship enclosures. (1982)	
19. W. Ziółkowski	Measurement of the complex modulus of viscoelastic materials with the beam point impedance method. (1984)	
20. S. Pogorzelski	Examination of processes connected with existing of oil-derivatives substances on the water surface with acoustical method. (1985)	
21. R. Wituła	Propagation of elastic waves in a thin uniaxial liquid crystal layer adjacent with solid. (1986)	
22. J. Szczucka	Using of acoustical methods for study inhomogenities in Baltic Sea. (1986)	

(+ 3 Thesis in other domains of physics).

D.Sc. Thesis

P. Kwiek, Near and far field in the phenomena of light diffraction by two ultrasonic beams. (1986)

Universities have various possibilities to couple acoustics with industry. One of the forms exemplified by the situation at the University of Gdańsk was erecting of the specialized acoustics laboratory in which a straight cooperation between university staff and industry people as well as students undertaking diploma projects is performed.

Important role in acoustics - industry integration is played by organizing postgraduate studies for training secondary school teachers as well as for acousticians from industry. The postgraduate education for Ph.D. and D.Sc. degrees in acoustics by proper choice of Thesis topics in research coupled with technological problems is also a field to obtain better cooperation between science and industry.

REFERENCES

1. Lundquist, D., The Academic - Industry Gap, Europhys. News, 11, 10-11 (1980).
2. University - Industry interaction, Europhys. News, 14, 8 (1983).
3. Śliwiński, A., Main trends of application of acoustics in industry (Abstract, Fizyka dla Przemysłu, Gdańsk, 20-22 September 1984) 51 (in Polish).
4. Since 1953 (Delft) and then 1956 (Cambridge), 1959 (Stuttgart), 1962 (Kopenhagen), 1965 (Liege), 1968 (Tokyo), 1971 (Budapest), 1974 (London), 1977 (Madrid), 1980 (Sydney), 1983 (Paris), 1986 (Toronto) Proceedings of the ICA Congresses organized by ICA (International Commission on Acoustics) has been regularly published.
5. François, P., 25 Years of International Standardization and the Measurements of Machinery Noise, Proc. of 31-st Open Seminar on Acoustics Błażejewko - Poznań, 1-13 (1984).
6. Trautman, A., Proposals and remarks on education and development of scientific staff, Postępy Fizyki, 35, 271-279 (1984) (in Polish).

THE SITUATION OF THE ACOUSTICAL EDUCATION IN HUNGARY

T. TARNOCZY
Scientific laboratories of Hung. Akad. of Sciences
Budapest, Hungary

We cannot discuss the problem of acoustical education without talking some words about the general situation of education. We have the feeling that education is being in revolt throughout the world, many questions of fundamental importance propound for answer, and further we are alive in a complicated era of process of changing. This is the reason why I should like to characterize the general situation in a few words.

I.

The science and technology developing very speedly resulted in a hard position of the education all over the world. It is out of question that we should teach in schools of various level the scientific bases of technological news which are integrated in practical life. The newspaper, radio and television talk about the new technical terms every day. They give reports in most natural tone about laser's applications, results of ultrasonic diagnostics, holography, X-ray tomography, isotopic chronology and mainly about computer technology. The cultured man has a requirement to know any idea about the bases of principle of these terminologies. But it is also important in the interest of preparing higher school education to give some knowledge to schoolchildren and students on new scientific results and their basic explanation. The enumerated examples are only a few ones among the several topics of up-to-date "technological revolution". Beside the mentioned subject matters many other technological or scientific results have penetrated the closed system of our classical knowledge, too.

The latest results must be taught without doubt because the man of the age needs them. On the other hand the length of learning hours and study time is not to be increased over a dayly eight hours and over 16-17 years. The "dose of study" cannot be augmented even in case of picking up the new knowledge in the syllabus of a course, and we may cherish hope about the "efficacy" of education only. It is clear that no miracles can be expected from pedagogy, therefore we have tb consider - beside the increased efficacy - the refashioning of the subject matter of instruction and the replacement less important positive knowledge by more important and up-to-date one.

This necessity has always been in action in the course of history but perhaps never such a basically and urgently than nowadays. All these concern the konowledge of science and technology, in particular. The situation is ridden with great difficulties; we almost may talk about critical period of education. New and new reform endeavours emerge year by year to changeing the method, the matter, the quantity and the profoundness of education. According to a group of experts, the most exciting method is to teach the fundamental or most essential laws and theories deeply; the detailed knowledge would be eitherway attained in practice.

According to another conception it would be enough to flash all interest and/or the most important results because the universal summary and the presentation of most interesting results awoke the creative instinct; saying, this will be enough for collecting deeper knowledge later. The awkward try to find a way out from this difficult situation was evidently caused by the rapid accumulation of new informations. And the exigence results in compulsory solutions.

Good results are to be expected when carrying through whenever of both mentioned conceptions but always made consequently and by correct pedagogy. In reality, however, it flows a struggle to the knife between the special subjects. Who should submit to the reduction or perhaps should give up his subject of instruction, and who should it not? Which are the most imporatnt subjects, which must be taught by all means?

To solve the difficult problem of equilibrium among the materials to be instructed, and to discuss it according to the new points of view, first of all it is necessary to find people who are experts in several scientific fields and who are in possession of proper review and summary, further who are able to make order "sine ira et studio" in the hierarchy of various topics. Such a person with the required qualification, degree of culture and with a wide intellectual horizon exists so far - or perhaps already - in very few numbers. And if he exists, it is problematical whether he is willing at all to deal with the problem of educational reform. I think this fact may be one of the reasons why we do not have a real solution till now, what is more, we are going toward a bad direction.

The "bad direction" can be experienced in our country in many respects. At the beginning the teaching time in schools were diminished to 45 min instead of 50 min. This measure was found prejudical by the school masters. The second step was to increase of the number of lessons from a dayly 5 to 6 sometimes even to 7. This is unfavourable both for teachers and students. The third fact was that the subject matter of instruction prescribed in the whole country is the same, regardless of the residential place and cultural background of pupils. Therefore the instructors have little individual freedom in their work. Recently the situation has changed in such sense that parallel textbooks are written with various conceptions, especially in physics and mathematics.

Last but not least, the authors of the textbooks carry out pedagogical experiments, e.g. they try to teach in public elementary schools theory of sets in the frame of mathematics or transport phenomen in biology; although these studies could be understood after proper preliminary training and only at adequate level of maturity. On the other hand, the same authors do not consider as important to learn such positive basic pieces of knowledge like descriptive geography or natural history, i.e. the names of mountains, rivers and cities or the morphological study of plants and animals.

A modern wish-dream seems to be the education of philosophers, world economists and computer programmers already in the secondary school although most students want to find employment in the normal life. In Hungary there are 1 300 000 children in the public elementary schools, of them 230 000 go on to secondary school and only about 40 000 attend a university of college (in our terminology high school). This means a proportion of 32:6:1.

I can hardly judge the situation in other countries; nevertheless I may conclude from many signs that similar problems emerge also in other countries and even world-wide, so the outlined conditions could be treated not only as a regional problem but as a general one.

II

Before a detailed discussion of acoustical education, I present a short survey of the Hungarian school system.

1. Between the ages of 6 to 14 practically all children attend the public elementary school. In fact there are few who keep away notoriously or must stay in a form a second year after fourteen. In spite of compulsory identical subject matter of instruction the various schools give very different education matters. It can be hardly made a comparison between the knowledge of a young people educated in the capital and of another one learned in a village in the backwoods. An equalization could be imagine by very good teaching staff but the good teachers were picked up in the big cities.

It is also to be mentioned that the textbooks were often rewritten and their contents are not already understandable even by the teachers.

2. The secondary school system gives education for the youth between 14th to 18th, and has a very large scale of types. The most general education will be given by the so

called general gymnasium where the traditional material subjects are in programme but not according to a traditional proportion. Physics e.g. will be taught all four years, in about 1.5 hours weekly. Some of these schools have special classes in which one or two subjects are accentuated and have a higher number of lessons on the timetable. There are e.g. classes in which the study of some languages are essential, and other ones where mathematics or biology could be learned thoroughly. In these classes the other subjects in a reduced time and with a reduced substance will be instructed.

Another type of secondary schools is the so called vocational (trade-, industrial-, sanitatary-, technological-, architectural-, agricultural- etc.) school with special programmes. In these schools also the basic cultural subjects (e.g. Hungarian literature and history, history of art etc.) are fundamentally diminished.

It is an important note that from a gymnasium it is possible to apply for admission to all universities or other high schools, whereas from vocational schools this possibility exists only for special colleges or special faculties or universities.

3. The universities of the country are also specialized in some extent; there are 4 scientific universities, 5 medical and 3 technological ones, further 6 colleges of university character for special tasks, e.g. veterinarial, agricultural etc. The 12 universities have the right to give out doctor degrees, the others have it not. Moreover there are many real high schools, e.g. teacher's training colleges, music academy, academies of fine arts, technicist schools, altogether 30 or 32. For admission to all these institutions an entrance examination is needed.

III.

Now we are coming to the education of acoustics in our country. Briefly we may not talk about acoustical education in the proper sense of the word because 1st:
before the highest studies one hears only a few words about acoustics, and 2nd: there are no acoustical chairs at the universities colleges and high schools. Therefore we were pressed to find the substitute possibilities for educating acoustical experts and scientists.

Let us begin with the bases. In the primary and secondary schools there are three courses in which the pupils could learn something about acoustics: physics, biology and music.

1. Physics will be taught in the 6th, 7th and 8th form of public elementary schools. The programme is mechanics, heat, electricity and optics. In the three textbooks of about 500 pages we find 5 pages for vibration and wave

motion, and 2 pages for sound, i.e. 4 per mille of the material. Because the teaching period amounts 1.5 hours per week, the acoustics will be taught in 2.2 minutes. Therefore it is understandable that not about decibels, hearing, noise will be heard. Only frequency, Doppler-effect and ultrasound are mentioned.

In biology one treats hearing at the 8th form in about 2 pages in which a short survey is given on the inner ear, frequency domain of hearing, hearing losses and 2 rows about the Nobel-prize winner Békésy. The half of the text is but written with petit characters which means, schoolchildren must not know these parts, they are pieces of reading only. In spite of this, this book offers the most data about acoustics in the primary school. I demonstrate the book before you. On the basis of teachers' information the working time for the subject amounts 10 minutes.

2. At the gymnasiums one learns physics in all four years. The book for the 3rd form have a part for vibrations and wave motions but mostly treated from point of view of optics and of electromagnetic field. The short mention about acoustics is obsolete, and if has something "modern", these are unimportant slangs, e.g. "high fidelity". In connection of Doppler-shift the author mentions the radar but not sonar. Now it is planned to write another parallel book for giving possibility for teachers to choose the better.

We have to make special mention about the teaching of physics in some vocational schools. In these schools oriented in technical direction acoustics with proper quality and quantity is lectured. Physics book of the 2nd year includes 160 pages, out of this wave motion amounts 20 pages and acoustics 10 pages. The text is - with few exceptions - correct. This book is also circulated here.

The book of biology for 2nd and 3rd classes of gymnasium is very good. Among others hearing and voice production are also treated. Altogether more than 3 pages in the book of 294 pages, i.e. about 1.2 %. This subject matter will be taught 1.5 hours in a week during two years. the book is also presented before the Conference.

In the sanitary vocational schools similar texts may be read but the contents are more directed into the medical treatment sense.

Unfortunately, according to the new course programmes nothing about acoustics is to be found in the present song teaching books.

3. The most important education about acoustics is going - of course - on the universities and colleges. In this level the fundamentals must be also taught in the frame of physics, but physical sciences are so much extent that the time is not enough to give profound study of all fields of physics. Mainly modern capitals will be lectured on classical ones' account; the latters are almost totally missing.

It is a fundamental problem that there are not

acoustical chairs or institutes at all. Filling in the gaps a five year course of the whole field of acoustics was given by myself in 2 hours per week between 1950 and 1975 five times, successively. During the whole period not more than 80-100 acousticians were educated.

In the Technical University Budapest a course of technological acoustics is given about for 30 years. This course of three semesters will be completed by one or two facultative lectures in the ninth semester for specialization. These courses of lectures were delivered in the frame of the Institute of Electronics and Telecommunication where some students specialize himself in acoustics. These people want to work with broadcast, television and film industry. About the half of all acousticians are educated in this way.

Another center for teaching acoustics is to be found in the faculty of architecture. Here in the frame of the Chair for Building Structures there is an intensive education of building acoustics. A third place in the same university where students receive acoustical education, is the Chair of Flow Sciences.

Recently the directing board of various colleges and/or high schools took note of the lack of fundamental knowledge on acoustics in general, and therefore these schools recently organize acoustical courses, too. A fair example about this endeavour are the lecture notes edited by the high school for Hydraulic Engineering which will be also demonstrated here. The essential treating of sound insulation between pages 205 to 246 is a very good introduction for all experts who work in planning buildings from acoustical point of view.

On the medical universities the foundamental subject of instruction is biophysics. Not more than 1.5 hour remains for lecturing something about acoustics at all. The situation is similar to the scientific universities, except the difference that here in the frame of otorhyno-laryngological studies one needs more basic knowledge on acoustics.

The sporadic and therefore not profound knowledge, can not substitute for a fundamental acoustical lecture, and results in a confused and obscure learning. This may be experienced especially with architects, engineers and sometimes also with physicists and physicians.

4. To eliminate this insufficiency we organize special courses after ending the university studies. Such courses are e.g. the refresher course for engineers and the education of special experts. It will be brought care to bear upon educating experts for environmental protection. A part of this a two year course deals with acoustics in the frame of which we teach physical acoustics, psychoacoustics, noise abatement, infra- and ultrasonics etc. In this year the third course is in progress, and till now about 40

persons are educated. All acousticians working in the country mount up to 300-350 but scientific activities are to be taken only 30-40 persons.

There are two social corporations outside the official teaching institutions in the country which organize courses of lectures, scientific meetings, conferences etc. One is the Acoustical Commission of the Hungarian Academy of Sciences, the other the Acoustical and the Noise Abatement Sections of the Optical, Acoustical and Film-technical Society. In proper sense the activities of the Acoustical Laboratory as a scientific and service institution, influence also fruitful the evolution of acoustical culture and civilization in Hungary.

Our last endeavour to compensate the missing basic education is the edition of books. In the last 5 years seven books and monographys have appeared, partly under the auspices of the Acoustical Commission, and a further five are being written.

ACOUSTICS FOR EVERYONE: A GENERAL COURSE IN ACOUSTICS

R. A. WALKLING

Department of Physics
University of Southern Maine
96 Falmouth Street, Portland, Maine 04103
USA

ABSTRACT

Acousticians in the academic profession have an excellent opportunity to help increase the public's awareness of acoustical matters. Since the field of acoustics touches on a great many other disciplines, it can catch the interest of students with many different backgrounds and experiences. Interest in the subject is also created by public concern about acoustics problems. A general course in acoustics can take advantage of this interest and serve a broad audience. A course of this type has been taught at the University of Southern Maine since 1974, usually on an alternate year basis, with a wide variety of students enrolled. The course presumes minimal mathematical background while providing an introduction to most of the major topics in acoustics. In addition to the usual final exam, a 5-to-10 page term paper is required. Student response has been positive.

Acousticians who teach at a university have an excellent opportunity to increase the general public's awareness of acoustical topics. Acoustics, which we know as the study of the production, propagation, and reception of sound, is a subject which incorporates a large number of disciplines and relates to many popular cultural activities. For this reason, a course in acoustics can capture the interest of many people. There are, for example, the musician and the concertgoer; the architect and the person who lives or works in the building; the audiologist and the person with a hearing problem; the speech therapist and the person with a speech defect; the legislator, the noise code enforcement officer, and the person bothered by noise; the factory manager and the worker; the sound engineer and the hi-fi enthusiast; and of course the individual who simply has a desire to know more about an important topic in physics. Through a course designed to serve this broad spectrum of people, it is possible to broaden public understanding of the science of sound, to pass on the excitement of the subject in its many facets, and to generate additional interest as students see the many ways acoustics touches their everyday lives.

To satisfy the needs of this broad audience, a course must of necessity be a survey, touching briefly but with some thoroughness on many areas of acoustics. It must not be too technical; only a minimum of mathematics should be required. Demonstrations and laboratory work should provide visual and auditory experiences, which can enhance the presentation. In all cases it must be remembered that the course is not intending to produce experts in any one branch of acoustics, but rather to pass on the interdisciplinary excitement inherent in this subject.

Such a general course in acoustics was introduced at the University of Southern Maine in 1974, during a period of rising public concern over noise pollution. The title ACOUSTICS AND NOISE was chosen partly to capitalize on that concern. The course has been taught approximately every other year since that time, with enrollments ranging from 9 to 61 students. The students come with varied backgrounds and interests, in large part due to the character and location of the University.

The University of Southern Maine is a part of the University of Maine system, which provides public higher education to the state of Maine. Maine is in the northeast corner of the United States, with an area not quite the size of Hungary and a population of about one million. The University of Southern Maine serves the southern part of the state and is within commuting distance of about half the state's population. Current enrollment is 9400 students: half of these are full-time and half part-time. The average age of undergraduate students is more than 26, indicating the presence of a large number of what we call "non-traditional" students, students who are returning to the academic setting after some time away. Classes are offered from 8 in the morning to 9:30 at night on two campuses and at several satellite locations.

One campus is located in Portland, Maine's largest city, the other in a rural community 12 miles (20 km) away. A bus provides shuttle service between the two campuses.

The course has primarily been given in lecture format one evening per week for 2-1/2 hours, for a 14-week semester. This scheduling allows people with jobs in business, industry, and teaching to take the course without having it conflict with their work schedule. It also reduces any necessary commuting time and doesn't occupy more than one day per week. No laboratory is currently offered with the course, due to scheduling difficulties, but demonstrations are an important component of the lectures. These demonstrations sometimes involve sophisticated apparatus, but generally use ordinary bits of equipment one can easily collect. Where appropriate, students themselves are involved in the demonstrations. For example, hollow paper tubes are handed out to the students when standing waves in air columns are being discussed, and students bring some of their own musical instruments when we discuss musical acoustics.

A typical outline for the course is shown in the first figure.

Figure 1.

Physical Acoustics — including vibrations, traveling waves,
    standing waves
Hearing — including the ear, decibels, loudness, masking, pitch
Speech — (included if the class has interest)
Music — including scales and the various categories of instruments
Sound Reproduction — covering all aspects of the reproduction
    chain, from original sound source to listening space
Room Acoustics and **Noise** — as time permits.

Since physical acoustics provides the foundation for all the other topics, more stress is placed on that subject than on any of the others. The mathematical ability of the students is quite varied, so it has been necessary to keep the mathematical requirements low, but certainly mathematics is not eliminated entirely.

Due to the large variety of topics covered, there are no restrictions on who can take the course. I feel that regardless of the person's background there is something to be learned. A sampling of the wide variety of backgrounds which students bring to the course is given in Figure 2.

Figure 2.

Art and Architecture	Professional Musician
Computer Science	Broadcast Industry
Business	Musical Instrument Manufacture
Philosophy	Speech Pathology and Audiology
Medicine	Hi-Fi Enthusiast
Geology	Industrial Noise and Vibration
Biology	School Teacher
Physics	Electronics Technician
Music Performance	Ultrasonography

Some students have expressed their interest in taking the course as simply "curiosity", "sounds interesting", and "I listen a lot". When the students are asked what course topics interest them most, they consistently answer "music" and "sound reproduction", with "architecture" and "hearing" mentioned slightly less often.

Although an attempt is made to attract music students, care is taken not to make musical acoustics the principle focus. This concern limits the choice of text, since several good texts are primarily texts in musical acoustics. Initially, I used several inexpensive paperbacks, by Benade, Pierce, and others, and other supplementary material. As these paperbacks began to go out of print, two very suitable books became available: one, by William Strong and George Plitnik, is entitled *Music, Speech, High-Fidelity*, and the other, Thomas Rossing's *The Science of Sound*.

The requirements for the course include a final examination and a term paper. The examination now consists of four parts: simple algebraic problems, true-false questions, multiple choice questions, and several brief essay questions. The term paper, 5 to 10 pages, permits the student to investigate a topic of his own interest and report on it. The student may choose to consider in greater depth a topic already covered in class, or may choose a subject which has not been discussed. Many interesting papers have been submitted. Some of the titles are given in Figure 3.

Figure 3.

The Acoustical Properties of Wood in Stringed Instruments
About Absolute Pitch: A Definition and Some of its Properties
Resonance and Diction as Applied to a Vocalist
Acoustics of the Baby Cry
Acoustical Elements of the Organ
Underwater Acoustics
Electronic Sound Generators
Acoustics and Auditorium Design
Tinnitus: An Auditory Abnormality
Pythagorean Tuning and Some History
Infrasound: A Short Survey
Music Therapy
Ear Hair
Thunder
The Phonograph
Speech Abnormalities

(Figure 3, continued)
- **The Acoustical Orientation of Bats and Men**
- **Acoustical Oceanography**
- **Loudspeakers**
- **Noise and Our World**
- **Concert Hall Design Criteria: An Overview**
- **Medical Ultrasonics**
- **Noise Monitoring and Area Noise Mapping**
- **Ultrasonic Welding**
- **Noise and Its Effects of the Human Body**
- **The Theory of Flute Design**
- **Personal Hearing Protectors**
- **A Survey of Three Different Styles of Guitar**

I think these titles demonstrate the range of interests the students bring with them to the class.

The students are also expected to fill out a form evaluating both the faculty member and the course itself. The evaluations have generally ranged from good to excellent.

In summary, I believe my experience shows that a general survey course in acoustics, intended for the average student, can reach an interested audience with the fun and excitement of sound in all its aspects. As acousticians in education, we have the unique opportunity of extending the knowledge of our subject to the general population, with the possibility of far-reaching benefits to scientific understanding and the quality of life.

THE IMPACT OF ACOUSTICS TO PHYSICS EDUCATION AND RESEARCH

WEI, RONGJUE
Nanjing University
P. R. C.

Dedicated to Professor Isader Rudnick for his 70th birthday and his distinguished contribution* to physics and acoustics, but revised.

ABSTRACT

The classical and highly interdisciplinary nature of the science of sound either occludes its being a branch of modern physics or makes its position in the latter somewhat dubious and mostly unjustified. The purpose of the paper is to explore with illustrative examples that from the unity of physics, the interaction of acoustics with other branch of physics, etc., acoustics is, as it has always been, an important and indispensable branch of physics though sometimes hidden. The relation between macroscopic acoustics and microscopic physics and other branches of modern physics are discussed in some length. It is a motivative force to the advance of physics for it is very heuristic to some concept and its impact to physics education and research cannot be overemphasized.

INTRODUCTION

Acoustics, as an old branch of physics, was developed through scientific study of one of the most conspicuous phenomena in our everyday life, and is advancing with all the rest of physical sciences. The negligence of the problem as is posed here arised from the fact that acoustics is not only dated but also a science of more interdisciplinary nature with everincreasing applications dispersed in engineering and technology, in life sciences and humanity, as shown in appendix 1 of this paper - a simplification and extension of Lindsay's wheel [1] proposed 21 years ago. The trend of modern physics concern more with the intrinsic of the physical world, with no direct bearings with human senses

---

* Partly introduced by the citation of the Gold Medal awarded to him at the 103rd A.S.A. meeting, Spring 1982(JASA) Supplement 1, vol 71, Spring 1982. The paper is based chiefly on a paper the author firstly reported at his visit to UCLA, Suny at Albery June 1979 (Preprint, Acta scientia, 1980) and then reported at 10 ICA, Sydney,1980 (appeared in its Proceedings and published in Science Progress 2,389(1982) in Chinese. A Guest comment appeared in Phys. today, Feb.,1985 emphasizing Acoustics and Education in Physics. See also[54].

e.g.,the particle physics dealing with the ultimate structure of matter. Acoustics seems to be off the main stream of modern physics. There are much indications that acoustics does not usually manifest itself as an independent chapter in an undergraduate text of general physics, though a young student of physics may know the name of Rayleigh from his law of black body radiation or elsewhere rather than from his classical work on theory of sound, from which he actually learns quite a lot. So do some authors [2] in summarizing the progress of modern physics in terms of energy scale. As claimed by P.M.Morse [3] and more recently by R.B. Lindsay [4,5], R.T.Beyer [6], and the present author [7] the extrinsic nature [8] of the science of sound never excludes the importance of its being an important branch of basic physics, and the oversight was exemplified [9] by the failure of some brilliant Chinese students with advanced training of theoretical physics to solve a fundamental problem on acoustics (App.2). To the present author, to make this point clear is not as much for those problems or publications already bearing the title of physical acoustics [10], or evidently physics-related, as for those subfields like psycho-acoustics, speech communications, are somewhat vague at first sight. The present author is attempting to clarify that the mutual relations dependence between modern acoustics and modern physics have not been lost in the age of increasing specialization in conformity with the viewpoint as was offered by the aforementioned authors some years ago. This paper will take up the following main themes:

(i) The parity and disparity of the trend of modern acoustics and that of modern physics;

(ii) acoustics as a branch of modern physics from unity of physics chiefly from the point of view of analogy;

(iii) acoustics as an indispensable part of modern physics. That acoustics has been and will always be one of the motive force for the progress of modern physics and that mutual dependence cannot be overlooked. The impact of acoustics physics education and research cannot be overemphasized.

PARITY AND DISPARITY OF THE TREND OF MODERN ACOUSTICS AND THAT OF MODERN PHYSICS

Physicists usually consider that the discoveries of cathode rays, radioactivity, etc., not explainable by classical theory, led to the quantum hypothesis by Planck and Einstein, and later the development of quantum mechanics marked a new epoch of modern physics. From then on, the main stream of physics seemed to have shifted to the exploration of microscopic world rather than its classical subjects as acoustics that had once reached its climax. It was really so for the edifice of classical phenomenological physics had been considered to be consolidated, finding its ultimate completion at the advent of the theory of relativity. Nevertheless, there were many special problems of hydrodynamics, the diffractions of light and sound, etc. Most of them have developed the theory of holography. As all physical phenomena are nonlinear with linearity as special cases, problems of nonlinearity have now dispersed in many current fields of sciences including acoustics and particle physics. All these problems once thought not new in their basic framework have had their renewed interests by both physicists and acousticians. For instance the monochromatic coherent sound waves had been discovered long before the advent of laser, however some others are enhanced by the development of quantum mechanics, field theory, etc. and also by the discovery of electrons and other nuclear particles. For instance, the scattering of light and sound (the analogy of which corresponds to the macroscopic relaxational problem of sound absorption in water fogs [32] and their microscopic counterpart [11] i.e., photons, or the scattering of waves and particles [12] cuts across the boundary of classical and modern physics. Specific problems thus from interaction of sound by sound [6,13] to phonon-phonon, electron-phonon and phonon--other "ones", [14] all these belong to problems of fundamental nature, which, of course, involves the study of superconductivity [15] and phonon physics [16]. In superfluidity, several sounds other than conventional first sound in ordinary fluids have been found and new modes are developing [18]. Strenuous efforts are being made including preliminary report on production of very high frequency coherent phonon beam[19] of the order 10^{12}-10^{13}Hz and for the study [20] of the condensed matter physics. No matter how these achievements are classified, they are at least partly acoustical and those admitted acousticians themselves are of actually participating such development.

The demarcation may seem to arise from some areas of sustaining or even increasing interests of acoustician, but due credits have been shifted to elsewhere but not modern physics. For instance, such pioneer work as was done by W.C.

Sabine in architectural acoustics at the beginning of the century has proved invaluable to human culture and yet it had not draw enough attention to the world of physics for the main trend then was atomic physics. One must not forget that in this branch of acoustics its merits have been shown by some eminent acoustical physicists [6] who had applied quantum mechanical method [21], modern mathematical computer techniques [22] etc. to a better solution of sound waves in enclosed space, in conformity with the criteria determined by subjective judgement. But if a physics teacher takes up such progress as typical example to illustrate the transition of ray theory to wave theory and to relate to modern computer technology, I believe the students will accept it more readily than any other example he has in mind.

Planck once remarked early in this century that one of the essential progresses made in the 20th century was that one could express physical world in quantities instead of qualities through human senses. In acoustics he mentioned expressing pitch on terms of frequencies. May be Planck had not given enough consideration of the complexity, if possible, of working out a physical theory for expressing sound quality of a concert hall or anything else of the sort. The hearing mechanisms are lest too intriguing that when Von Bekesy essentially a physicist, won his Nobel prize his theory of hearing is rather rudimentary indeed. Many capable acousticians have worked for years for the combination of physical theory with engineering practice, and finally many a concert hall with good acoustics is now in existence. Others are working on cochlear mechanics. But either Bekesy's or these ablest acousticians' achievements have been ascribed to elsewhere in spite of the fact that it is essentially physics that leads to such successes. The situation might be quite similar in many other fields of modern acoustics.

So far we may classify modern acoustics into two main categories, one is microscopic and intrinsic which may be more accessible to the trend of modern physics, and the other chiefly macroscopic and extrinsic, which is by no means deviating too far from this trend. In what follows in this paper we shall lay greater emphasis on the second in order to show that both categories obey certain fundamental principles of modern physics.

ACOUSTICS AS A BRANCH OF MODERN PHYSICS
FROM UNITY OF PHYSICS

In general the development of quantum mechanics is an extension rather than a cut-off from the classical physics. Some branches of classical physics are much related to acoustics, as was declared by Pauli and others. The revolution that brought physics from Newtonian to quantum mechanics, left the thermohydrodynamics unchanged and for this reason the quantum mechanics has been regarded as micromechanics [23]. N.Bohr advocated his corresponding principle with classical physics as the limit of quantum mechanics although his principle must be handled with care. This sort of analogy sometimes engenders new ideas also. Without previous knowledge of acoustical waves, Clerk Maxwell could hardly give the guess that light waves might be just like a sort of mechanical waves in "jelly-like" ether, which finally led to his famous and elegant theory and equations of electromagnetic field. Appendix 3 shows the unity of waves equations. Here only linear equations are considered. The detailed derivations can be found in some standard text of mathematical physics. The generalized equation /8/ actually laid the basis of field quantization by Heisenberg and Pauli with Appendix 4 showing how it was started.

Appendix 4 also shows the generality of continuity equation or of mass conservation proposed by De Broglie which holds true for classical acoustical and electromagnetic waves in connection with quantum mechanics, gravitational waves in connection with general theory of relativity and a number of coupled waves, etc. De Broglie defined density "ρ" and flux "ρV" as the very important properties of all waves. The equation also holds true for particle picture based on his wave-particle dualism.

Appendix 5 shows the similarity of uncertainty principle in quantum mechanics and that in acoustics. Heisenberg's uncertainty principle is a very basic one which states the impossibility of determining the position and momentum or time and energy of a particle (say, an electron) at the same time. There are many ways to state such affairs, what we adopt here [24] is a convenient way by comparing the corresponding well-known problems in acoustics [25]. The comparison shows remarkable resemblance in mathematical formulation. The acoustical "transients" play an important role both in speech and music and in vibration, so do the sound recording and reproduction systems. In this connection more extentive works were reported and confirmed by communication theory [26], and recently by auditory analysis [27].

Appendix 6 shows the similarity of tunneling effect of quantum mechanics (such as α-emission through a potential barrier, etc.) as was much discussed in almost every elementary text of quantum mechanics with transmission of plane sound waves through layered media solved by the matrix method [28] in mathematical expressions of transmissivity.

There are much more examples for illustration. It is very interesting to mention that P.J. Westervelt [13] had once identified his result obtained from the intensity of a scattered sound with that of the number of scattered particles of Rutherford's scattering of α-particles with thin foil after quantum mechanical treatment similar in mathematical expression, and recently more interesting to see that he had correlated his result of transient performance of parametric sound end-fire arrays with emission of radar or light pulses emitted from the earth to a planet and thus confirmed the gravitational waves predicted by Einstein's general theory of relativity. Putterman [29] suggested the ultimate emergence of acoustics and classical hydrodynamics to modern physics from classical nonlinear theory.

All examples thus mentioned show the unity of physics. Similarity or analogy in physical concept or in mathematical formulation is by no means accidental. The analogy usually happens in very remote and unexpected places.

I should like to mention as one more concrete example to illuminate this argument, i.e., the problem of mutual actions between aerosols in a sound field of practical importance to sonic agglomeration and precipitation already found applications in industry but not so assured to atmospheric science. So far nine different kinds of interactions have been found [30,31], varying in some 17 orders of magnitudes with some attractive some repulsive, some depending on the property of sound fields, etc. Compared with the four interactions of the publicized corresponding problems in particle physics which is of course more intrinsic, it is little known and its progress is very slow. I am sure there will be a breakthrough if more brilliant physicists do not bother to take some time attacking this problem of "classical" nature. An analogy was already shown by the intrasonic absorption of water fogs[32], a purely macroscopic problem, to ultrasonic absorption of molecular level [33,34], both showing relaxational peaks (by considering the fog as an ideal gas with internal degree of freedom) as was confirmed by many authors (Appendix 7).

ACOUSTICS AS AN INDISPENSABLE PART OF MODERN PHYSICS

In retrospecting the development of the modern physics one should not forget its historical merit to the advance of concept of optical and electromagnetic waves, and hence to that of all waves and wave fields hitherto not fully investigated. One may deny Rayleigh's contribution to Planck's quantum theory but not his contribution to quantum mechanics, such as partial wave solution for scattering problem, and to Bohr's liquid drop model in nuclear physics, the idea of which might be derived from his early work on surface tension of water [34] in addition to theory of mechanical vibration. Enrico Fermi who was well-versed in hydrodynamics [35] and preferred thinking in terms of concret things [36], started his theoretical derivation of "quantum theory of radiation" [35,1] from the oscillation of a simple pendulum and that of a string as analogy, and elucidating his beautiful theory of beta decay [37] by resorting normal modes of vibrations in enclosed space as an important step. Now acoustics has entered the area of space physics [5].

Since the microscopic part of acoustics belongs to modern physics itself, as already stated above the investigations of physics of fluids and of condensed matter by means of ultrasonics and practersonics would naturally emerge. By combining efforts of physicists and acousticians, high energy coherent phonon beam [39] thereby to obtain the highest lattice frequencies will soon be realized, and then a new epoch will be ushered in with acoustics as an unsurpassed means for the study of atomic forces, electronic structure, and a variety of collective excitations in solids while combining itself with X-ray and neutron diffractions. What we want to bring to notice is that acoustics has been applied frequently by some physicists unconsciously; the development from electron-phonon interaction to a successful BCS theory of superconductivity furnishes one of many examples. In addition, the ultrasonic experiment [40] already predicted a successful verification of this theory.

Let us talk a little more about an old problem of non-linearity concerning solitary waves now called solitons [41] discovered by Scott Russel from a narrow channel in 1834. It turns out to be a fashionable subject of diversified interest from applied mathematicians to particle theorists in

more or less abstract ways. But their existence in acoustics
[42] has been shown experimentally, and the discovery of non-
propagating soliton [43] in small rectangular tray of water
by vibrating vertically has brought this phenomenon in nature
back to the laboratory and classroom demonstration. We can
then study the physics underneath more intuitively. Up to the
present, we just can tell how much physics and mathematics
and technical applications this unique soliton will involve
[44,45,46]. Another more important and universal phenomena
which are deserving greatest interests to physicists,
chemists,mathematicians,biologists and ecologists named
"chaos" could be traced back to the time of Faraday and
Rayleigh. We now already found out route to chaos can be
realized in several acoustic system [47-50].

The combination of acoustics, optics and electronbeam
gave birth to the discovery of acoustic microscope, photo-
microscope and electron acoustic microscope---these have
brought a revolution to conventional microscopy.

CONCLUSION

What has been discussed so far might have shown the
close relationship between acoustics and modern physics.
"Acoustics can't get on very far without the rest of physics
and vice versa---the rest of physics can't get very far
without acoustics" was what P.M.Morse [3] remarked some 32
years ago. It seems that as diversity goes farther, the demarcation between acoustics and modern physics becomes even
vaguer [51] in those overlapped areas where mutual under-
standings seem to be more pertinent. I should like to quote
the paper entitled "Unity of Concept in the Structure of
Matter" by John Bardeen [52], in which he elucidated with
convincing illustrations that progress made in the understanding in one area may often be applied in many other fields.
In problems of the structure of matter, he mentioned that the
method of elementary excitations should be one of great generality. This certainly holds true in modern physics and
acoustics for there are always some linkages between both. In
retrospect the author recalls he talked the problem of the
importance of acoustics in physics with professor L.M.Brebo-
vskikh, the director of Acoustical Institute of the USSR
Academy of Sciences in 1956, and more recently with more
American and European physicists and acousticians, they all
shared the same viewpoint of mine. From the above mentioned
arguments and examples, the author gives the name of this

lecture as entitled in order to call attention to educators and researchers who think otherwise. As a matter of fact, a concise article but by no means less convincing has already been published [53-54]. The present author also demonstrated by video tapes in various occasions to verify the topic as entitled, the latest one [55] showed that out of a simple experiment [43], there will be even more intriguing phenomena worthy of further investigations in so far as the term "impact" is concerned.

APPENDIX 1
EXTRINSIC NATURE OF ACOUSTICS

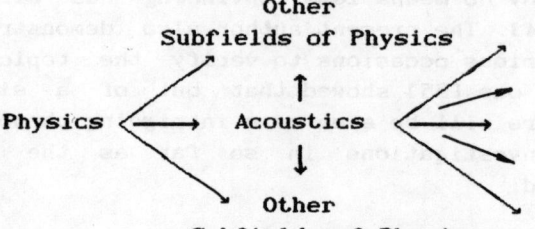

APPENDIX 2
UNITY IN WAVE EQUATION

L - Lagrangian density, a function of field variable;
$\mathbb{L}$ - Lagrangian function;
Q_α - field variables.

Hamilton's principle:

$$\sigma \int_{t_o}^{t} (T - V) \, dt = \delta \int_{t_o}^{t} L \, dt = 0 \quad \text{(for any conservative system);} \quad /1/$$

$$\mathbb{L} = \int L \, dt; \quad /2/$$

$$L = L(Q, \frac{\partial Q_\alpha}{\partial x_f}, \frac{\partial Q_\alpha}{\partial t}, t) \quad /3/$$

If x_β is the coordinate of a certain point in the field, we can prove

$$\frac{\partial L}{\partial Q_\alpha} - \sum \frac{\partial}{\partial x_\beta}\left(\frac{\partial L}{\partial \left(\frac{\partial Q_\alpha}{\partial x_\beta}\right)}\right) - \frac{\partial}{\partial t}\left(\frac{\partial L}{\partial \left(\frac{\partial Q_\alpha}{\partial t}\right)}\right) = 0 \quad /4/$$

Eq. /4/ includes wave equation, heat diffusion equation etc., e.g., let

$$L = \frac{K}{2} (\nabla \phi)^2 - \frac{\rho}{2} \left(\frac{\partial \phi}{\partial t}\right)^2,$$

and define sound velocity $C = \sqrt{K/\rho}$, then we obtain the wave equation of sound as

$$\nabla^2 \phi = \frac{1}{C^2} \frac{\partial^2 \phi}{\partial t^2} \quad /5/$$

Further, let

$$L = \frac{\hbar}{2m} [\nabla \phi^* \nabla \phi] + \phi^* V \phi + \frac{\hbar}{2m} \left(\phi^* \frac{\partial \phi}{\partial t} - \phi \frac{\partial \phi^*}{\partial t}\right) \quad /5.1/$$

The Schrodinger equation
$$\frac{\hbar}{2m}\nabla^2\phi + V\phi = \frac{\hbar}{i}\frac{\partial\phi}{\partial t} \qquad /6/$$
can show /4/—the Klein-Gordon equation etc., and
/5/—the Hamilton-Jacobian equation as geometric approximation.

APPENDIX 3
SIMPLE ACOUSTICAL PROBLEM UNSOLVED BY BRILLIANT PHYSICS STUDENTS

On the 1983 CUSPEA exam, students either omitted the unfamiliar problem (for they could choose six problems out of seven or gave wrong answers. The problem runs thus:

> Two immiscible fluids are in planar contact as shown below Planar acoustic waves of pressure amplitude A and frequency f are generated in medium (1), directed toward medium (2).  Take A and f as given quantities and assume the wave propagation is normal to the interface. Medium (1) has density ρ_1 and sound velocity c_1, while medium (2) has density ρ_2 and sound velocity c_2.
> (a) What are the appropriate boundary conditions at the interface?
> (b) Apply these boundary conditions to derive the pressure amplitude A_r, of the wave reflected back into the medium (1), and the pressure amplitue B, of the wave transmitted into the medium (2).

This belongs to the simplest case of wave transmission phenomena, whether the wave is mechanical or electromagnetic in nature. Anyone with a modest training in acoustics could give the right solution right away. One assumes a wave of the form $e^{i(\omega t \pm kx)}$ with the plane of contact at $x = 0$; the continuity of acoustic pressure at the interface then gives

$$A + A_r = B \qquad /7/$$

with $k = 2\Pi f/c_1$, and $k = 2\Pi f/c_2$; c_1 and c_2 are the phase velocities of sound waves in medium (1) and medium (2), respectively. The continuity of particle velocity (not velocity as given in the official version of the solution that accompanies the problem) yields

$$U_A + U_{A_r} = U_B \qquad /8/$$

The specific acoustic impedance is the ratio of sound pressure to particle velocity, or $P/U = \pm \rho c$. The sign is taken as positive for the incident plane wave and for the transmitted wave, which propagates in the same direction; it is negative for the reflected wave. One can thus express the velocity-continuity relation as

$$\frac{A}{\rho_1 c_1} - \frac{A_r}{\rho_1 c_1} = \frac{B}{\rho_2 c_2} \qquad /9/$$

Equations 7 and 9 are the required answer in part a. From them, one obtains

$$A_r = A \frac{\rho_2 c_2 - \rho_1 c_1}{\rho_1 c_1 + \rho_2 c_2}$$

$$B = A \frac{2\rho_2 c_2}{\rho_1 c_1 + \rho_2 c_2}$$

which is the answer for part b. No calculus is required.

Since the acoustic impedance that characterized a given medium in which the sound wave travels is a concept never considered on an equal footing with the refractive index in optics. That may be why the official version of the examined solution answered the second part of the problem by deriving the relation between the sound pressure and particle displacement (actually $-\partial \mathcal{K}/\partial \lambda$ is the condensation, usually expressed by s), which only made the calculations much more complicated, and solved part b of the problem in a round-about way.

APPENDIX 4
CONSERVATION OF PARTICLES AND IDEAL FLUIDS

We also see that if we drop the last term of /4/, then it reduces to Lagrangian form of equation of motion for classical field. By assuming the field quantities to be expressed by complex Q_α, and its conjugate Q_α^*, we have

$$\left. \begin{array}{l} \dfrac{\partial L}{\partial Q_\alpha} - \dfrac{\partial}{\partial x_\beta}\left(\dfrac{\partial L}{\partial \left(\dfrac{\partial Q_\alpha}{\partial x_\beta}\right)}\right) = 0 \\[2em] \dfrac{\partial L}{\partial Q_\alpha^*} - \dfrac{\partial}{\partial x_\beta}\left(\dfrac{\partial L}{\partial \left(\dfrac{\partial Q_\alpha^*}{\partial x_\beta}\right)}\right) = 0 \end{array} \right\} \qquad /10/$$

Pauli introduced the transformation $\phi_\alpha \rightarrow e^{i\alpha} Q_\alpha$, $\phi_\alpha^* \rightarrow e^{-i\alpha} Q_\alpha^*$, ($\alpha$ real and small) or the first gauge transformation. By some manipulation, an equation closely connected with conservation of particles was obtained,

$$i\frac{\partial}{\partial x_\beta}\left[\frac{\partial L}{\partial\left(\frac{\partial \phi_\alpha}{\partial x_\beta}\right)}\phi_\alpha - \frac{\partial L}{\partial\left(\frac{\partial \phi_\alpha^*}{\partial x_\beta}\right)}\phi_\alpha^*\right] = 0 \quad /11/$$

De Broglie defined density "ρ" and flux "ρV". All waves obey the continuity equation of "conservation of fluids",

$$\frac{\partial \rho}{\partial t} + \nabla\cdot(\rho V) = 0 \quad /12/$$

which is applicable to both classical and quantum physics, and $\int \rho dt$ is independent of time and equal to 1.

In quantum mechanics:

$$\rho = \phi\phi^* = |\phi|^2 \quad /13/$$

With the Schrodinger equation multiplied by ϕ^* and its conjugate equation by ϕ, we get

$$\frac{\partial}{\partial t}(\phi\phi^*) + \nabla\cdot\left[\frac{\hbar}{2im}(\phi^*\nabla\phi - \phi\nabla\phi^*)\right] = 0 \quad /14/$$

which is another way of expressing /9/ with

$$\rho V = \frac{\hbar}{2im}(\phi^*\nabla\phi - \phi\nabla\phi^*) \quad /15/$$

We can likewise get Eqs. /10/, /12/ for equations of Klein-Gorden and Dirac. Therefore an analogy exists both in physical concepts and in mathematical formulation.

APPENDIX 5

UNCERTAINTY PRINCIPLE IN QUANTUM MECHANICS AND IN ACOUSTICS

In quantum mechanics:

x — position coordinate,

$$\bar{x} = \int x|\phi|^2 dx, \quad \phi(x) = \delta(x-x_o); \quad /16/$$

$P = \dfrac{\hbar}{i}\dfrac{d}{dx}$ momentum operator; $f(x)$, its eigenfunction; p', eigenvalue

$$P f(x) = P'f(x), \quad f(x) = e^{\frac{i}{\hbar}P'x} \quad /17/$$

Let x vary in $\pm a$, i.e., $\delta r = a$, we can prove the probability distribution of P',

$$P' \sim \left[\int_{-a}^{a} e^{-\frac{i}{\hbar}P'x} e^{i\varkappa x} dx\right]^2 = \left[\int_{-a}^{a} e^{i(\varkappa-\frac{P'}{\hbar})x} dx\right] -$$

$$= \frac{\sin^2\left[(P' - \hbar\varkappa)\frac{a}{\hbar}\right]}{(P' - \hbar\varkappa)^2}, \quad /18/$$

with maximum at $\hbar k$, and a main peak varies in the interval "$2\pi\hbar/a$" as shown in the figure, i.e.,

$$\delta p' = \frac{\pi\hbar}{a}, \qquad /19/$$

$(14),(16) \rightarrow \delta p' \delta x \sim \hbar$ /20/

similarly $\quad \delta t \cdot \delta E \sim \hbar$ /21/

$E = \hbar\omega, \quad \delta t \cdot \delta\omega \sim 1$ /22/

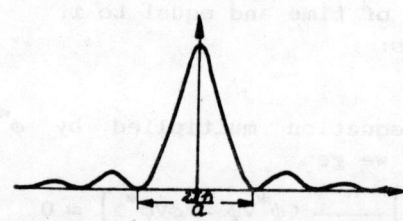

In acoustics:

Let

$$P(t) = \begin{cases} 0 & (t < -\frac{\Delta}{2}) \\ p\cos(\omega_0 t) & t \in (-\frac{\Delta}{2}; +\frac{\Delta}{2}) \\ 0 & (t > +\frac{\Delta}{2}) \end{cases} \qquad /23/$$

By Fourier transform, the amplitude distribution with ω is

$$A(\omega) \sim \frac{1}{\pi} \frac{\sin(\frac{\Delta t}{2}(\omega - \omega_0))}{(\omega - \omega_0)},$$

which shows $A(\omega) = 0$ at $\omega = \omega_0 \pm \frac{2n\pi}{\Delta t}$ $(n = \pm 1, \pm 2 \ldots)$.

Hence $\quad \Delta\omega = \omega - \omega_0 = \frac{2\pi}{\Delta t}$ /24/

Since only the order of magnitude is considered, therefore,

$\Delta t \cdot \Delta\omega \sim 1$ /25/

is in agreement with /19/ and the sound intensity is

$$E \sim A^2 \sim \frac{\sin^2(\frac{\Delta t}{2}(\omega - \omega_0))}{(\omega - \omega_0)^2} \qquad /26/$$

which is similar to /15/, all are of the form $\sin^2 x/x^2$.

APPENDIX 6
α—EMISSION IN QUANTUM MECHANICS AND TRANSMISSION OF SOUND IN LAYERED MEDIA OR IN PARTITIONS.

In quantum mechanics:

α — decay,

$$\phi \sim \alpha\, e^{ikx} + \beta\, e^{-ikx}$$

$$\phi_{II} \simeq \varkappa\, e^{\sqrt{\frac{2m}{\hbar}(V_o - h)x}} + L\, e^{-\sqrt{\frac{2m}{\hbar}(V_o - E)x}}$$

$$\phi_{III} = r \cdot e^{ikx}$$

where the transmissivity

$$T \sim e^{-\sqrt{\frac{2m}{\hbar}(V_o - E)d}} = e^{-2\delta d} \quad II(1)$$

The solution depends on $E \lessgtr V$ and the shape of the potential barrier. The classical limit, i.e., $\hbar \to 0$, $II(1) \to 0$.

Solution in quantum mechanics:

$$T = \frac{\text{particle transmitted}}{\text{particles incident}} = \frac{|\phi_{III}|^2}{|\phi|^2} =$$

$$= \frac{4}{4\cosh^2 \delta d + (\delta/\varkappa - \varkappa/\delta)\sinh^2 \delta d} \quad /27/$$

In acoustics:

The corresponding problem is the transmission of plane waves trough layered media, which can be solved easily by matrix method with

$$T = \frac{4 r_{13}^2}{(r_{13}+1)^2 \cos^2 \varkappa_2 d + (r_{12} + r_{23})^2 \sin^2 \varkappa_2 d} \quad II(3)$$

where $K_2 = \omega/c_2$,

r_{13} etc. are impedance ratios,

$$r_{12} = \frac{z_2}{z_1} = \frac{\rho_2 c_2}{\rho_1 c_1},$$

$$r_{13} = \frac{z_3}{z_1} = \frac{\rho_3 c_3}{\rho_1 c_1},$$

$$r_{23} = \frac{z_3}{z_2} = \frac{\rho_3 c_3}{\rho_2 c_2},$$

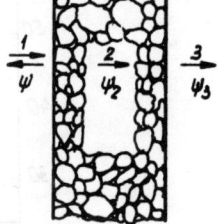

by assuming that all media are with real z's.

$r_{13} = 1$ in $II(2)$, for $z_1 = z_3$; $\delta = i\varkappa_2$, $II(2) = II(3)$ in mathematical form.

APPENDIX 7.1

Infrasonic absorption per wavelength due to mass transfer (denoted by M, solid and dotted curves are results from two different authors; C other mechanisms not considered in this example. See Ref.[36]).

μ — dimensionless absorption coefficient,
ω — circular sonic frequency,
τ — relaxation time, determined by fog constants, etc.

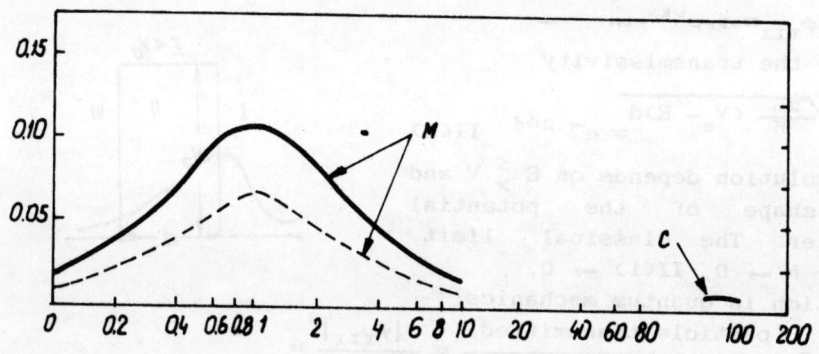

APPENDIX 7.2.

Ultrasonic absorption per wavelength arising from rotational isomeric relaxation in ethyl acetate (See Ref. [33]).

ω — sonic frequency,
τ — relaxation time.

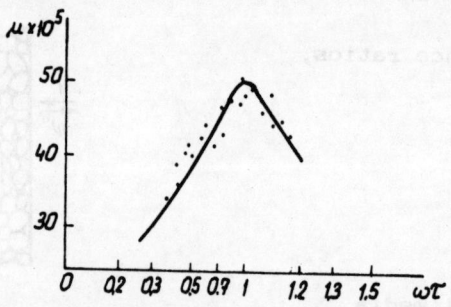

APPENDIX 8
Acoustics solitons are to be demonstrated by video tapes

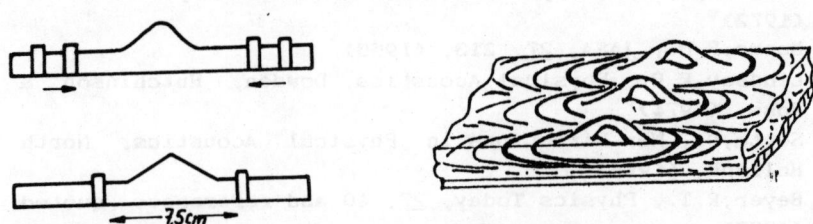

Fig.1   Fig.3

Fig.1. shows the standing soliton remains unchange in shape at very short l_x.

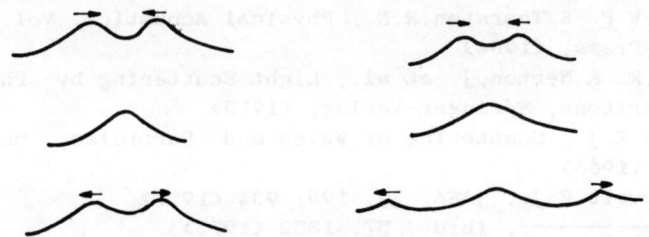

Fig.2.1 l_λ-20 cm   Fig.2.2 l_λ-26 cm

Fig.2. show that 2 standing solitons at likephase attract and pass over each other, the process is repeated indefinitely however the manner differs for different l_x.

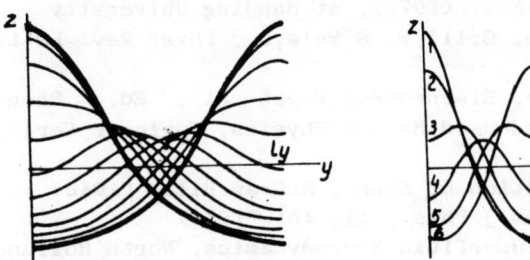 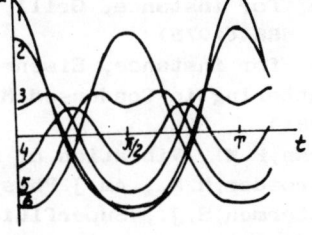

Fig.4. The y-section of water surface for mode (0,1) at different times (computer graph)

Fig.5. Evolution of water surface for different positions along y-axis (computer graph).

REFERENCES

[1] Lindsay,R.B., JASA, **37**, 361 (1965)
[2] Weisskopf,V.F., Physics of the 20th Century, M.I.T. (1972)
[3] Morse,P.M., JASA, **27**, 213, (1955)
[4] Lindsay,R.B., Physical Acoustics, Dowden, Hutchinson & Ross, (1974)
[5] Sette,D., New Directions in Physical Acoustics, North Holland, (1976)
[6] Beyer,R.T., Physics Today, **27**, 40 and references quoted (1973)
[7] Wei,R.J., Proceedings in I.C.A., Sydney, (1980); Prog. Phys., **2**, 389 (1982), in chinese
[8] Physics in Perspective, Vol. **1**, Phys. Survey Committee, National Acad. Sc., Washington D.C., (1972)
[9] Wei,R.J., Physics Today, **2**, Feb (1985)
[10] Mason,W.P. & Thurston,R.M., Physical Acoustics, Vol. **1A**, Acad. Press, (1964)
[11] Claus,R. & Merten,J. et al., Light Scattering by Phonon - Polaritons, Springer-Verlag, (1975)
[12] Newton,R.J., Scattering of Waves and Particles, McGraw Hill, (1966)
[13] Westervelt,P.J., JASA, **29**, 199; 934 (1957)
[14] ---------------, ibid., **57**, 1352 (1975)
[15] Saenz,H. et al., Science, **198**, 295 (1977)
[16] See, for instance, Blatt,J.M., Theory of Superconductivity, Acad. Press, (1964)
[17] See, for instance, Reissland,J.A., The Physics of Phonons, John Wiley, (1973)
[18] Rudnick,I., a chapter of Ref. **10**; his Lecture Notes of Low Temperature, Nov. (1979), at Nanjing University
[19] See, for instance, Grill,W. & Weis,O., Phys. Rev. Lett., **35**, 588 (1975)
[20] See, for instance, Eisenmenger,W. et al., Ed., Phonon Scattering in Condensed Matter Physics, Springer-Verlag, (1984)
[21] Morse,P.M., Vibration of Sound, McGraw Hill, (1948)
[22] Schroeder,M.R., Am.J.Phys., **41**, 461 (1973)
[23] Putterman,S.J., Superfluid Hydrodynamics, North Holland, (1974)
[24] De Broglie,L., The Current Interpretation of Wave Mechanics, Elsiver, (1963)
[25] Fermi,E., Notes on Quantum Mechanics, Univ. of Chicago

Press, (1961)
[26] Malecki,I., Acoustical Condition of Spoken Program Transmission Reprint, (1958)
[27] Majernik,I. et al., Auditory Uncertainty Principle, Acoustics, $\underline{43}$, 132 (1979)
[28] Wei,Y.T., Res. Report at Int. Rm. and Arch. Acoustics, Sept. (1957), Elektro-Technik und Elekt-Akustik, (1959), in German
[29] Putterman,S., Acoustics, Classical Hydrodynamics and Modern Physics, (1984), Leture notes at Nanjing University, to be published in chinese
[30] Hueter,T.F. & Bolt,R.H., Sonics, John Wiley, (1955)
[31] Wei,Y.T. et al., J. Univ. of Nanjing (Natural Sci. Ed.) $\underline{8}$, 249 (1964)
[32] Wei,Y.T.(Wei Rong-jue), Acta Scientia Sinica, $\underline{2}$,245, (1953); Report at ASA*50, Boston (1979); with Wu,J., JASA, $\underline{70}$,1213, (1981)
[33] Wei,Y.T. & Zhang Shu-yi, Acta Physica Sinica, $\underline{18}$, 298, (1962)
[34] Matheson,A.J., Molecular Acoustics, John Wiley, (1971)
[35] Bohr,N., Roy. Soc. (London) Phil. Trans., $\underline{A209}$, 281, (1909)
[36] Segie,E., Biographical Introduction, Collected Papers of Erico Fermi, Univ. of Chicago Press, (1962)
[37] Fermi,E., Atoms in the Family, George Allen & Unwin (1953)
[38] Fermi,E., Quantum Theory of Radiation, Rev. Mod. Phys., $\underline{4}$, 100 (1932)
[39] Fermi,E., Nuclear Physics, Univ. of Chicago, (1950)
[40] Morse,R.W., Bohn,H.B., Phys. Rev., $\underline{108}$, 1094 (1957)
[41] Scott,A.C. et al., Proc. IEEE, $\underline{61}$, 1443 (1973)
[42] Rudenko,O.V. & Soluyan,S.I., Theoretical Foundation of Nonlinear Acoustics, N.Y., (1977)
[43] Wu Jun-ru, Rudnick,I. et al., Phys. Rev. Lett. $\underline{52}$, 1421, (1984)
[44] Putterman,S.J. et al., See Acoustics and Education in Physics
[45] Wei,R.J. et al., Technical Papers, Westpac II, HongKong, 196 (1985)
[46] Wei,R.J. et al., Chinese Phys. Lett., $\underline{3}$, 213 (1986)
[47a] Hao,B.L., Progress in Phys., $\underline{3}$, 329 (1983)
[47b] Feshbach,H. Phys. Today, April, $\underline{7}$, 17 (1986)
[48] Ni,W.S., Acta Physica Sinica, $\underline{34}$, 504 (1985)

[49] Wang,B.R., Miao,G.Q., Chinese Physics, 5, 326 (1985)
[50] Wei,R.J., Nonlinear Acoustics-Soliton and Chao, reported at 12th Int. Congress of Acoustics, July 24-31,1986, Toronto.
[51] Beyer,R.T., Phys. Today, Nov. (1981)
[52] Bardeen,J.,Annual Rev. of Material Sc., 10, 1 (1980)
[53] Wei,R.J., Wuli(Physics), 5, 265 (1986), in chinese.
[54] Ibid., Invited lecture, Int.Copt on Phys. Education, sponsored by IUPAP,Wanjing, August 31 Sept 5 (1986)
[55] Ibid, Contemporary Nonlinear Acoustics, Joint Conf. China-Japan on ultrasonic, Nonjing, Nov 11-14, Nonjing.

THE PHASE VELOCITY OF SOUND IN THE ACOUSTICAL FIELD OF A GENERAL SPHERICAL SOURCE

BARBARA WYRZYKOWSKA, ROMAN WYRZYKOWSKI
Institute of Physics
Pedagogical University
Rzeszów, Poland

ABSTRACT

This paper concerns the waves having non-constant phase velocity, depending on the position in the sound field. The velocity of both pressure and acoustical velocity waves produced by a dipol were calculated - they have a singularity at the point r=0. The dipol can not radiate relatively low frequencies -it has a cut-off frequency. That frequency occurs at the value of the so called diffraction parameter ka (a-radius of a vibrating sphere) equal $\sqrt{2}$. Both velocities tend to constant velocity of sound wave in the medium when kr (r - the distance from the center) tends to infinity.

INTRODUCTION

The solution of the wave equation in spherical coordinates we have in a general form [5]:

$$F(r,\vartheta,t) = P(\vartheta)F(r,t) = P(\vartheta) [A(kr) - iB(kr)]e^{i\omega t} \qquad /0.1/$$

where: $F(r,\vartheta,t)$ — denotes any physical quantity
$P(\vartheta)$ — function depending on the angle - will play no role in our considerations
$\omega = 2\Pi\nu$ — circular frequency
ν — frequency of vibrations
$k = \omega/c_o$ — wave number
c_o — constant phase velocity
t — time
r — radius determining the position

In order to obtain the physical interpretation of the formula /0.1/ we must rewrite it in the exponential form:

$$F(r,t) = F_o(kr) e^{i[\omega t - f(kr)]} \qquad /0.2/$$

where: $F_o(kr) = \sqrt{A^2(kr) + B^2(kr)}$ — amplitude $\qquad /0.3/$

$f(kr) = tg^{-1}\dfrac{B(kr)}{A(kr)}$ — the function of the wave front $\qquad /0.4/$

We obtain the formula for the phase velocity of a wave in dependence of the variable "r" c=dr/dt differentiating with respect to time phase of the wave of any choosen constant value:

$$\omega t - f(kr) = \text{const} \implies \omega = \frac{df}{d(kr)} k \frac{dr}{dt} \qquad /0.5/$$

From /0.5/ by means of the above definition of c we get:

$$\frac{c}{c_o} = \left[\frac{df(kr)}{d(kr)}\right]^{-1} \qquad /0.6/$$

Differentiating f(kr) in the form /0.4/ we obtain a formula we will use in specific cases:

$$\frac{c}{c_o} = \frac{A^2 + B^2}{B^2 A + A^2 B} \qquad /0.7/$$

THE VELOCITY OF THE PRESSURE WAVE

The formula for the acoustical pressure of the spherical wave of the order was given by Stenzel [3,5]. We write here that formula in a condensed form to visualize only the part depending on the variable r. The remaining part, immaterial for our considerations, we denote by P_n.

$$p_n = P_n \frac{1}{\sqrt{kr}} \left[I_{n+1/2}(kr) - iN_{n+1/2}(kr) \right] \qquad /1.1/$$

where: $I_{n+1/2}(kr)$ - Bessel's function
$N_{n+1/2}(kr)$ - Neumann's function

Denoting by c_{Pn} the velocity of the pressure wave we calculate it from the above formula /0.7/.

The function A(kr) and B(kr) from the formula /1.1/ we will write in the form:

$$A(kr) = \frac{I_{n+1/2}(kr)}{(kr)^{n+1/2}} \qquad B(kr) = \frac{N_{n+1/2}(kr)}{(kr)^{n+1/2}} \qquad /1.2/$$

which can not change the value B/N appearing in /0.4/ but considerably facilitates farther calculations because the derivative of the function in the form /1.2/ has a very simple form [4]:

$$\frac{d}{dz}\left(\frac{Z_\nu(z)}{z^\nu}\right) = -\frac{Z_{\nu+1}(z)}{z^\nu} \qquad /1.3/$$

where: $Z_\nu(z)$ - any cylindrical function of z.

According to the above:

$$\frac{c_{Pn}}{c_o} = \frac{I^2_{n+1/2}(kr) + N^2_{n+1/2}(kr)}{N_{n+1/2}(kr) I_{n+1/2+1}(kr) - N_{n+1/2+1}(kr) I_{n+1/2}(kr)} \qquad /1.4/$$

In the denominator of /1.4/ we have a known expression: it is a Wronskian of the Bessel's and Neumann's functions in a general form represented by the formula /1.2/:

$$I_{\nu+1}(z) N_\nu(z) - I_\nu(z) N_{\nu+1}(z) = \frac{2}{\Pi z} \qquad \text{for any } z \qquad /1.5/$$

On the other hand the sum of squares of Bessel's and Neumann's Functions of the order n+1/2 appearing in the numerator of /1.4/ is known in the form of a polynomial[1,2]:

$$I^2_{n+1/2}(z) + N^2_{n+1/2}(z) = \frac{2}{\Pi z} \sum_{m=0}^{n} \binom{2m}{m} \frac{(n+m)!}{(n-m)!(2z)^{2m}} \qquad /1.6/$$

Accordingly the formula for the looked for velocity c_{Pn} takes of a multinomial with "n+1" terms:

$$\boxed{\frac{c_{Pn}}{c_o} = \sum_{m=0}^{n} \binom{2m}{m} \frac{(n+m)!}{(n-m)!(2kr)^{2m}}} \qquad /1.7/$$

We excerpt some specific cases:

for n = 0 $\qquad \frac{c_{Po}}{c_o} = 1 \quad \Longrightarrow \quad c_{Po} = c_o \qquad /1.8/$

for n = 1 $\qquad \frac{c_{P1}}{c_o} = 1 + \frac{1}{(kr)^2} \qquad /1.9/$

for n = 2 $\qquad \frac{c_{P2}}{c_o} = 1 + \frac{3}{(kr)^2} + \frac{9}{(kr)^4} \qquad /1.10/$

for n = 3 $\qquad \frac{c_{P3}}{c_o} = 1 + \frac{6}{(kr)^2} + \frac{45}{(kr)^4} + \frac{225}{(kr)^6} \qquad /1.11/$

We see from the formula /1.7/ and cited examples /1.8/... /1.11/ that:

$$c_{Pn} > c_o \quad \text{for } n>0 \qquad /1.12/$$

In the limit when $kr \rightarrow \infty$ the terms of the polynomial tend to zero excepting the first one (m=0) which always equals unity, so we have:

$$\lim_{kr \rightarrow \infty} c_{Pn} = c_o \qquad /1.13/$$

That implies that for a given value of the frequency ν $\left(k = \frac{2\Pi\nu}{c_o}\right)$ only at a distance respectively large from the source we may approximately take $c_{P_n} \simeq c_o$. But when $kr \longrightarrow ka$ (a - the radius of the sphere) for a given value of ν, in the near field of the vibrating sphere the velocity c_{P_n} may take values considerably greater than c_o. That depends on both ν and a.

Comparing succesive expressions for c_{P_n} /1.8/... /1.11/ we notice that:
$$c_{P_{n+1}} > c_{P_n} \qquad /1.14/$$

That means that a wave of higher order must out run waves of lower orders.

THE VELOCITY OF THE ACOUSTICAL VELOCITY WAVE

The formula for the velocity of so called acoustical wave (the velocity of vibrations) we obtain from the Euler's equation [5]. We are interested in the propagation of the velocity of harmonic vibrations in the direction of the radius r - in that case the formula takes the form:
$$u_n = \frac{i}{\rho c_o} \frac{\partial p_n(kr)}{\partial (kr)} \qquad /2.1/$$

The differentiation with respect to kr we carry off adapting the formula /1.3/ with suitable changing of p_n /1.1/. As a result we obtain:
$$u_n = \frac{i}{\rho c_o} P_n \frac{1}{(kr)^{3/2}} \left\{ \left[nI_{n+1/2}(kr) - krI_{n+3/2}(kr) \right] + \right.$$
$$\left. - i \left[nN_{n+1/2}(kr) - krN_{n+3/2}(kr) \right] \right\} \qquad /2.2/$$

We denote by c_{U_n} the velocity of the acoustical velocity wave and by means of the formula /0.7/ we can begin to calculate the formula for c_{U_n}.

The functions A(kr) and B(kr) are respectively equal:
$$A(kr) = nI_{n+1/2}(kr) - krI_{n+3/2}(kr)$$
$$B(kr) = nN_{n+1/2}(kr) - krN_{n+3/2}(kr) \qquad /2.3/$$

Let us begin with the calculation of the denominator of /0.7/. The derivative A' we obtain from /1.3/:
$$A' = \frac{d}{d(kr)} \left[n(kr)^{n+1/2} \frac{I_{n+1/2}(kr)}{(kr)^{n+1/2}} - (kr)^{n+5/2} \frac{I_{n+3/2}(kr)}{(kr)^{n+3/2}} \right] =$$
$$= n(n+1/2) \frac{I_{n+1/2}(kr)}{kr} - (2n+5/2)I_{n+3/2}(kr) + krI_{n+5/2}(kr) \qquad /2.4/$$

To reduce the number of terms we apply the recurrence formula for cylindrical functions [1,2,4]:
$$Z_{\nu+1}(z) + Z_{\nu-1}(z) = \frac{2\nu}{z} Z_\nu(z) \qquad /2.5/$$

Substituting $\nu = n+3/2$ we get the respective expression for $I_{n+5/2}(kr)$ we can introduce into the formula /2.4/. As a result we obtain:

$$A' = \left[\frac{n(n+1/2)}{kr} - kr\right] I_{n+1/2}(kr) + 1/2\, I_{n+3/2}(kr)$$

and by analogy /2.6/

$$B' = \left[\frac{n(n+1/2)}{kr} - kr\right] N_{n+1/2}(kr) + 1/2\, N_{n+3/2}(kr)$$

By means of the formulae /2.3/ and /2.6/ we may calculate the denominator in /0.7/: $M = B'A - A'B$. After multiplication and reduction of similar terms and application the formula for Wronskian /1.5/ we obtain a very simple expression:

$$M = \frac{2}{\pi kr} \cdot \left[(kr)^2 - n(n+1)\right] \qquad /2.7/$$

The calculation of the numerator $L = A^2 + B^2$ appearing in the formula /0.7/ is more complicated. We substitute A and B according the formula /2.3/. After squaring and setting in order we get:

$$L = n^2 \left[I^2_{n+1/2}(kr) + N^2_{n+1/2}(kr)\right] + (kr)^2 \left[I^2_{n+3/2}(kr) + N^2_{n+3/2}(kr)\right] +$$
$$- 2nkr \left[I_{n+3/2}(kr) I_{n+1/2}(kr) + N_{n+3/2}(kr) N_{n+3/2}(kr)\right] \quad /2.8/$$

To shorter the right side we introduce the following denotations:

$$W_\nu(kr) = I^2_\nu(kr) + N^2_\nu(kr) \qquad /2.9/$$

$$W_{\nu,\mu}(kr) = I_\nu(kr) I_\mu(kr) + N_\nu(kr) N_\mu(kr) \qquad /2.10/$$

Then:

$$L = n^2 W_{n+1/2}(kr) + (kr)^2 W_{n+3/2}(kr) - 2nkr\, W_{n+1/2,\,n+3/2}(kr) \quad /2.11/$$

$W_{n+1/2}(kr)$ and $W_{n+3/2}(kr)$ are the polynomials of the form already known /1.6/. To simplify the formula /2.11/ we express $W_{n+1/2,\,n+3/2}(kr)$ by means of these polynomials. For that purpose we use the recurrence formula for cylindrical functions /2.5/. We write that formula for $Z_\nu(z) = I_{n+1/2}(kr)$ and $Z_\nu(z) = N_{n+1/2}(kr)$ in the following order:

$$I_{n-1/2}(kr) = \frac{2n+1}{kr} I_{n+1/2}(kr) - I_{n+3/2}(kr) \qquad /2.12/$$

$$N_{n-1/2}(kr) = \frac{2n+1}{kr} N_{n+1/2}(kr) - N_{n+3/2}(kr)$$

We square both formulae and add side by side introducing simultaneously foreshortening forms /2.9/ /2.10/. In that way we obtain the looked – for formula:

$$W_{n+\frac{1}{2},\,n+\frac{3}{2}}(kr) = \frac{kr}{2(2n+1)}\left[\frac{(2n+1)^2}{(kr)^2}W_{n+\frac{1}{2}}(kr) + W_{n+\frac{3}{2}}(kr) - W_{n-\frac{1}{2}}(kr)\right] \quad /2.13/$$

Introducing /2.13/ to /2.11/ we get:

$$L = \frac{n+1}{2n+1}(kr)^2 W_{n+\frac{3}{2}}(kr) - n(n+1)W_{n+\frac{1}{2}}(kr) + \frac{n}{2n+1}(kr)^2 W_{n-\frac{1}{2}}(kr) \quad /2.14/$$

We replace the shortening forms by the respective polynomials

$$L = \frac{1}{\Pi}\frac{n+1}{2n+1}\cdot\sum_{m=0}^{n+1}\binom{2m}{m}\frac{(n+1+m)!}{(n+1-m)!(2kr)^{2m-1}} + \quad /2.15a/$$

$$-\frac{1}{\Pi}2^2 n(n+1)\sum_{m=0}^{n}\binom{2m}{m}\frac{(n+m)!}{(n-m)!(2kr)^{2m+1}} + \quad /2.15b/$$

$$+\frac{1}{\Pi}\frac{n}{2n+1}\sum_{m=0}^{n-1}\binom{2m}{m}\frac{(n-1+m)!}{(n-1-m)!(2kr)^{2m-1}} \quad /2.15c/$$

The obtained formula containing three polynomials we can shorten to one. For the purpose we must change particular polynomials in such a way to be able to add them together. Let us take into account that in the sum /2.15c/ from the formal point of view we may shift the summing of "m" to "n+1" because in the denominator of the expressions under the sigma will appear (-1)! with m=n and (-2)! with m=n+1 and for that reason both terms must equal zero. Then in the sum /2.15b/ we shall shift the summation index by "1" to realize the summation from m=1 to m=n+1. In that case we must, under the sigma sign, substract "1" from every "m". For instance the Newton's symbol will take now the form $\binom{2m-2}{m-1}$. We transform that symbol in such a way to get under sigma the same form as before. Then:

$$\binom{2m-2}{m-1} = \frac{(2m-2)!}{(m-1)!(m-1)!} = \frac{(2m)!}{m!m!}\frac{mm}{2m(2m-1)} = \binom{2m}{m}\frac{m}{2(2m-1)} \quad /2.16/$$

We note immediately from /2.16/ that for m=0 the term of the sum with the shifted index must equal 0 and we may formally begin the summation from m=0.

After the above mentioned changes the formula for L /2.15/ may be rewritten with one common symbol of the $\sum_{m=0}^{n+1}$:

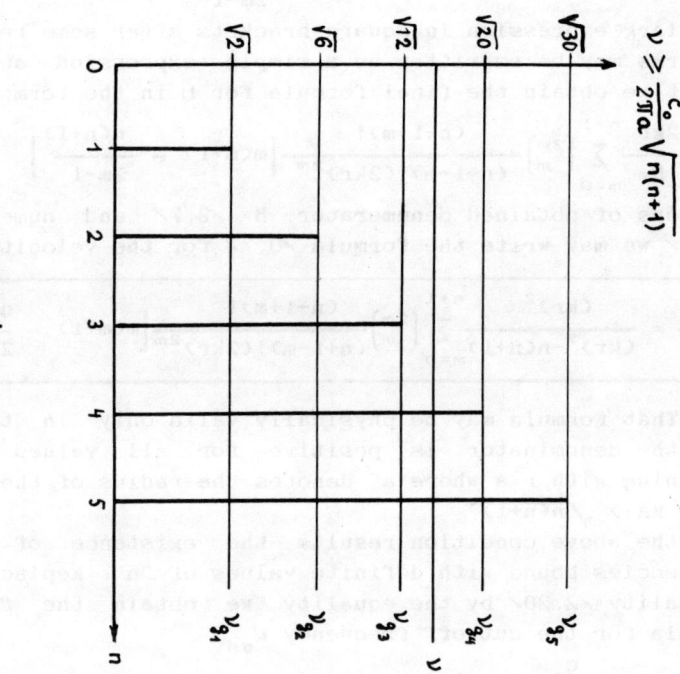

Fig. 1

$$L = \frac{1}{\Pi} \sum_{m=0}^{n+1} \binom{2m}{m} \frac{(n-1+m)!}{(n+1-m)!(2kr)^{2m-1}} \left[\frac{(n+1)(n+1+m)(n+m)+n(n+1-m)(n-m)}{2n+1} + \right.$$

$$\left. - \frac{2n(n+1)m}{2m-1} \right] \qquad /2.17/$$

The large expression in square brackets after some regrouping of terms may be rewritten as a simple expression and as a result we obtain the final formula for L in the form:

$$L = \frac{2kr}{\Pi} \sum_{m=0}^{n+1} \binom{2m}{m} \frac{(n-1+m)!}{(n+1-m)!(2kr)^{2m}} \left[m(m+1) - \frac{n(n+1)}{2m-1} \right] \qquad /2.18/$$

By means of obtained denumerator M /2.7/ and numerator L /2.18/ we may write the formula /0.7/ for the velocity c_{U_n}:

$$\boxed{\frac{c_{U_n}}{c_o} = \frac{(kr)^2}{(kr)^2 - n(n+1)} \sum_{m=0}^{n+1} \binom{2m}{m} \frac{(n-1+m)!}{(n+1-m)!(2kr)^{2m}} \left[m(m+1) - \frac{n(n+1)}{2m-1} \right]}$$

$$/2.19/$$

That formula may be physically valid only in the case when the denominator is positive for all values of "r" beginning with r=a where "a" denotes the radius of the sphere:

$$ka > \sqrt{n(n+1)} \qquad /2.20/$$

From the above condition results the existence of cut-off frequencies bound with definite values of "n". Replacing the inequality /2.20/ by the equality we obtain the following formula for the cut-off frequency ν_{gn}:

$$\nu_{gn} = \frac{c_o}{2\Pi a} \sqrt{n(n+1)} \qquad /2.21/$$

If a sphere with a given radius "a" is vibrating with a frequency ν enclosed between the cut-off values for n=s and n=s+1 it will radiate only the waves of the order n=0,1,2...s - the waves of higher order (n > s) shall not be radiates. Especially when $\nu = \nu_{gn}$ the sphere may radiate only the waves of the order n=0,1,2...(s-1). We can explain it by means of the fig.1. On the horizontal axis we place in turn the numbers n=0,1,2... indicating the order of the wave, on the vertical axis we put the values of the left and right side of the inequality $\nu \geq \frac{c_o}{2\Pi a}\sqrt{n(n+1)}$. The vertical segments corresponding to the values of n represent the values of the right side, and the horizontal lines the left side, that is the values of ν. The broken lines represent the cut-off frequencies ν_{gn}. As an example we have shown a continuous lines corresponding to the frequency ν between ν_{g3} and ν_{g4}.

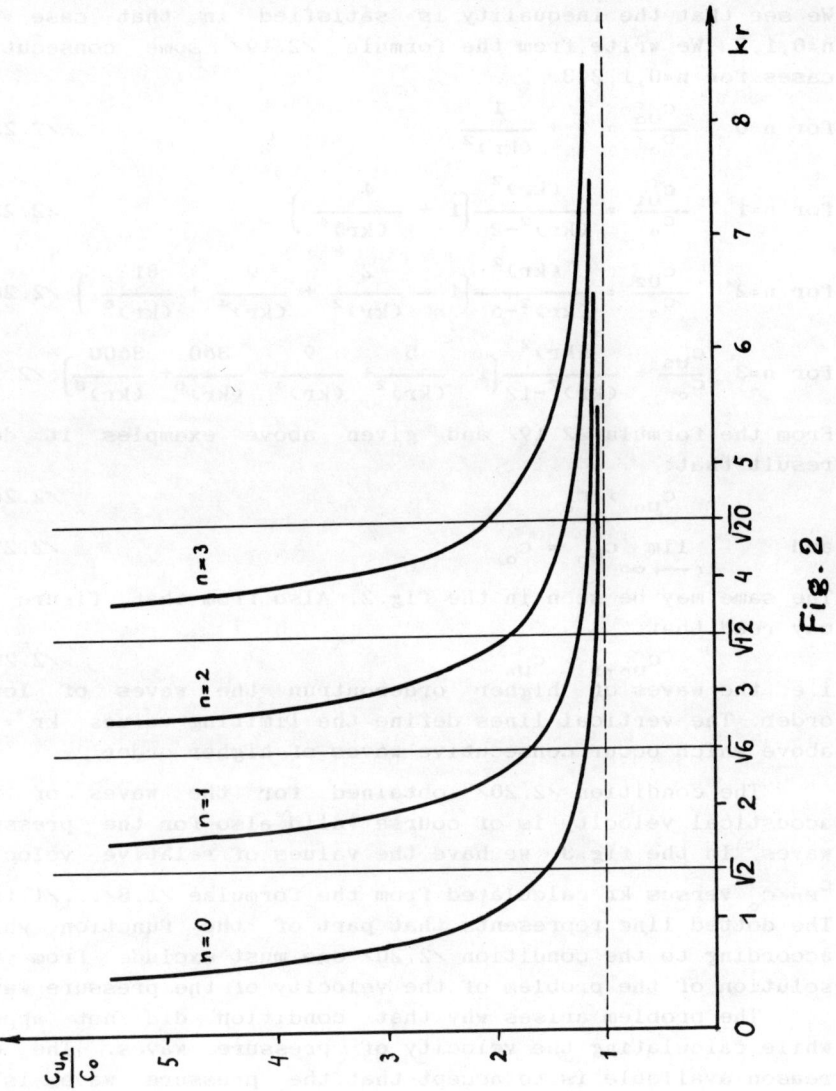

Fig.2

We see that the inequality is satisfied in that case for n=0,1,2. We write from the formula /2.19/ some consecutive cases for n=0,1,2,3.

for n=0 $\quad \dfrac{c_{U_0}}{c_0} = 1 + \dfrac{1}{(kr)^2}$ /2.22/

for n=1 $\quad \dfrac{c_{U_1}}{c_0} = \dfrac{(kr)^2}{(kr)^2-2}\left\{1 + \dfrac{4}{(kr)^4}\right\}$ /2.23/

for n=2 $\quad \dfrac{c_{U_2}}{c_0} = \dfrac{(kr)^2}{(kr)^2-6}\left\{1 - \dfrac{2}{(kr)^2} + \dfrac{9}{(kr)^4} + \dfrac{81}{(kr)^6}\right\}$ /2.24/

for n=3 $\quad \dfrac{c_{U_3}}{c_0} = \dfrac{(kr)^2}{(kr)^2-12}\left\{1 - \dfrac{5}{(kr)^2} + \dfrac{9}{(kr)^4} + \dfrac{360}{(kr)^6} + \dfrac{3600}{(kr)^8}\right\}$ /2.25/

From the formula /2.19/ and given above examples it does result that:

$$c_{U_n} > c_0 \qquad /2.26/$$

and $\quad \lim\limits_{kr \to \infty} c_{U_n} = c_0$ /2.27/

The same may be seen in the fig.2. Also from that figure we may read that:

$$c_{U_{n+1}} > c_{U_n} \qquad /2.28/$$

i.e. the waves of higher order outrun the waves of lower order. The vertical lines define the limiting values kr = ka above which occur consecutive waves of higher order.

The condition /2.20/ obtained for the waves of the acoustical velocity is of course valid also for the pressure waves. In the fig.3. we have the values of relative velocity c_{P_n}/c_0 versus kr calculated from the formulae /1.8/.../1.11/. The dotted line represents that part of the function which according to the condition /2.20/ one must exclude from the solution of the problem of the velocity of the pressure wave.

The problem arises why that condition did not appear while calculating the velocity of pressure waves. The one reason available is to accept that the pressure wave is a secondary phenomenon. The propagation of an acoustical wave consists the transmission of the vibrations, described by the acoustical velocity. The pressure is the secondary phenomenon describing the physical state of the medium.

Comparing the values of c_{U_n} and c_{P_n} we see that velocity wave of the order n always outruns the pressure wave of the same order.

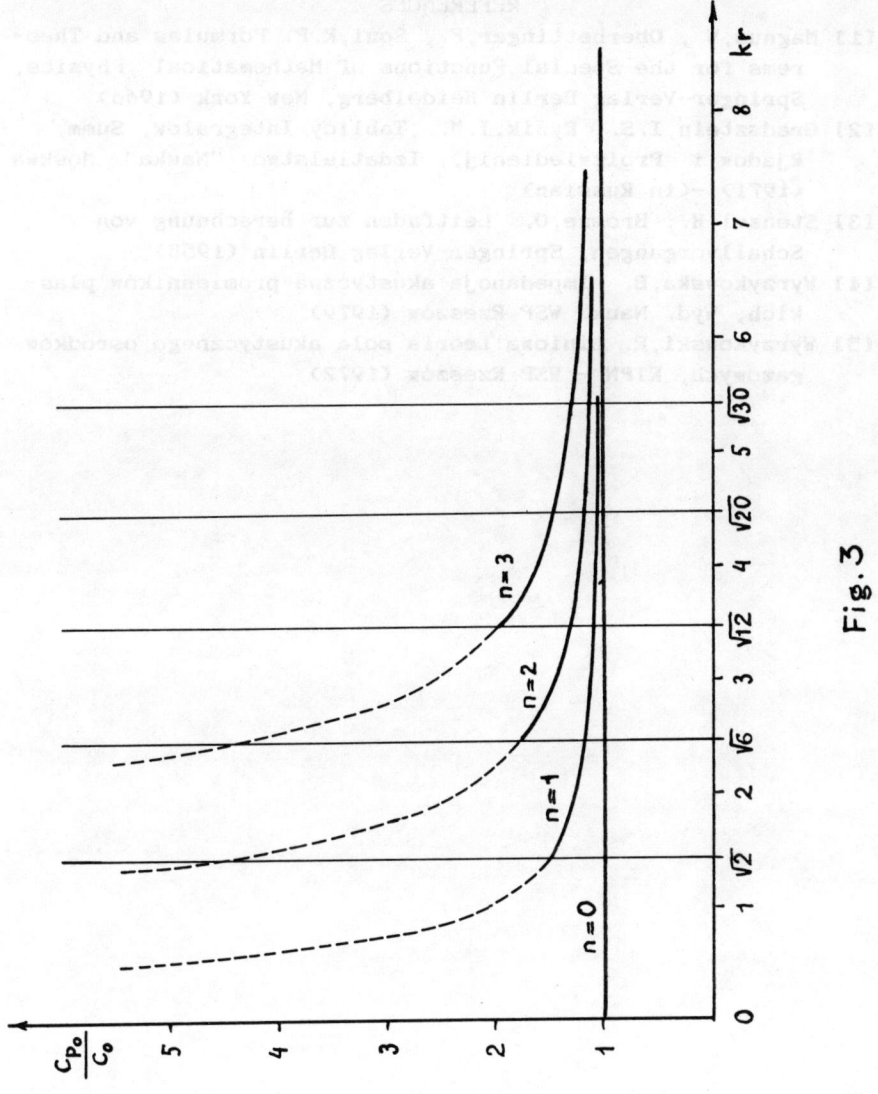

Fig. 3

REFERENCES

[1] Magnus,W., Oberhettinger,F., Soni,R.P. Formulas and Theorems for the Special Functions of Mathematical Physics, Springer-Verlag Berlin Heidelberg, New York (1966)

[2] Gradsztein,I.S., Ryżik,I.M. Tablicy Integrałow, Summ, Rjadow i Proizwiedienij, Izdatielstwo "Nauka" Moskwa (1971) -(in Russian)

[3] Stenzel,H., Brosze,O. Leitfaden zur Berechnung von Schallvorgangen, Springer-Verlag Berlin (1958)

[4] Wyrzykowska,B. Impedancja akustyczna promienników płaskich, Wyd. Nauk. WSP Rzeszów (1979)

[5] Wyrzykowski,R. Liniowa teoria pola akustycznego ośrodków gazowych, RTPN - WSP Rzeszów (1972)

THE USE OF ANALOGIES IN THE TEACHING OF ULTRASONICS.
Concepts of "old physics" still active in the modern development of acoustics.

A. ZAREMBOWITCH

Laboratoire Dynamique Cristalline et Ultrasons - D.R.P.
Université Pierre et Marie Curie
4, place Jussieu
75252 PARIS CEDEX 05
FRANCE

ABSTRACT

The use of analogies is fundamental for the elaboration of concepts ; consequently it plays an important didactic role. This statement is illustrated by examples which emphasize mainly the analogies between optics and ultrasonic acoustics. In this paper we try to explain that "old" as well as "new" physics are parts of the modern development of acoustics.

Introduction

When studying the history of science it is surprising to note how most of the eminent scientists used analogies in their scientific activities. This extensive and often fruitful use of analogies raises several questions whose most fundamental one is probably the understanding of the mechanisms which, in the brains' work, explain the elaboration of concepts and consequently the prominent part played by analogies ; this question, however, goes beyond the subject of this paper.
Other questions are fundamentally epistemologic :
 - are analogies always heuristic ? (we know they are not : for example, the concept of "caloric fluid" has delayed the development

of thermodynamics ; it is likely that the use of analogies drawn from mechanics has sometimes been a handicap to the development of other branches of physics...)

- can analogies be heuristic when they are based upon wrong concepts ? (we know they can : for example, in the case of DESCARTES' formula for the reflection and refraction angles of light beams at the boundary of two dielectrics as well as in the case of FRESNEL's formula for the amplitudes, where mechanical analogies have been successful though founded on wrong basis.

In the same way, we can also wonder whether the analogic approach is not a timorous one where we try to erect a new building by fixing it, instead of completely reconsidering the whole structure.

It is legitimate too, to question oneself upon the didactic interest of analogies ; indeed, analogies allow us a "spare of thoughts" but consequently they are a little bit simplifing.

In spite of these reservations analogies are in general very useful ; in that paper, we intend to review some of the aspects of analogies mentionned previously and to illustrate some statements by using examples mainly borrowed to analogies between optics and ultrasonic acoustics.

1 - The Didactic Interest of Analogies.

People interested in acoustical imaging know that ultrasonic images can be obtained by working out an ultrasonic camera |1| in which an acoustical lens replace an optical lens (fig. 1). Such a device, very interesting from the technical point of view, is rather poor concerning its conceptual content and its didatic message (it is noteworthy however that a convergent optical lens is a divergent acoustic lens and conversely).

On the contrary, powerful analogies between optics and acoustics can be used at a more abstract level ; as an example we will report a pedagogical experience carried out at Paris University |2|; this experience is intended for first level students in order to make clear the connection between the basic principles of geometrical optics and

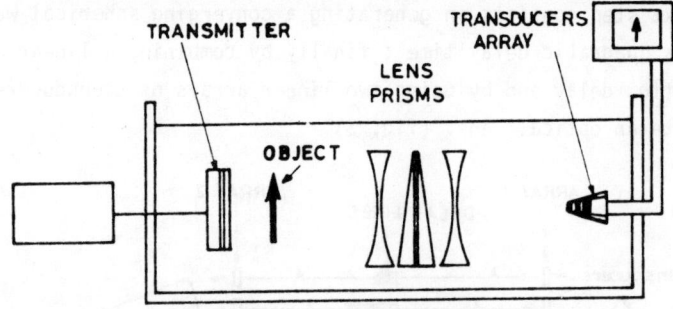

Fig. 1 - Real time acoustical imaging with a linear array and mechanical scanning of the focused image using rotative prisms. (from |3|).

the HUYGENS - FRESNEL wave approach. This connection which is not obvious in the field of optics can be understood more profoundly in ultrasonics. |3|

The starting point is to demonstrate that one can form and steer an acoustic beam with the help of a linear array of transducers. If an electric signal excite the transducers in phase, each transducer acts as a point source and according to the HUYGENS - FRESNEL principle the envelope of the spherical waves will be a plane wave ; now if each transducer is excited with a delay, a delay time varying linearly along the linear array will produce a plane wave with an angle θ (fig. 2).

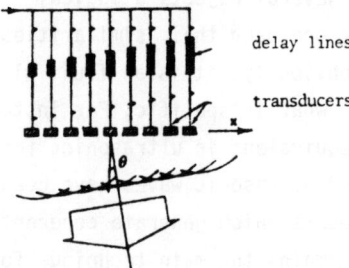

Fig. 2 - Acoustical beam forming.

The next step consists in generating a converging spherical wave by using a quadratic delay time ; finally by combining a linear and a quadratic delay and by using two linear arrays of transducers one simulate an optical lens. (fig. 3).

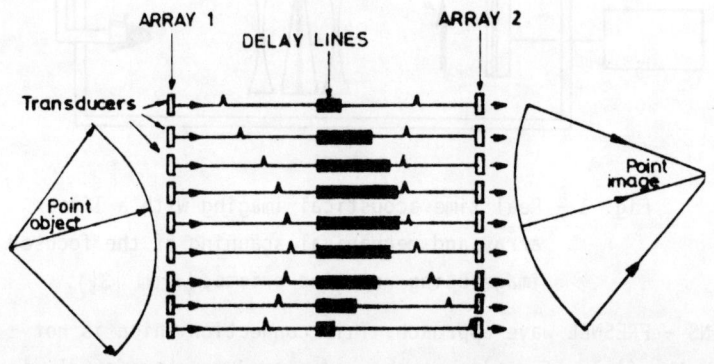

Fig. 3 - Simulation of an optical lens.

This approach enables us to illustrate the basic principles of wave physics and to demonstrate easily many formulas of geometrical optics :
- the conjugaison formula for simple lens immersed in two externe identical media.
- the formula for magnification
- the LAGRANGE - HELMOLTZ formula...

2 - Analogies and Differences.

When studying analogies between two or several objects a logical attitude consists in identifying on the one hand their similarities, on the other hand their differences ; obviously, it is as fruitful to emphasize what is general, as to stress what is specific. For instance, the laser revolution on optics has no equivalent in ultrasonics for a simple reason : from the early beginning,ultrasonic waves have been obtained by using piezoelectric transducers which generate coherent waves and even now, piezo-electricity remains the main technique for generating ultrasounds from frequencies as low as 20.000 Hertz up to

20 Gigahertz ; thus the requirement for spatial coherence is met automatically.

To analyse some specificities of acoustical waves in comparaison with optical waves we will examine more in detail their nonlinear behaviour. Indeed, many branches of physics are debitors to acoustics from the point of view of nonlinearity. It is usual to find in acoustics striking examples to illustrate nonlinear physics : the aolien harp, the wind whistling through tall grass, the singing telephone wire...

Acoustical waves are nonlinear for two reasons :

- the basic equations of acoustics are nonlinear ; for instance (1) is the wave equation for the adiabatic propagation of a plane wave in a non-viscous fluid.

$$\frac{\delta^2 \chi}{\delta t^2} = \frac{c^2}{1 + (\frac{\delta \chi}{\delta a})^2} \frac{\delta^2 \chi}{\delta a^2}$$

- the constitutive equation connecting stress to strain in a medium shows important departure from linearity even for small strain (that is very different from optics).

As a consequence, in acoustics we can observe specific behaviour that we will examine in connection with two phenomena, dispersion and absorption.

a. Influence of dispersion

In optics, the refractive index depends on frequency ; as a result, in nonlinear optics, phase matching problems are very important for harmonic generation. On the contrary dispersion is very often negligible in acoustics ; for example the velocity of longitudinal waves propagating in NaCl is constant with an accuracy of 10^{-3} from 10^4 to 10^{10} Hertz. Consequently, when the wave propagates into the medium, the amplitude of generated harmonics can increase at the expense of the fundamental mode (cumulative effect) since phase matching conditions are satisfied. Because of harmonic growth the wave front becomes more straight and instabilities can occur if this growth is not limited by attenuation (fig. 4).

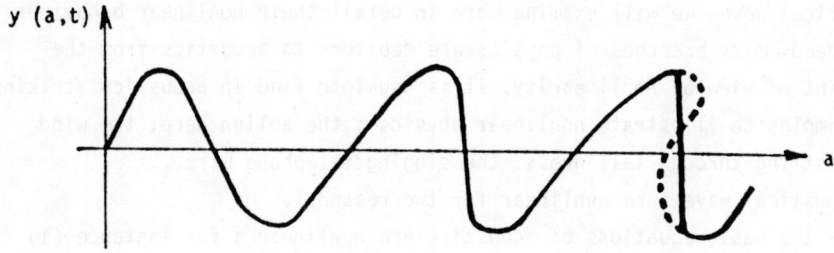

b. Influence of attenuation

Many different processes are responsible for the attenuation of elastic waves in solids and fluids ; however, a large number of these processes leads to a regime in which the attenuation coefficient is proportional to the square of the frequency ; thus, high frequencies are strongly damped and the distorsion of the wave is usually limited by the attenuation. A very interesting, though particular situation, occurs in the case of an exact balance between harmonic generation and attenuation effects leading to the propagation of solitons |4|.
Another interesting aspect of nonlinearity in ultrasonics is a parametric interaction used in the so-called "parametric arrays".
A basic problem in underwater acoustics is to find a compromise between directivity and attenuation. If high frequencies are used the ultrasonic beam will be directive but strongly damped ; at low frequencies directivity is lost. This severe limitation in the use of high frequencies can be overcome thanks to nonlinearity. Two high frequency ultrasonic beams whose frequencies are f and $f + \Delta f$ are generated by a transducer ; the nonlinear behaviour of water, though weak, is sufficient to allow the generation of a low frequency beam (frequency Δf) which is directive since it results from the beating of two directive beams. This technique (parametric arrays) is an important progress in ultrasonics within the last years |5| |6|.

3 - Concepts of "Old Physics" still active in the modern Developments of Acoustics.

To draw the border between "modern" and "old" physics is not an easy task. Of course, technological progress now allow the achievement of modern and remarkable devices but the concepts and analogies on which they are founded are often originated from "old physics" as shown in that paper concerning for instance beam forming, beam steering, parametric arrays...

However, one cannot deny that "modern" physics is also characterized by the use of new and powerful concepts. To illustrate these statements we will examine successively some analogies coming from old physics and still active in the modern development of ultrasonics, then traditional concepts revisited in a modern approach.

A large number of classical optical effects can be transposed into the ultrasonic field. Some transpositions are trivial, for instance as seen previously the realization of ultrasonics lenses ; some are sophisticated as the evidence for a "GOOS - HANSCHEN" effect in acoustics (fig. 5).

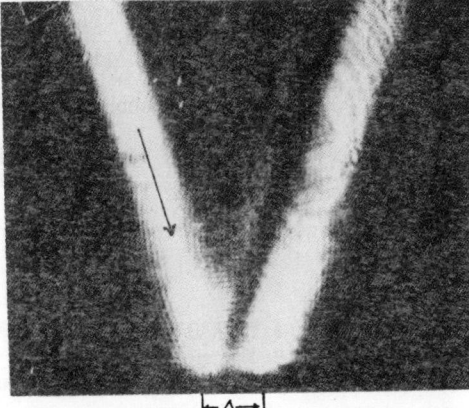

Fig. 5 - Ultrasonic beam reflected from a water-glass interface. Δ is the lateral displacement of the reflected beam. |7|

Obviously, we can expect fruitful analogies to be based on concepts rather than on devices ; a typical illustration is the analogy between optical and acoustical holography ; another is the close similarity between pure shear waves propagating in solids and electromagnetic waves, a similarity which allows us to build an ultrasonic polarimetry similar to the optic polarimetry.
Starting from this idea it is possible in ultrasonics.

a. To demonstrate the MALUS law

Two Y-cut quartz transducers are used to generate and to receive transverse acoustic/waves in an isotropic medium ; they act as polarizers and analysers.
The amplitude of the electric signal received through the analyser is shown to be proportional to $\cos^2 \alpha$, where α is the angle between the directions of vibration of the two Y-cut quartz transducers |8|.

b. To generate circularly polarized waves (FRESNEL's parallelepiped)

The phase difference between the reflected and the incident transverse elastic waves at the boundary between an isotropic solid and a gas can be chosen equal to $\frac{\pi}{2}$. As in optics, this property is used to generate circularly polarized waves |9| |10|.

c. To demonstrate a "POCKELS effect" (acoustical birefringence)

A uniaxial stress is applied to an isotropic solid medium ; an acoustical birefringence occurs since the velocity of transverse waves polarized along or perpendicular to the stress is no longer the same |11|.

d. To give the evidence of a "FARADAY effect"

The polarization plane of the acoustical transverse wave rotates under the influence of an external applied field in magnetic crystals (ferrimagnetic |12|), paramagnetic |13|, antiferromagnetic |14|.

e. To find out a natural acoustical activity (ARAGO effect)

Using high frequency elastic waves |15| and BRILLOUIN scattering |16| the polarization plane has been found to rotate without applied magnetic field in quartz crystals.

f. To realize an acousto-elasticimeter comparable to a photoelasticimeter |17|.

From these examples we can see that, in this field, analogies between optics and acoustics are very stimulating.

4 - Traditional Concepts revisited in a modern Approach

Technological progress allow today the achievement of devices able to produce outstanding performances ; as by-product, they also stimulate the elaboration of new concepts or the reconsideration of old ones. The concept of "coherence" is a good illustration for that statement. Before the second world war opticians were the most advanced people in the understanding of this concept ; they had inheritate from the 19th century's physicists beautiful experiments illustrating coherence partial coherence, incoherence. After the second world war radar people became also very expert in the use of coherence. With the discovery of lasers the concept of coherence came back to the opticians communauty and, it seems, for a long time. Indeed, they have been able through a clever analysis to enlarge the concept along several lines :

- Statistical physics ; it is necessary to reconcile photons with optical waves. For instance an HEISENBERG-like relation $\Delta N . \Delta \varphi \geq \frac{1}{2}$ can be found |18|. This relation plays also a major role for superconducting media. (in a superconductor the phase of the wave function has a macroscopic extent).

- Symmetry properties : time reversal(reverse t in -t) or equivalently phase conjugaison is an operation that opticians know how to realize and to apply in modern devices.

- Signal processing : for example speckle which can be an unpleasant by-product of coherence can however be used in an heuristic way for example to measure the angular distance of double stars |19|.

Conclusion.

In old as well as in modern physics analogies play a major role. The "free trade" of experience between acoustics, optics, statistical physics, quantum mechanics... is a necessary condition for the elaboration of powerful concepts.

REFERENCES.

|1| Green, P.S et al., Acoustical Holography, Plenum 5, 493-503, (1974).

|2| Maneval, J.P., Bulletin Union Phys 692, 379-392 (1987).

|3| Fink, M., Lecture Notes in Physics, Springer 112, 438-452 (1979).

|4| Libchaber, J. and Toulouse, G., La Recherche 73, (1976).

|5| Westerwelt, P.J., J. Acout. Soc. Amer 32, 339 (1960).

|6| Beyer, R., Nonlinear Acoustics, U.S. Navy, 299 (1974).

|7| Breazeale, M.A. and Bjorno, L., Ulrasonics 440-443 (1977).

|8| Pauthier, S., Ann. Phys. Paris 14, 195 (1966).

|9| Plicque, F., Feupier, J., Zarembowitch, A., J. Acoust. Soc. Amer. 47 168 (1970).

|10| Fischer, M., Zarembowitch A., Acoustica 28, 259 (1973).

|11| Fischer, M., C.R. Acad. Sc. 274, 1115 (1972).

|12| Matthews, H., Lecraw, R.C., Phys. Rev. Let. 8, 397 (1962).

|13| Guermeur, R., Joffrin, J., Levelut, A., Penne, J., Sol. St. Com. 6, 519 (1968).

|14| Boiteux, M., and al., Phys. Rev. B (6) 2752 (1972).

|15| Joffrin, J., Levelut, A., Solid St. Com. 8, 1573 (1970).

|16| Pine, A.S., Phys. Rev. B.2 2049 (1970).

|17| Zarembowitch, A., and Khalifa, E., Revue Phys Appl. 20, 359 (1985).

|18| Glauber, R.J., Phys. Rev. 131, 2766 (1963).

|19| Françon, M., Granularite laser-Speckle. Masson (1978).

4

CONTRIBUTED PAPERS

Phonon Propagation in Disordered Media

K.G. BREITSCHWERDT

Institut für Angewandte Physik
Universität Heidelberg
Heidelberg
F.R. Germany

ABSTRACT

Phonon propagation properties and phonon lifetimes are shown to be a valuable tool in the investigation of the structural dynamics of disordered materials. As an example, quantitative results are presented for liquid and solid ionic solutions in the frequency range $10^5 - 5 \cdot 10^{12}$ Hz at temperatures between 100 and 300 K.

1. INTRODUCTION

Many physical properties of disordered or amorphous materials, such as thermal conductivity, specific heat, electrical conductivity, dielectric function, optical properties, ultrasonic propagation etc., differ from those in the crystalline state. This indicates that these properties strongly depend on the microscopic structure and on the structural dynamics of the material. While the ultrasonic propagation properties and the phonon characteristics are well established for crystalline materials, much less is known about the "quasi-phonon" propagation in disordered media, the phonon-phonon interactions in these systems, and the phonon coupling to structural modes, especially at very high frequencies.

According to the hydrodynamic approach the ultrasonic absorption α is proportional to the square of the frequency f, and the sound velocity v is independent of frequency.[1-4] If interactions of the ultrasonic waves with special structural properties take place, which may be present in disordered media but not in ideal crystalline materials, deviations of the ultrasonic absorption and of the sound velocity from the hydrodynamic predictions are observed (Fig. 1).

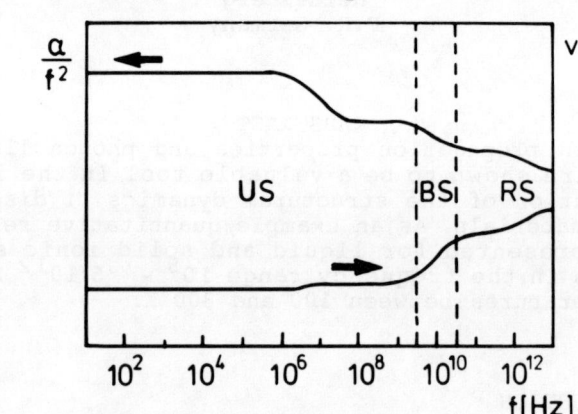

Fig. 1. Ultrasonic absorption and velocity of sound versus frequency

Because of the strong frequency dependence of the ultrasonic absorption a number of different experimental techniques have to be employed: Ultrasonic methods include resonator techniques[5] in the lower frequency range up to about 10^6 Hz, conventional pulse and Debye-Sears techniques[6] in the frequency range $10^6 - 10^8$ Hz, and surface excitation[7] of a quartz delay rod in the frequency range $10^8 - 5 \cdot 10^9$ Hz. Optical methods include Brillouin light scattering,[8] with which the frequency range may be extended to about $3 \cdot 10^{10}$ Hz, and Raman spectroscopy.[9] If an acoustic-mode interpretation is adopted in the analysis of

Raman spectra taken from disordered media information may be obtained about the propagation properties of these acoustic modes up to frequencies of several 10^{12} Hz.[10-12]

A great variety of materials have been investigated in the different frequency ranges. The present discussion will be restricted to a number of materials that show a characteristic ultrasonic absorption behavior in certain frequency ranges and thus are typical examples of materials with structural or chemical relaxations which couple to the ultrasonic waves or to the acoustic modes in the phonon regime:

1. Aqueous solutions of 2-2 electrolytes, especially those containing transition metal ions which show a pronounced ultrasonic relaxation in the frequency range 10^5-10^9 Hz.
2. Aqueous solutions of 1-1 electrolytes, in particular alkali halides, some of which exhibit a transition into the glassy state at lower temperatures and thus are particularly useful for the investigation of the ultrasonic propagation in the solid state, in the liquid state, and in the glass transition region.

Since they are optically transparent these systems are well suited for the optical investigations, Brillouin and Raman scattering, too.

2. CONVENTIONAL FREQUENCY RANGE

The ultrasonic absorption in aqueous solutions of many 2-2 electrolytes, including the sulfates of the bivalent transition metal ions, deviates considerably from the hydrodynamic prediction α/f^2 = const. Generally, several relaxations[13] can be separated in the frequency range 10^5-10^9 Hz if the structural relaxation of the electrolytic solution is properly taken into account. The experimental results for a number of these 2-2 electrolytes are shown in Fig. 2.

A so-called chemical relaxation was proposed[14] in order to explain the excess ultrasonic absorption in aqueous solutions of 2-2 electrolytes involving the stepwise removal of water molecules from between the cation and anion in ion pairs.

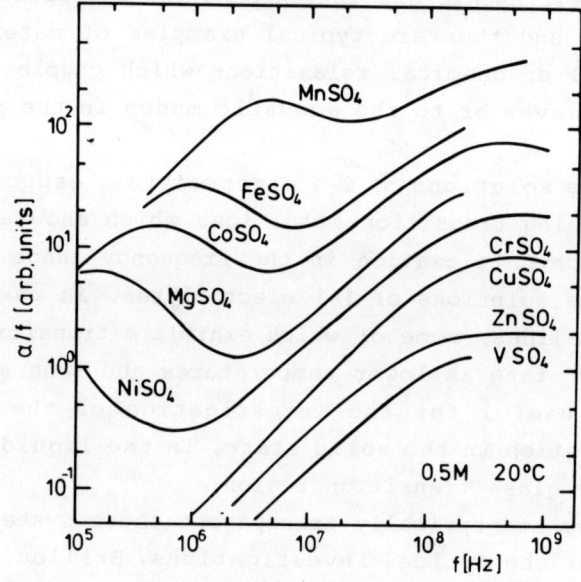

Fig. 2. Frequency dependence of α/f for various 2-2 electrolytes

This interpretation is supported by the fact that in the series of the transition metal ions, although they have comparable charge density, the relaxation times depend strongly on the number of electrons in their d shell. A theory[15] in which the activation energy for the exchange of water molecules in the inner hydration sphere is calculated quantitatively on the basis of ligand field stabilization is in good agreement with the experimental results.

The ultrasonic absorption in aqueous solutions of alkali halides[16] does not show any relaxation in the conventional frequency range. However, the quantitative explanation[17] of the ultrasonic absorption data leads to relaxation frequencies of about 10^{11} Hz. Lower as well as higher ultrasonic absorption than in pure water is found in these solutions depending on type and concentration of the ions present.[17]

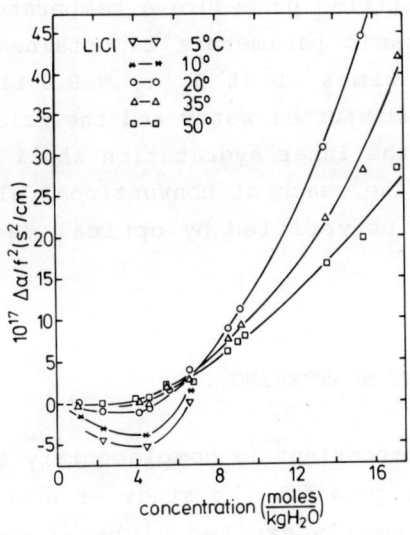

Fig. 3. $\Delta\alpha/f^2$ versus concentration for LiCl

The ultrasonic absorption versus concentration in LiCl and KBr solutions is typical for two types of behavior: α/f^2 values which, after a slight decrease at lower concentrations, increase with concentration, such as in LiCl solutions, and α/f^2 values which decrease with concentration in the total range of solubility, such as in KBr solutions. The experimental results for LiCl solutions are presented in Fig. 3 as difference $\Delta\alpha/f^2$ between the α/f^2 value of the solution and that of pure water versus ion concentration.

The excess ultrasonic absorption may be ascribed formally to structural relaxations between states assumed to be in equilibrium with one another. The states M_{ik}, k = 1,2,...,n, which represent different structural forms k in different ionic environments i are characterized by their free enthalpies and molar volumes. The transition between the states are governed by free activation enthalpies.

In a curve fitting procedure a temperature-independent set of thermodynamic parameters is obtained.[17] The resulting relaxation times at 20°C $\tau_1 = 0.6 \cdot 11^{-12}$s and $\tau_2 = 1.9 \cdot 10^{-12}$s for undisturbed water and the relaxation time $\tau_{Li} = 4.1 \cdot 10^{-12}$s for the inner hydratation shell of the Li^+ ion are well beyond the reach of conventional ultrasonic techniques. They may be verified by optical means (Section 3 and 4).

3. BRILLOUIN LIGHT SCATTERING

Brillouin scattering[8] is complementary to conventional ultrasonics techniques for the study of acoustical properties because thermally excited higher-frequency phonons with higher attenuations are usually accessible by this method. The velocity of sound is determined directly from the Brillouin shift. The spectral width of the scattered light, examined with high resolution, yields information about acoustic attenuation arising from anharmonicity, structural relaxation or other acoustic interactions with structural excitations.

The dynamic structure factor is usually employed to analyze Brillouin light scattering spectra. Using the hydrodynamic equations the following expressions for the center line and the two Brillouin lines are obtained:

$$S(q,\omega) = \frac{1}{\pi} V\rho_o^2 kT \kappa_T \left\{ (1-\frac{1}{\gamma}) \frac{2D_T q^2}{\omega^2 + [D_T q]^2} \right.$$
$$\left. + \frac{1}{\gamma} \left(\frac{\Gamma q^2}{(\omega-\omega_q)^2 + (\Gamma q^2)^2} + \frac{\Gamma q^2}{(\omega+\omega_q)^2 + (\Gamma q^2)^2} \right) \right\}, \quad (1)$$

where

$$\omega_q = v_q q = 2\pi f_q \quad (2)$$

is the shift in angular frequency, and

$$\Gamma q^2 = \pi \delta f_q, \quad (3)$$

where δf_q is the Brillouin line width.
The approximate value of the sound velocity is then

$$v_q = \frac{\lambda f_q}{2n_i \sin(\theta/2)}, \quad (4)$$

where λ is the wavelength of the incoming laser line and n_i the refractive index. The ultrasonic absorption α/f^2, which is frequency independent at frequencies well above and below a relaxation region, is approximately

$$\frac{\alpha}{f^2} \simeq \frac{\pi \delta f_q}{v_q f_q^2}. \quad (5)$$

The frequency dependence of the sound velocity v and of the ultrasonic absorption α/f^2 in the investigated system indicate the existence of relaxations in the Brillouin scattering range[18] (Figs. 4 and 5). The dispersion and absorption data yield relaxation times (Fig. 6) which are, within the accuracy of the measurement, identical in the temperature range between +20°C and -40°C if the light scattering data are properly transferred into the acoustic regime.[19]

The relaxation times obtained from dispersion and from absorption data with the single rate process analysis differ more strongly at temperatures below -40°C. This is the temperature range where the viscosity of the solutions approaches 1 P, and cooperative effects of the structural re-

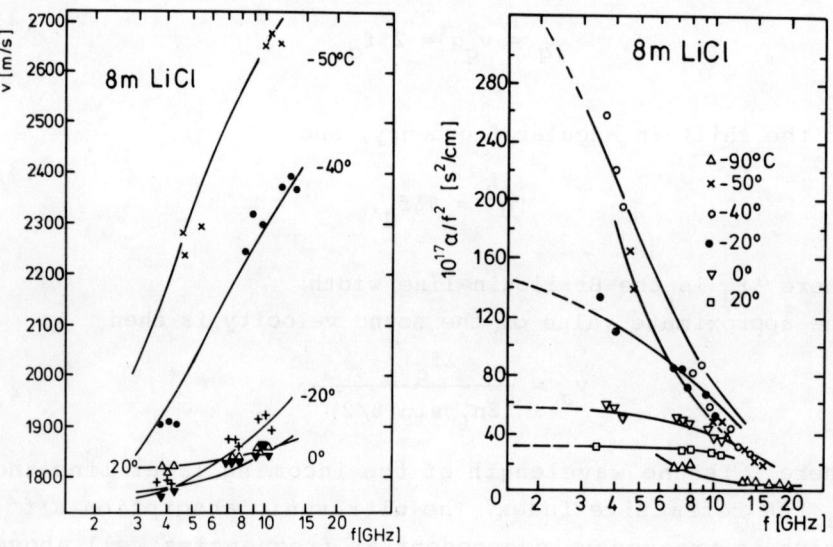

Fig. 4. Velocity of sound versus frequency

Fig. 5. α/f^2 versus frequency

arrangement have to be taken into account. Data analysis on the basis of the viscoelastic theory with a distribution of relaxation times is probably more appropriate in this case.

The activation energy for the structural rearrangement in 8 m LiCl solutions is 0.175 eV or 16.9 kJ/mole at temperatures above -40°C. The higher activation energy of 0.51 eV at -90°C is in agreement with the activation energy obtained from the viscose properties of the solution.[20]

The experimental results may be compared with the relaxation times and the activation energies which were predicted from the ultrasonic absorption at lower frequencies using the dynamic n-state model[17] (Table 1).

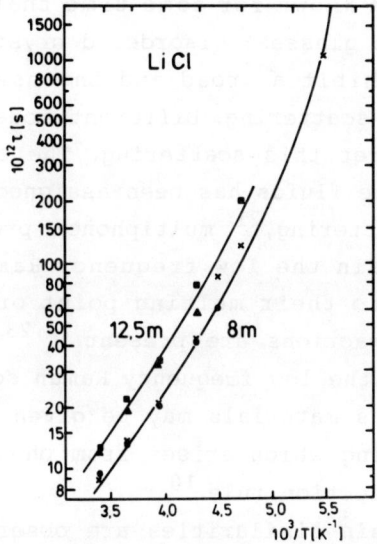

Fig. 6. Relaxation time versus reciprocal temperature

4. DEPOLARIZED RAMAN SCATTERING

High-frequency acoustic modes in amorphous solids are characterized by a very strong attenuation. The propagation properties of these modes may be studied in the frequency range around 10^{12} Hz by means of depolarized Raman scattering techniques since Raman spectra are produced by the modulation of the electrical polarizability due to fluctuations and thus contain information about the dynamical properties of the scattering system. Again certain aqueous ionic solutions are particularly useful in these investigations because they exhibit a transition into the glassy state at lower temperatures, and their Raman spectra can be obtained in the glass transition region as well as in the liquid and in the solid phase.

It has been known for some time that many disordered systems, such as glasses, disordered crystals, liquids, and dense fluids exhibit a broad and intense depolarized low frequency Raman scattering. Different models have been proposed to interpret this scattering. The so-called Rayleigh wing in all dense fluids has been assigned to collision induced light scattering,[21] multiphonon processes have been invoked to explain the low frequency Raman scattering in crystals close to their melting point or whenever strong anharmonic interactions are present.[22,23] A successful interpretation of the low frequency Raman scattering in glasses and amorphous materials may be given assuming disorder induced scattering which arises from phonons that violate the $\vec{k}$ vector selection rule.[10]

Since certain similarities are observed for different kinds of disordered systems, together with a continuous trend going from one phase to another, it is reasonable to assume that in these cases the same mechanism is responsible for the scattering process. Accordingly, attempts have been made to interpret the scattering[10-12] in terms of disorder induced one phonon processes in which strong anharmonic effects arising from phonon-phonon interactions and structural relaxations are included. With the Green's function formalism no restrictions are required on the degree of anharmonicity and therefore the approach is applicable also to liquids.

The depolarized Raman spectra may be taken in the standard HV configuration, i.e., the incident laser light is polarized in the horizontal scattering plane and the scattered light intensity with the polarization vector perpendicular to the scattering plane is recorded as a function of wave number $\omega/2\pi c$.

The depolarized Stokes intensity for an 8 m aqueous LiCl solution is depicted in Fig. 7. At room temperature the spectral distribution consists of a center line and a wing with relatively low intensity extending up to frequencies of about 300 cm^{-1}. With decreasing temperature the

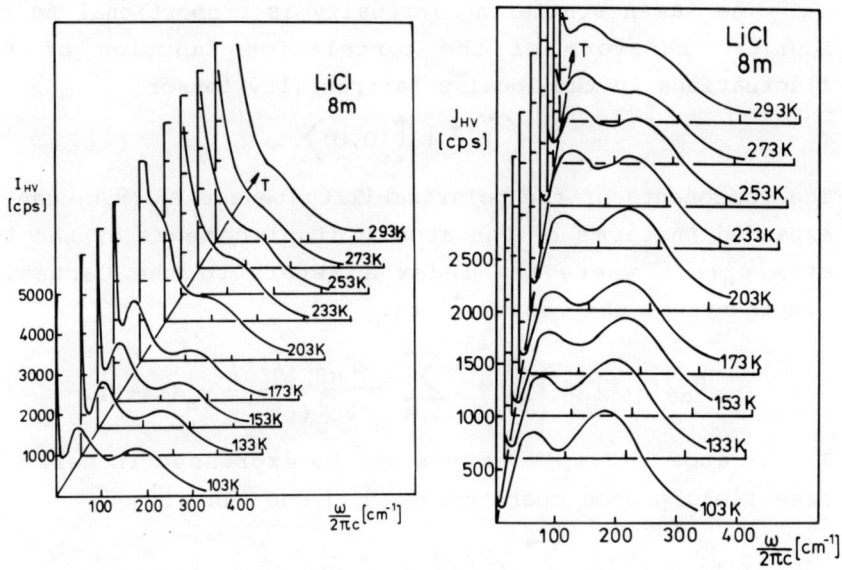

Fig. 7. Stokes intensity versus wave number

Fig. 8. Reduced Stokes intensity versus wave number

distribution exhibits more structural details. At 160 K, still in the liquid phase, the spectrum shows two maxima, one at 60 cm^{-1} and one at 180 cm^{-1}. The glass transition temperature of the system under investigation is around 150

K. In the solid phase at the lowest measuring temperatures of about 100 K two overlapping bands may be clearly distinguished. It is this low-temperature behavior in the solid phase that suggests an analysis of the Raman spectra in terms of phononlike excitations. With decreasing temperature the center line gets narrower and more intense. Below 170 K the width of the center line lies within the instrumental width of 0.5 cm^{-1}.

The Raman scattering intensity is proportional to the Fourier transform of the correlation function of the fluctuations in the local polarizability tensor

$$\left\langle \overleftrightarrow{P}(\vec{r},t)\overleftrightarrow{P}(0,0) \right\rangle . \tag{6}$$

The components of the polarizability tensor $P_{\alpha\beta}(\vec{r},t)$ can be expanded in terms of the atomic displacements of the ith atom $u_a^i(t)$, where the index a refers to the Cartesian components of the vector $\vec{u}^i(t)$:

$$P_{\alpha\beta}(\vec{r},t) = P_{\alpha\beta}^o + \sum_{i,a} \frac{\partial P_{\alpha\beta}(\vec{r},t)}{\partial u_a^i(t)} u_a^i(t) + \ldots \tag{7}$$

If the atomic displacements can be expressed in terms of free field phonon operators $A_b(\vec{q},t)$ one can write

$$P_{\alpha\beta}(\vec{r},t) = P_{\alpha\beta}^o + \sum_{b,\vec{q}} C_b^{\alpha\beta}(\vec{q}) R_b(\vec{q},\vec{r}) A_b(\vec{q},t) \left[2\omega_b(\vec{q}) \right]^{-1/2}, \tag{8}$$

where b is the phonon branch index, $C_b^{\alpha\beta}(\vec{q})$ are the electromagnetic-phonon coupling constants, and $\omega_b(\vec{q})$ the phonon eigenfrequencies. For the Stokes component the spectral distribution of the Raman scattering at a frequency shift $\omega = \omega_i - \omega_s$ is given by

$$I_{is}(\omega) = \left(\frac{\omega_s}{\omega_i}\right)^4 \left[n(\omega,T)\right] \sum_{b,\vec{q}} C_b^{is}(\vec{q}) U_b(\vec{q},\vec{k}) \operatorname{Im}\left[D_b(\vec{q},\omega)\right], \tag{9}$$

where $\vec{k}$ is the transferred wave vector,

$$n(\omega,T) = \left[1 - e^{-\hbar\omega/k_B T}\right]^{-1} \qquad (10)$$

the Bose factor, and $D_b(\vec{q},\omega)$ the phonon propagator, while

$$U_b(\vec{q},\vec{k}) = \int \left\langle R_b(\vec{q},\vec{r}) R_b(\vec{q},0) \right\rangle e^{i\vec{k}\cdot\vec{r}} d\vec{r} \qquad (11)$$

is the space correlation Fourier transform of the fluctuation in the polarizability due to a phonon of wave vector $\vec{q}$.
The sum over b and $\vec{q}$ may be replaced by

$$\sum_b \int \rho_b(\omega_b(\vec{q})) d\vec{q}, \qquad (12)$$

where $\rho_b(\omega_b(\vec{q}))$ is the density of states in the phonon branch b.

For a meaningful analysis of the Raman spectra the spectral distribution obtained experimentally must be reduced by a factor which is related to the population of the vibrational states and by a factor which takes into account the characteristic frequency dependence of dipole radiation. For the Stokes component the reduced spectral distribution (Fig. 8) is

$$J_{is}(\omega) = \left(\frac{\omega_s}{\omega_i}\right)^{-4} \left[n(\omega,T)\right]^{-1} I_{is}(\omega). \qquad (13)$$

$J_{is}(\omega)$ is, in the simple case,[12] given by a linear combination of the densities of states of those phonon branches which are coupled to the electromagnetic radiation. In the more general case the Raman intensity can still be interpreted as due to disorder induced one phonon scattering yielding a "Raman effective density of states".

It is generally assumed that vibrational excitations in disordered systems are localized due to lack of long-range order.[24] Nevertheless, for all kinds of disorder rather extended modes have been observed.[25] Furthermore, neutron scattering experiments have indicated coherence effects in noncrystalline solids and in liquids. Therefore, phononlike excitations may be used as an approximate description of vibrational excitations in disordered materials. These plane wave excitations are, however, not eigenstates of disordered systems since they are damped by scattering on structural irregularities and by structural relaxations due to rearrangements of the atoms or molecules.

In a simple model[10] the space correlation function is represented using a cutoff length $\Lambda_b(\vec{q})$:

$$\langle R_b(\vec{q},\vec{r}) R_b(\vec{q},0) \rangle = e^{\left[i\vec{q}\cdot\vec{r} - |\vec{r}|\Lambda_b^{-1}(\vec{q})\right]}. \quad (14)$$

This will remove the $\vec{q} = \vec{k} \approx 0$ selection rule since the momentum is no longer a good quantum number, and the scattering cross section is related to the density of all the vibrational states of the system.

The reduced Stokes intensity is then given by an integral over the phonon density of states and a sum over individual bands b:

$$J_{is}(\omega) = \sum_b \int_0^{q_b} dq\, C_b^{is}(\omega_b(q)) \rho_b(\omega_b(q)) \operatorname{Im}\left[D(\omega_b(q),\omega)\right], \quad (15)$$

where the superscripts i and s in the coupling constants denote the polarizations of the incident and scattered light, respectively.

A general form for the phonon propagator in the case

where phonons of different branches do not interact is

$$D_b(\vec{q},\omega) = \left[\omega^2 - \omega_b^2(\vec{q}) + G_b^2(\vec{q},\omega) + iF_b^2(\vec{q},\omega)\right]^{-1}, \quad (16)$$

where

$$\Sigma_b(\vec{q},\omega) = G_b^2(\vec{q},\omega) + iF_b^2(\vec{q},\omega) \quad (17)$$

is the phonon self-energy with both $G_b^2(\vec{q},\omega)$ and $F_b^2(\vec{q},\omega)$ being real positive quantities.

The renormalized phonon eigenfrequencies are

$$\bar{\omega}_b^2(\vec{q}) = \omega_b^2(\vec{q}) + G_b^2(\vec{q},\omega). \quad (18)$$

Whenever the phonon self-energy has a consistent real or imaginary contribution, the reduced Raman spectrum reflects the convolution of the vibrational density of states with the phonon propagator. Therefore, the knowledge of both the density of states and the phonon self-energies is needed to interpret the Raman spectra.

In the present case both anharmonicity effects and particle motions must be taken into account. Consequently, the propagator will contain in its self-energy phonon-phonon interactions and structural relaxations. It may be written in the following form:

$$D(\omega_b(q),\omega) = \left[\omega_b^2(q) - \omega^2 + i\omega M(\omega_b(q),\omega)\right]^{-1}, \quad (19)$$

where

$$M(\omega_b(q),\omega) = \Gamma_b(\omega_b(q)) + \frac{g_b(\omega_b(q))\tau}{1+i\omega\tau}. \quad (20)$$

The real term $\Gamma_b'(\omega_b(q))$ describes the attenuation due to phonon-phonon interaction. The relaxation term, written as a single Debye-type process with a relaxation time τ, takes into account the damping due to structural rearrangements that couple to the elastic distortions.

Neither the dispersion $\omega_b(q)$ nor the electromagnetic coupling constants $C_b(\omega_b(q))$ are known for large values of q. Since no detailed theoretical model specific for the amorphous state exists at present, the high-momentum excitations are approximately described by extending the hydrodynamic approach. A Debye-type density of states and a frequency dependence of the phonon damping

$$\Gamma_b(\omega_b(q)) = \Gamma_b^o \omega_b^2(q) \tag{21}$$

and of the relaxation strengths

$$g_b(\omega_b(q)) = g_b^o \omega_b^2(q) \tag{22}$$

are employed.

The reduced experimental Raman spectra are shown in Fig. 8 for the total temperature range. Starting from the solid phase the reduced spectral distribution exhibits essentially two overlapping bands. With increasing temperature these bands broaden and their maxima shift to lower frequencies. Furthermore, a quasi-elastic scattering contribution appears. The spectral distribution changes steadily across the glass transition region. This indicates that the Raman spectra can be analyzed in terms of phonon-like excitations in the liquid as well as in the solid phase.

Using Eqs.(15), (19), (21), and (22) the experimental curves of the reduced Raman spectra (Fig. 8) may be fitted quantitatively in the total temperature range. Two bands, b=1 and b=2, with temperature dependent damping constants $\Gamma_1^0(T)$ and $\Gamma_2^0(T)$ and a single Debye-type relaxation with a temperature dependent relaxation time $\tau(T)$ and a temperature dependent relaxation strength $g^0(T)$ suffice for a satisfactory agreement between experimental and theoretical spectra.

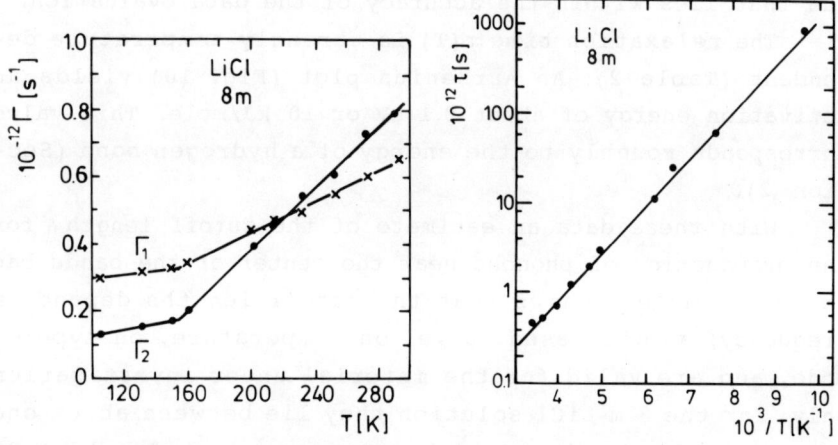

Fig. 9. Damping constants versus temperature

Fig. 10. Relaxation time versus reciprocal temperature

It is found that the relaxational term in the phonon propagator, which is responsible for the quasi-elastic scattering, also causes the shift of the band maxima to lower frequencies with increasing temperature. The broadening of the bands with increasing temperature is due to the increasing phonon damping and the decreasing relaxation time.

The values of the damping constants $\Gamma_1(\omega,T)$ and $\Gamma_2(\omega,T)$ for frequencies near the maxima of the bands, obtained in the curve fitting procedure, are given in Table 2, and their temperature dependences between 100 and 300 K are depicted in Fig. 9. A linear relationship is found for both bands in the liquid state. The transition into the glassy state around 150 K is marked by a change in the slope of the temperature dependence to a relatively small value that lies within the accuracy of the data evaluation.

The relaxation time $\tau(T)$ is strongly temperature dependent (Table 2). An Arrhenius plot (Fig. 10) yields an activation energy of about 0.1 eV or 10 kJ/mole. This value corresponds roughly to the energy of a hydrogen bond (Section 2).

With these data an estimate of the cutoff length Λ for the propagation of phonons near the center of the bands can be made. Table 2 shows that the cutoff lengths depend on frequency, as discussed above, on temperature, on type of mode, and are valid for the material under investigation only. For the 8 m LiCl solution they lie between about one wavelength for band 1 at higher temperatures and about 20 wavelengths for band 2 at lower temperatures. These cutoff lengths of 50-300 Å are about an order of magnitude smaller than those observed with Brillouin light scattering in the frequency range around 10^{10} Hz.

REFERENCES

1. Herzfeld, K.F. and Litovitz, T.A., Absorption and Dispersion of Ultrasonic Waves, Academic Press, New York, 1959.
2. Tamm, K., in Handbuch der Physik, S. Flügge, Ed., Springer-Verlag, Berlin, 1961; Vol. 11/1, p. 202.

3. Bhatia, A.B., Ultrasonic Absorption, Oxford, 1967.
4. Matheson, A.J., Molecular Acoustics, Wiley, New York, 1971.
5. Eggers, F., Acustica 19, 323 (1968).
6. Tamm, K., in Handbuch der Physik, S. Flügge, Ed., Springer-Verlag, Berlin, 1961; Vol. 11/1, p. 226.
7. Boemmel, H.E. and Dransfeld, K., Phys. Rev. 3, 83 (1959).
8. Berne, B.J. and Pecora, R., Dynamic Light Scattering, Wiley, New York, 1976.
9. Brodsky, M.H., in Light Scattering in Solids, M. Cardona, Ed., Springer-Verlag, Berlin, 1975.
   Gut, S., Thesis, Universität Heidelberg, to be published.
10. Shuker, R. and Gammon, R.W., Phys.Rev.Letters 25, 222 (1970).
11. Winterling, G., Phys.Rev.B 12, 2432 (1975).
12. Mazzacurati, V., Nardone, M., and Signorelli, G., J. Chem.Phys. 66, 5380 (1977).
13. Bechtler, A., Breitschwerdt, K.G., and Tamm, K., J. Chem.Phys. 52, 2975 (1970).
14. Eigen, M., and Tamm, K., Z.Elektrochem. 66, 93, 107 (1962).
15. Breitschwerdt, K.G., Ber.Bunsenges.Physik.Chem. 72, 1046 (1968).
16. Breitschwerdt, K.G., Kistenmacher, H., and Tamm, K., Phys. Letters 24 A, 550 (1967).
17. Breitschwerdt, K.G. and Kistenmacher, H., J.Chem.Phys. 56, 4800 (1972).
18. Breitschwerdt, K.G. and Polke, E., Ber.Bunsenges.Physik. Chem. 85, 1059 (1981).
19. Deguent, P. and Boon, J.P., J.Chem.Phys. 54, 4443 (1971).

20. Moynihan, C.T., Balitactac, N., Boone, L., and Litovitz, T.A., J.Chem.Phys. 55, 3013 (1971).
21. Bucaro, J.A. and Litovitz, T.A., J.Chem.Phys. 54, 3846 (1971).
22. Hardy, J.R. and Karo, A.M., in Light Scattering Spectra of Solids, G.B. Wright, Ed., Springer-Verlag, Berlin, 1969, p. 99.
23. Fleury, P.A., Worlock, J.M., and Carter, H.L., Phys. Rev.Letters 30, 591 (1973).
24. Dean, P., Proc.Phys.Soc. London 84, 727 (1964).
25. Bell, R.J., J.Phys. C5, L315 (1972).

Table 1. Relaxation times: τ_a from absorption data, τ_v from velocity data, τ_{th} from n-state model.

t [°C]	τ_a	τ_v [10^{-12}s]	τ_{th}
20	9	9.5	6.2
0	13.5	13	11.8
-20	20	20	24.8
-40	42	41	58.5
-60	120	-	160
-90	1000	-	1045

Table 2. Damping constants Γ, cutoff lenghts Λ, and relation times τ for the two observed bands. Sound velocity taken from Brillouin scattering (Ref.18).

T [K]	Band 1: $f_1=10^{12}$Hz, $\lambda_1=40$Å		Band 2: $f_2=3\cdot 10^{12}$Hz, $\lambda_2=13$Å			
	Γ_1 [s^{-1}]	Λ_1 [Å]	Γ_2 [s^{-1}]	Λ_2 [Å]	τ [s]	
292	$6.5\cdot 10^{11}$	62	$7.8\cdot 10^{11}$	51	$4.6\cdot 10^{-13}$	
273	$6.0\cdot 10^{11}$	67	$7.2\cdot 10^{11}$	55	$5.5\cdot 10^{-13}$	
253	$5.6\cdot 10^{11}$	71	$6.0\cdot 10^{11}$	67	$7.6\cdot 10^{-13}$	
233	$5.1\cdot 10^{11}$	78	$5.4\cdot 10^{11}$	74	$1.3\cdot 10^{-12}$	
213	$4.7\cdot 10^{11}$	85	$4.4\cdot 10^{11}$	91	$2.0\cdot 10^{-12}$	
202	$4.5\cdot 10^{11}$	89	$3.9\cdot 10^{11}$	103	$3.2\cdot 10^{-12}$	
161	$3.5\cdot 10^{11}$	114	$2.0\cdot 10^{11}$	200	$1.2\cdot 10^{-11}$	
150	$3.3\cdot 10^{11}$	121	$1.7\cdot 10^{11}$	235	$2.7\cdot 10^{-11}$	
132	$3.2\cdot 10^{11}$	125	$1.5\cdot 10^{11}$	267	$6.3\cdot 10^{-11}$	
105	$3.0\cdot 10^{11}$	133	$1.3\cdot 10^{11}$	308	$8.3\cdot 10^{-10}$	

THE TEACHING OF ACOUSTICS AT THE FACULTY OF ELECTRICAL ENGINEERING

V.CHALUPOVÁ, K.SOBOTKOVÁ

DEPARTMENT OF PHYSICS, ČVUT FEL, PRAGUE, CZECHOSLOVAKIA

ABSTRACT

In the presented paper the current state of teaching of acoustics in regular, interdisciplinary and post-graduate study at the Faculty of Electrical Engineering is shown. The questions of curriculum of study and its forms are also discussed.

Teaching of acoustics at the Faculty of Electrical Engineering has a long tradition. It was founded by professor Slavík. He was the head of the Department of Physics at the Faculty of Electrical Engineering from 1950 to 1964. His main specialization was physical and technical acoustics. His scientific activity was very extensive. He published seventy works especially from the branch of physical and technical acoustics. Professor Slavík's contribution to the education of new acoustic specialists was also important. Besides his own scientific and pedagogic work he took part in the solution of problems in dealing with development of the faculty.

He was the chairman of Prague branch of the Union of Czechoslovak mathematicians and physicist for many years. He emphasized basic problems of acoustics, that is generation of sound and conditions for propagation of sound. His premature death in full work meant a big loss for acoustics.

New generation of students of professor Slavík started its work. The orientation of the work of the Department of

Physics was changed, but foundation of branch of physical acoustics remained firmly connected with this department. Further special centers of acoustics were founded; VÚZORT is specialized in the architectural and room acoustics, SVÚSS in the branch of acoustic sources, VUPS in the branch of architectural acoustics and ÚVMV in the branch of noise of motor cars.

There was a great engagement in standartization. Professor Kolmer, who was a disciple of professor Slavík, became after the professor's death the chairman of acoustic group of the Czechoslovak Academy of Sciences. The important role was played and still is by the Czechoslovak Society for Science and Technology. Dr.Němec also a disciple of professor Slavík,was a chairman of special group for noise and enviromental acoustics for a long time.

In the seveties there was some reduction in the volume of teaching acoustics in the basic course of physics. Acoustics was partially replaced by new developing branches of physics, mainly by quantum physics. That's why recommended lectures of physical acoustics were started. The posgraduate study of physical acoustics was organized. 120 students were educated in three repeated postgraduate courses. In connection with these courses similar courses were organized at the Faculty of Engineering and at the Faculty of Civil Engineering.

There was further specialization at the faculties in following years. The Faculty of Electrical Engineering deals mainly with the physical problems of the branch, e.g. measuring methods, measuring instruments, processing of acoustic signals and technical diagnostics.Problems of physical acoustics and electroacoustics were solved in postgraduate courses.

Now we come back to the undergraduate study. In those days when the environmental problems had become compeliing the necesserity of an education in the environmental engineering was needed. Since 1980 the courses of environmental engeneering have been integrated into curriculum. There is one lesson per week in the last term of study of every specialization. Enviromental study is carried out within the Department of Physics.

In the cources of environmental engineering there are also lectures of the basic problems of sound, noise and vibration in the connection with the environment and people. The environmental engineering courses include both the complex conception of the problem and the evaluation of a quality of an environment according to several points of view. This courses are closed by special homework and by exam. Students now are more interested in problems of acoustics of the environment. Some students take up again with these problems in the form of student´s scientific activity or diploma work. One course of postgraduate study of environmental acoustics was organized. Now another course is prepared by other faculties of the Czech Technical University. There is a special group of acoustic consultants at the Faculty of Electrical Engeneering sponzored by the Czechoslovak Society for Science and Technology, which is very often consulted not only by students, but also by many people from industry.

There is also a new interdisciplinary branch-Metrology. This branch arose in connection with the inovation of study at the Faculty of Electrical Engineering. In this course lectures of subject Metrology of acoustic and optic quantities are read. There are two hours of lectures and two hours of practices per week in the seventh term. Students of this interdisciplinary branch get acquainted in this subject with metrology of basic quantities of sound field, with modern methods of measurment technology and also with standartization and testing. Graduate of this branch work especially in measurment practice, in standartization and in supervising.

Further education in acoustics for graduate is organized by the Czechoslovak Society for Science and Technology. These courses do not last for a long but they deal with special problems (for example traffic noise, aerodynamics etc.). Postgraduate and graduate education at the Czech Technical University lays special emphasis on the relevance of science and technology to the complex problems of society.

Graduate study has two possible forms. Internal graduate study requires three academic years beyond an undergraduate study. The external graduate student fulfils his duties of the graduate study besides his work. This study lasts five years.

Graduate student has to pass prescribed examinations and defend thesis, then he receives the degree the candidate of sciences. There are also other educational centers, e.g. other universities, the institutes of the Czechoslovak Academy of Sciences and some research institutes. Acoustic study and the research at the Czech Technical University are carried out within the Department of Physics in specialization acoustics and within the Department of Radioelectronics in specialization electroacoustics.

There are also updated methods of teaching at the Faculty of Electrical Engineering. Audiovisual equipment is used at every lecture. Instructive films are also shown and the utilization of computers is extended. The greatest emphasis is laid on understanding of basic physical problems in every form of study, but we pay also attention to the new development of technology. These were some options we use for an extension of acoustical education of the undergraduate, graduate, and postgraduate students working in science and industry.

REFERENCES:
1. Studijní plány FEL, ČVUT, Praha
2. Práce ČVUT, č.4, 1965, Praha

SELECTED PROBLEMS OF TEACHING ACOUSTICS TO STUDENTS OF THE LUTERY FACULTY IN SECONDARY AND HIGHER SCHOOLS

H. HARAJDA

Institute of Experimental Physics
Pedagogical University (WSP)
Zielona Góra Pl. Słowiański 6
POLAND

ABSTRACT

In the Polish system of art and music education, lutery has been a unique specialist faculty in the State Lycee of Music and at the Poznań Academy of Music. In these two schools the subject called lutery acoustics have been introduced and the curricula comprise selected problems from different branches of acoustics thought to be indispensable for a violin-making master when forming the sound quality of the instrument being manufactured. In the analysis of the most important problems concerning the selection of curriculum contents special stress has been laid upon the correlation to be made between the curriculum of acoustics and the curriculum of physics.

INTRODUCTION

Since 1973 in the Polish system of music education lutery has been a unique specialist faculty at which students are trained within two stages. The first stage, the education on the secondary level at the State Lycee of Music in Poznań, lasts for six years and those who leave this school pass their final examinations and get a certificate. The graduates from that lycee are formally prepared to join the Union of Polish Lutery - Masters or to apply for a job of violin-making artist. Since 1978 it has been made possible for them to continue their education on the higher level, the second stage, at the Poznan Academy of Music, Instrumental Department with lutery as a specialist subject. First graduates with a degree of magister in lutery completed their in 1984. In 1985 the first persons successfully completed the post-graduate research stage to hold the position of adiunct of that faculty.

LUTERY ACOUSTIC

In the above mentioned schools, a separate subject called "lutery acoustics" has been introduced, one hour a week at the lycee, and 60 hours of lectures for the second-year students at the Academy. Acoustics is one of the obligatory professional courses for the students to take. In accordance with the educational objectives, those who leave the secondary level school ought to get acquainted with "general knowledge of ... lutery acoustics and ought to be capable of putting the theoretical knowledge of lutery into practice, when manufacturing an instrument or improving its quality" [1]. Lectures are meant "to broaden and deepen the knowledge of lutery acoustics and physics to the level indispensable for a lutery artist in his professional work" [2]. Those who acquire the knowledge of lutery get in touch with acoustical problems practically not only when forming or evaluating the tone quality of the instrument which has already been made but also at the very beginning of their creative work. The first decision they make is the selection of material for the instrument to be manufactured. Therefore ,a wide variety of acoustical problems must be included in the syllabus of acoustics at the lycee level. Acoustics introduced at this stage of schooling creates a precendential situation what secures specific conditions for the transfer of inforrmation leading inevitably to experimenting in order to determine the range of knowledge for the pupil - violin - maker - to acquire. It also points to the necessity of wide application of the knowledge of other subjects as well, physics first place, but also mathematics and biology. Frequent references to professional luter problems are also called for.

Curricula of lectures lay stress upon
1. the problems which broaden the knowledge and practical skills indispensable for individual experimenting on forming and improving the sound qualities of the instruments being manufactured,
2. the development of individual abilities to make use of the world acoustical literature, of journals in particular (Journal of the Catgut Acoustical Society, Das Musikinstrument, Hudebni Nastroje).

THEMATIC RANGE AND INTERDISCIPLINARY CORRELATION

Taking into consideration the objectives of teaching lutery acoustics and practical observations made during the first five years of schooling, the following curriculum of teaching has been worked out.

Table 1 presents the syllabus meant to be applied at the first stage of training violin-maker specialists.

Table II presents the syllabus of lectures at the second stage of schooling (at the Academy).

The current physics syllabus in primary and secondary schools allows for the correlation to be made between sylabus contents in physics and acoustics, with the possible exception of wave phenomena and the acoustical field. They appear in the physics syllabus later than they are needed for the purpose of acoustics. Therefore, these problems are introduced at acoustics lessons prior to their appearance in the physics syllabus though the discussion of them can be left out when speaking about the acoustics of material or about vibrations. Considering the correlation between acoustics and mathematics, logarithms seem to cause most trouble. They prove to be useful in acoustics on several occasions. It has been found, however, that the

TABLE I

Number	General topics	Form semester	Number of hours
I	Acoustics of the material	V,1	8
II	Elements of physical acoustics		
	A. Vibrations	V,1,2	10
	B. Wave-sound	V,2	8
	Revision of the material - tests		4 hours
III	Elements of psychophysiological acoustic	VI,1	8
IV	Elements of electroacoustics	VI,1,2	6
V	Recording and audio monitoring of a sound	VI,2	2
	Revision of the material - tests		4 hours

knowledge of logarithms, in general, is hardly satisfactory even among higher school students. Although, at the lessons in the science of man, the structure of ear is discussed superficially, the information the pupils get is fairly sufficient to be further developed at acoustics lessons. It occurs, however, that pupils are not provided— with a logical coherent picture of the way of transmitting musical information in the process of hearing when they get acquainted with the structure and function of the nervous system.

TABLE II

Number	General topics	Number of hours
I	Lutery instruments as the complex source of sound – –String vibrations, sound board vibrations, sound board tuning, physical and acoustical problems when manufacturing an instrument, physical foundations of bow construction and bow functioning.	10
II	Elements of audiology and psychoacoustics – –The process of hearing as a physical process, the audibility field, equal loudness contours	6
III	Sound of lutery instruments – –A. Bow instruments –B. Guitar sound and attempts at its forming	10 6
IV	Changes in the construction and their effects on the tone quality of an instrument- –Changes in the scheme of transmission, –in the structure of wood, the field of liberty of pitch, dynamic, ton quality, form.	6
V	Acoustical properties of wood in standards and in lutery practice	6
VI	Acoustical measurments – –Foundations of oscillographic analysis,–spectroscopic analysis, –intonographic analysis	6
VII	Methods of measurment organisation in view of the research literature	6
VIII	Elements of room acoustics – the effect of acoustical climate on the reception of sound sensation	6
	Lectures in the form of discussion	4 hours

Therefore, this problem has to be clearly explained at acoustics lessons.

CONCLUSIONS

Since the number of students, whether at the lycee or at academy, is rather small (1 - 6 persons), there are very favourable conditions for individual treatment of any student. Visual aids available in the teaching process such as 1. the instrument itself (violin, guitar), 2. prints of sound structures and their various courses, 3. a basic set of electroacoustical equipment-well serve the purpose of obtaining better results in schooling. There are, however, shortcomings which prevent the results from being optimum ones:
1. very low level of understanding physical terminology which, as it can be presumed, is not introduced at the lessons on the grounds of word etymology,
2. pupils acquire the knowledge of physics almost exclusively by heart and not on the way of logical and semantic operations,
3. huge gaps in the knowledge of certain notions such as vibration, acceleration, or field are in evidence (these remarks come from the experience in teaching acoustics to university students of musicology).
Observations on the interest in acoustics made when teaching young people at the lutery faculty lead to the conclusion that their interests develop when they enter inte a close contract with analysis and measurment equipment or when they are provided with visual aids of acoustical courses dealing with auditory sensations. To sum up, my appeal is to secure pupils with wider access to the basic measurement and analysis equipment in school laboratories.

REFERENCES

1) Organization of the teaching process, MKiS Warsaw (1973).
2) Curriculum of lectures in acoustics at the lutery faculty, the Poznań Academy of Music, Poznań (1978).

355

THE RAY METHOD AS A TEACHING TOOL

Andrzej Kulowski

Sound Engineering Department
Institute of Telecommunication
Technical University of Gdansk
80-952 Gdansk, Poland

ABSTRACT

The ray method as a part of the teaching course on room acoustics has been described. The principle and the teaching advantages of the method have been discussed. The topics of the teaching program and the M.Sc. theses prepared in the field of the ray method have been presented.

INTRODUCTION

Students graduating from the Technical University of Gdansk at the Sound Engineering Department take the two-semester course on room acoustics. The course consists of 30 hours of the lectures, 15 hours of the laboratory exercises and 15 hours of the design exercises. 2-3 students per year prepare also their M.Sc. theses in the field of room acoustics.

Among the various topics of the room acoustics course, the numerical methods of the sound field modelling play an important role in the teaching program. The importance of this particular topic arises from its high teaching efficiency, mainly due to the systematic, algorithmic form of the numerical methods presented at the lecture. Another reason of the significance of this subject is the increasing importance of the computer methods of the sound field modelling in the profession of the room acoustician.

The ray method is the one of the numerical methods which are discussed at the room acoustics course.

THE PRINCIPLE OF THE RAY METHOD

The ray method is based on partitioning the energy of the spherical sound wave, which is emitted from the given point in the room, into a number of elements. The elements, called the sound rays, are assumed to be the punctual, weightless objects running with the speed of sound and obeying the laws of geometrical acoustics. The directional sound wave radiation from the source and the diffuse wave reflection by the wall can be also represented using the ray method. The ray energy decreases due to the absorption of the room walls and the absorption of the air. The ray is traced as long as its energy exceeds the given threshold value.

If it is assumed that all the rays are emitted at the same moment of time, the energy decrease observed in the room can be interpreted as the room impulse response.

The sketches often met in the books on room acoustics, which represent the directions of sound propagation in sections of a room, are nothing more but a graphical form of the ray method employed in the two-dimensional space. The three-dimensional ray tracing and calculation of the ray running energy can be also made by hand, though this is a task which requires some patience.

The ray method is practically implemented in a form of a computer program. The dramatic increase of both the number of rays traced and the number of reflections taken into account, facility of considering the complex room shapes, the radical improvement of accuracy etc. are the results of using the new tool, i.e. the computer, but the rule of the method remains the same. So the ray method is easy to be imagined in a visual form, which is an important teaching advantage of the method.

THE PRESENTATION OF THE RAY METHOD AT THE LECTURE

The presentation of the ray method at the lecture takes about 3 - 3.5 hours. The following topics are presented:
- the assumptions of the ray method,
- the phaseless ray and the phase carrying ray tracing,
- the algorithm of the ray tracing in an enclosure [1],
- the algorithm of the impulse response processing [2],
- the error resulting from the substitution the continuous structure of sound field with the granular structure of sound rays field [3] (see fig. 1),
- the room acoustical parameters provided by the ray method, for instance: the reverberation time (see fig. 2), the steady state sound level, the sound energy diffusion coefficient, the lateral reflection coefficient etc. [4],[5],
- the state of the art in the literature [6]-[9],
- the examples of the halls designed with the use of the ray method [10],[11] (see fig. 3),
- the applications of the ray method in other fields of acoustics (for instance geoacoustics [12], underwater acoustics [13]).

THE EMPLOYMENT OF THE RAY METHOD AT THE DESIGN EXERCISES

The students prepare 2-3 design works during 15 hours of the design exercises. The works concern the acoustics of both the theatre and concert halls, auditoria, recording studios etc. and the industrial halls.

The student s task is to consider the room shape and the acoustical coverage of walls, as well as to fix the sound sources positions and to determine the acoustical properties of the sources. From these data, the steady state sound level and the reverberation curve at the given points of the room are computed. From the reverberation curve the required acoustical parameters of the room are determined. After the data correction and the result re-calculation, the final values of the parameters are obtained.

The graphical subroutines of the computer program are the significant aid at the design work by facilitating the design drawings preparation. This is the typical facility for every computer-aided design work. However, for the student exercises this is particularly advantageous due to the considerable time savings which enable the student to spend more time on the design job.

THE EMPLOYMENT OF THE RAY METHOD AT THE M.Sc. THESES

There are three groups of the M.Sc. theses concernig the ray method:
1. Acoustical design and investigations of the concert and theatre halls, auditoria, recording studios etc.
2. Acoustical design and investigations of the industrial halls.
3. Optimization of the numerical representation of the ray method, investigation of the granularity error of the ray method, new algorithms of the modelling result processing and presentation etc.

The following M.Sc. theses have been completed in the above mentioned groups, respectively:

1. "Computer investigation of the TV studio acoustical properties". (1980)
   "Numerical method of the speech intelligibility determination in the room". (1982)
   "Investigation of the acoustical properties of the Musical Theatre in Gdynia". (1983)
   "Investigation of the acoustical properties of a concert hall by the analysis of its impulse response". (1985)

2. "The analysis of the ray method usefulness for the industrial designing of industrial halls". (1983)
   "Prediction of the sound level distribution in the industrial hall". (1986)

3. "Computer investigation of the room acoustics using the Monte Carlo Method". (1978)

"Numerical modelling of the sound source directional pattern using the ray method". (1980)
"Investigation of the statistical properties of the ray model of sound field in the room". (1982)

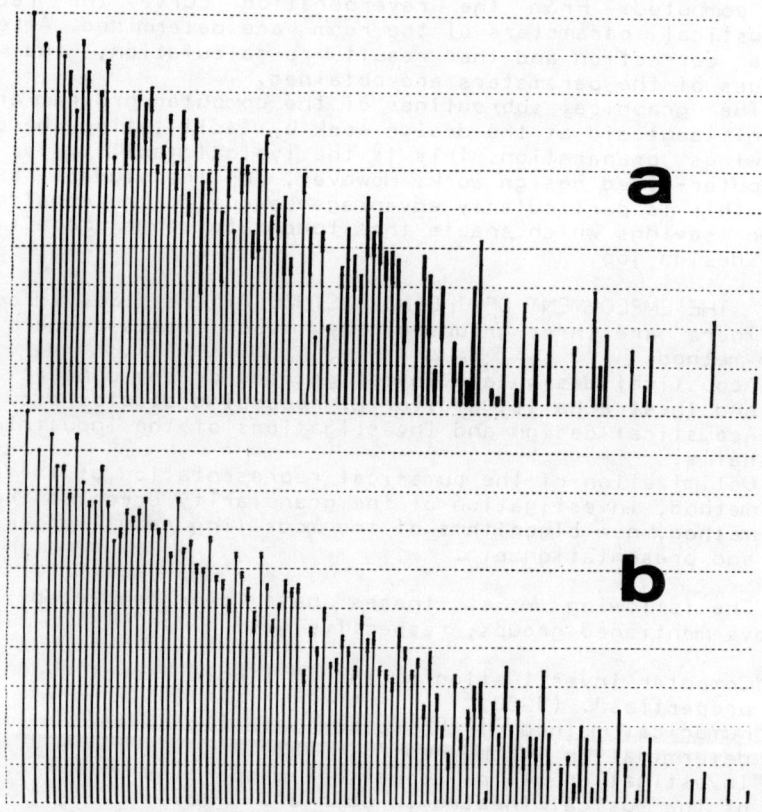

Fig. 1. The example of the impulse response at the selected point of the room, calculated using the ray method (the level of space energy density versus time). For each interval of the histogram the granularity error range is indicated. The dynamics and the duration of the modelled phenomenon: 60 dB and approximately 1.5 s; the number of the rays traced: 2000 (a) and 20 000 (b), respectively.

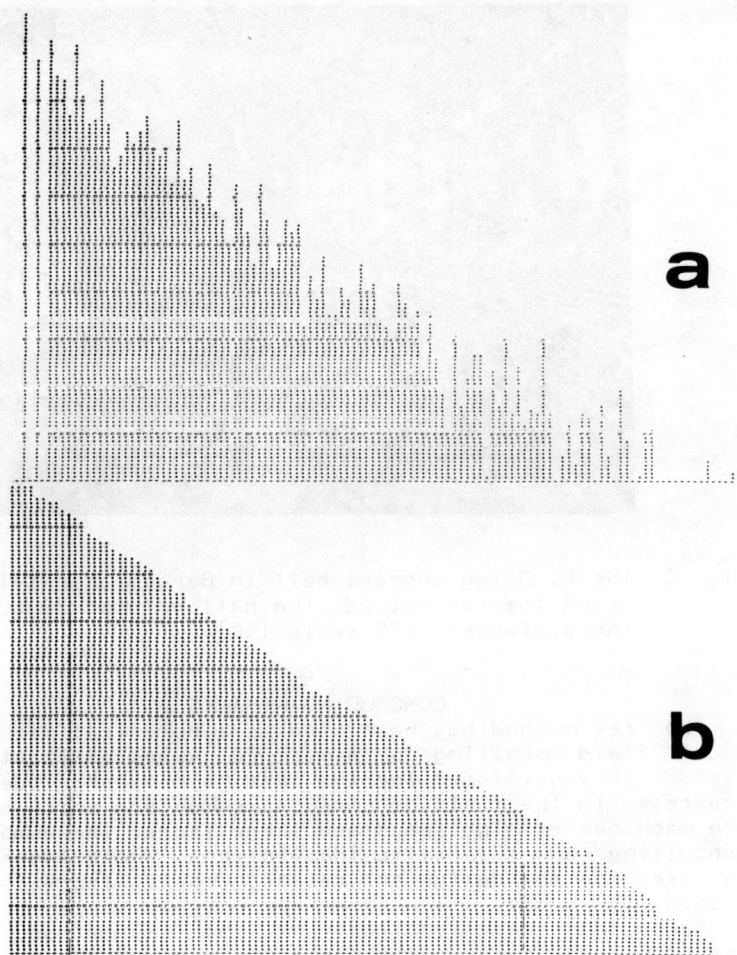

Fig. 2. The example of the reverberant response at the selected point of the room (b) determined from its impulse response (a) (the level of spatial energy density versus time). The impulse response is calculated using the ray method. The dynamics and the duration of the modelled phenomena: 60 dB and approximately 1.5 s.

Fig. 3. The E. Grieg concert hall in Bergen (Norway) designed using the ray method. The hall volume: 20 000 m3, the audience: 1520 seats [10].

CONCLUDING REMARKS

The ray method has been used as a numerical method of the sound field modelling for almost 20 years [14]. It has been particularly developed for the last 4-6 years due to the progress in the field of personal computers which has become the machines of high speed and large store. The number of the consulting and scientific institutions, where the ray method is used for the design and research works, increases year by year.

Apart from these applications, the ray method has great teaching advantages which enable to employ it as an efficient teaching tool. In such use of the method, both the room shape can be simplified and the number of sound sources and observation points can be limited. It is also acceptable that the granularity error may be greater than at the design or research applications, due to the tracing the limited number of rays. In these cases, also the small personal computers can be used for the teaching purposes due to the limitation of both the computation time and the required store capacity.

REFERENCES

[1] A. Kulowski: "Algorithmic representation of the ray tracing technique", Appl. Acoust., vol.18, p.449-469, 1985.
[2] A. Kulowski: "Relationship between impulse response and other types of room acoustical responses", Appl. Acoust., vol.15, p.3-10, 1982.
[3] A. Kulowski: "Error investigation for the ray tracing technique", Appl. Acoust., Vol.15, No.4, p.263-274, 1982.
[4] A. Kulowski: "Calculation of acoustical parameters of concert and theatre halls using ray method", Proceedings of 6th FASE Symposium, p.107-110, Sopron (Hungary) 1986.
[5] A. Kulowski: "Calculation of sound diffusion in industrial halls using the ray method", INTER-NOISE 87, Pekin 1987 (in print).
[6] M.R. Schroeder: "Digital simulation of sound transmission in reverberant spaces", J.A.S.A., vol.47, No.2, p.425-431, 1970.
[7] H. Kuttruff: "Simulierte Nachhallkurven in Rechteckraumen mit diffusem Schallfeld", Acustica, vol.25, H.6, S.334-342, 1971.
[8] S. Santon: "Numerical prediction of echograms and of the intelligibility of speech in rooms", J.A.S.A., vol.56, No.6, p.1399-1405, 1986.
[9] J. Wayman: "Computer simulation of sound fields using ray method", Ph.D. Thesis, University of California, Santa Barbara, 1980.
[10] A.Krokstad et al.: "Fifteen year s experience with computerized ray tracing", Appl. Acoust., vol.16, p.291-312, 1983.
[11] S.Strøm et al.: "Acoustical design of the Grieg Memorial Hall in Bergen", Appl. Acoust., vol.18, p.127-142, 1985.
[12] P. M. Shah: "Ray tracing in three dimensions", Geophysics, vol.38, No.3, p.600-601, 1973.
[13] H.-G. Schneider: "Rough boundary scattering in ray-tracing computations", Acustica, vol.35, p.18-25, 1976.
[14] A. Krokstad et al. : "Calculating the acoustical room response by the use of a ray tracing technique", J. of Sound and Vibr. vol.8, No.1, p.118-125, 1968.

Diffraction as time-space phenomenon: Educational aspects[*)]

Henryk LASOTA
Institute of Telecommunications
Technical University of Gdańsk
80 952 Gdańsk, Poland

The paper discusses particular features of the time-domain impulse description of acoustic diffraction phenomena, underlining educational aspects of this new method. When compared with the traditional harmonic description that operates in the domain of sine waveforms, the impulse approach appears to be particularly advantageous from the didactic point of view, as it constitutes a very efficient help in understanding general principles and physical mechanisms of the field effects, and, simultaneously, it provides a simple and economic tool for technical calculations of acoustic fields.

1. INTRODUCTION

When teaching acoustics to specialists of various branches of knowledge and technology, the first question appearing to the lecturer is: "What to teach?" - it concerns the problem of the choice of the m a t t e r to be taught. The proper choice means a reasonable compromise between the quantum of general knowledge of basic principles and the amount of specialized information dealing with the applications in the domain of interest. The present paper, leaving apart this problem, puts the main stress onto another one of no less importance - "How to teach?". This question concerns the choice of the m e t h o d o f a n a l y s i s when teaching a specific problem. When dealing with space phenomena, such a choice exists between the traditional harmonic approach that uses the complex exponential notation and works in the spectral domain, and the new impulse approach - using the distribution notation and working directly in the time domain.

[*)] *This work was performed under the grant CPBP 02.16.4.1*

In the author's opinion, when the time-domain impulse approach is used, students reach deeper understanding of basic principles, obtaining in the same time practical means for everyday computations of acoustic fields, both without any need of extending the lecture hours. The efficiency of this approach results from the notation that replaces the notion of phase shift by the notion of time delay having a more general value.

The time-domain approach has been adopted and is being developped in the Institute of Telecommunications, Faculty of Electronic Enigineering, Technical University of Gdańsk, and is now broadly used in both the research and education activity for studying space phenomena in acoustics related to such field and technology problems of acoustic antennas as diffraction of acoustic waves, on one hand, and multilayer structures of broadband piezoceramic transducers, on the other hand.[1,2]

The present paper discusses general and practical reasons that make of the time-domain approach a very interesting and efficient tool, better suited for analysing diffraction problems than the harmonic approach. The didactic advantages of the approach is presented at the example of a sequence of lectures on acoustic antenna fields given to students of Electronic Engineering specializing in Electroacoustic Systems. The sequence concerns the broadly understood diffraction phenomena, covering such problems as radiation of surface sources, wave propagation in the free field, diffraction on weighed apertures and space sensitivity of receiving transducers.

2. CALCULATION OF ARBITRARY FIELDS

Calculating the pressure field of acoustic antenna is one of the practical problems belonging to a more general question of the plane wave diffraction. The question is classical when the excitation is harmonic, the solutions being broadly known for cases when the excitation is uniformly distributed over an aperture of a simple shape. However, in more complex situations, when the excitation is arbitrary and the aperture is weighed, the calculations of the pressure waveform in the field become quickly involved, needing not only a great calculation power - easily attainable nowadays - but, first of all, rather sofisticated knowledge of mathematics, covering such domains like, for instance, integral series of the Bessel functions – not very popular among engineers. The rescue to this problem comes from the new approach to the diffraction, that uses the concept of the impulse field.[3,4]

The main difference between the harmonic method of analysis (denoted here as H) and the impulse one (denoted as I) lies in the way of decomposition of the exciting signal and the resulting field into elementary components. In H, the field $p(\vec{x},t)$ of a composed excitation $s(t)$ is seen as the superposition of elementary harmonic fields $P_\omega(\vec{x}) \exp(j\omega t)$, each being related to one spectral component S_ω of the exciting signal, and is calculated in the frequency spectrum domain with the use of the Fourier transformation (denoted below symbolically as $F\{\ \}$). In I, the field is treated as the superposition of elementary impulse fields $p_\delta(\vec{x},t)$ related to elementary impulse excitation $\delta(t)$, and the calculations are performed in the domain of time waveforms with the application of the operation of convolution (denoted by an *). The two approaches can be presented in the following symbolic way:

H: $\qquad p(\vec{x},t) = F^{-1}\{ P(\vec{x},\omega) \}$

I: $\qquad p(\vec{x},t) = s(t) * p(\vec{x},t)$

As it is claimed by all textbooks, both approaches are mathematically e q u i v a l e n t (that means: H = I). This fact is broadly made use of in the practice of field analysis. However, it should be firmly stressed here that from the point of view of physical phenomenology they are n o t i d e n t i c a l (H $\neq$ I). The goal of the present paper is to discuss educational consequences of the latter fact.

It is interesting to examine and compare the main features of both approaches from various points of view, taking into account such important aspects as their tradition and textbook "environment", calculation availability for everyday engineer computations, and matching of respective notations to the nature of the phenomena.

3. TRADITIONAL ANALYSIS

The classical approach has a long and worthy tradition in both research and education. It is deeply rooted in the rich environment of very complete textbooks covering various domains of physics, and multiple branches of engineering. The mathematical notation has an evident elegance and beauty remembering its baroc origins. Being familiar to every scientific worker, it has given to all of us the impression of being perfectly natural and well matched to resolve practical problems of any kind.

In fact, this impression is not really founded. The harmonic analysis is merely matched to the s i n e w a v e excitation - S_ω exp($j\omega t$). In the practice of acoustic engineering it is related to steady states of sinewave pulses and to spectral components of audible sound. However, its straightforward use can be problematic in other cases, especially when dealing with the excitations that have a simple time waveform $s(t)$ and, simultaneously, an involved form of related frequency spectrum $S(\omega)$. This is the case of many transients in sound engineering, of short pulses used in ultrasonic echography, and of special broadband pulses designed for the pulse compression technique. In such situations the field computations, although feasible, become rather difficult for an engineer - usually not very friendly to the idea of performing the integration of functions more involved than the exponential ones. Obviously, this drawback of the method is not the most important and could be neglected. The point is, however, that the traditional notation is n o t n a t u r a l ! It replaces a pulse wave - strictly localised in both time and space - by its spectral components, each occupying the whole space and existing forever. Being based on the use of alternative functions, the analysis puts an excessive stress on the second-order effects of interference, masking the first-order space effects the field phenomena are ruled by.

4. IMPULSE ANALYSIS

The "tradition" of the impulse approach in acoustics reaches 1971, the year of the publication of the papers by Stepanishen. The method has been developped since that date by numerous authors as a very efficient calculation tool for broadband field analysis.[5] Its textbook environment is that of the linear communication system theory - rather rich and very dynamic nowadays. There are, for the moment, no books on acoustics relating directly to the impulse method. Nevertheless, a good theoretical basis for operating in the time domain can be found in the book of Jessel.[6]

Although not as generally used as the traditional complex amplitude notation, the impulse response concept and the related distribution notation are known to a growing number of specialists of various domains of physics and engineering. However, it is usually treated as a mere help in calculations and, possibly, as an interesting, alternative point of view onto some specific phenomena related to signal processing. In the author's opinion, the features of the impulse method are not fully made use of, the time-domain approach being undervalued as a potential tool of fundamental analysis of space phenomena. When dealing with diffraction that is ruled by the delay effects related to mutual distances, the impulse method − disposing of the notation that expresses delays in the most direct way − is the right one, perfectly matched for treating theoretical and practical problems as well. It is not only simple and very powerful in computations of the field, but is also, and first of all, extremely precise in detailed analysis of the broadly understood diffraction mechanisms.[3,4] It can be seen − and should be taught − as the natural approach. Moreover, the method is autonomous, absolutely independent of the traditional one. As its solutions are general and are easily applicable to any type of excitation waveforms, the field proper to the harmonic excitation can be derived of them as a particular case.

The impulse analysis concerns, first of all, the Dirac impulse excitation − $\delta(t)$ and the related elementary impulse field − $p_\delta(x,t)$. It is worth noting that the analysis of this field is generally more easy for a given aperture than that of the related harmonic field. There are numerous cases when the impulse field can be calculated analytically in the whole space, whereas the corresponding harmonic one has only approximate solutions restricted to the far field. It results from the fact that diffraction integrals, when formulated in the time domain, have the form of a convolution integral involving the Dirac delta function of composed argument. The simple change of the integration variable from the space one − that of the aperture element position, to the temporal one − describing the time distance between the aperture element and the observation point, changes the form of the integral to that of a regular time-domain convolution with the delta function. Hence the solution is a straightforward one and, finally, the only "serious" operation to be performed when calculating the diffracted impulse field is a derivation at the stage of changing variables.[3,4]

In the general case of an arbitrary excitation s(t), a direct use is made of the superposition principle in its mathematical form of the convolution integral involving the elementary impulse field and the excitation waveform. In any specific case, even when there are no analytic solutions, the calculations can be easily performed by everyone. In fact, calculating the convolution is as simple as the multiplication of two long numbers, and to perform this operation should be taught more generally, for it is the most frequent and natural operation taking place in the physical world.[7]

The above mentioned multiple advantages of the time-domain approach can be explained as a direct consequence of the fact that the mathematical notation of the approach is absolutely natural. It describes in the most straightforward way the space phenomena as they pass in reality. The field calculated as the superposition of elementary impulse fields is always localised in space and time, and fulfills the principle of casuality. The direct relation between the notation and the intuitive image of the delay mechanisms the diffraction phenomena are ruled by, enables a "real-time" physical interpretation to be performed at all stages of analysis and computations. It is not necessary to point out the first-order importance of the latter advantage for the educational utility of the method.

5. LECTURES ON ACOUSTIC ANTENNA FIELDS

The time-domain approach has been developped for educational purposes and has been successfully applied in the lectures on fundamentals of acoustics given in our Insititute to the third-year students specializing in Electroacoustic Systems. The didactic advantages obtained thanks to the application of this method are illustrated below by the example of the sequence of lectures concerning acoustic antenna fields. The sequence takes the Helmholtz-Kirchhoff integral formula as the basis for calculating pressure waveforms in the field of arbitrarily weighed plane sources excited by arbitrary signals.

First, the features of primary point sources and secondary surface ones are analysed in details, and the Helmholtz-Kirchhoff integral is formulated on the basis of the superposition principle as a direct consequence of the physical action of the Huyghens' surface sources.

Second, the Huyghens' principle is directly made use of for calculating the wave propagation in the free field. Here the actual wave front is taken for the source of the subsequent wave field. The fact that the plane wave does not change its form while propagating is obtained in the issue of the impulse field analysis as a natural result of the secondary radiation from the wavefront and not as the classicalheuristic solution of the wave equation. The performed analysis is simple in form and gives a close insight into a very interesting mechanism of dynamic equilibrium that exists at the wavefront - the mechanism ruling the forward-only propagation of waves.[3]

Third, the diffraction is studied as the propagation through modulated apertures, starting from the Huyghens-Fresnel principle. The impulse analysis shows clearly up the creation of an aperture wave on each aperture nonuniformity. In the region of the geometrical "light" this wave follows the original one that, having had illuminated the aperture, has passed through it. In the region of the "shadow", this wave is the only one. Being a generalized form of the Young-Rubinowicz edge wave, the aperture wave represents in a pure form the phenomenon of the wave field diffraction. In this way the time-domain analysis enables the two traditional views of the diffraction to be linked in the same elementary evaluation, giving to both of them a clear and straightforward interpretation of two aspects of the same dynamic space phenomenon.[3,8]

Finally, the field distribution of acoustic antennas is calculated starting from the time-domain form of the Rayleigh formula with the use of the notion of the aperture impulse response that expresses geometrical aspects of diffraction. The impulse responses being easy to be found, the pressure waveforms in the field and related signals in acoustic systems are calculated, in otherwise complicated cases, with the use of pocket calculators.[1,5]

6. CONCLUSIONS

In the light of the discussion presented in the paper it is clear that the time-domain impulse description of space diffraction phenomena ruled by the effects of time delays, is better matched to the problem than the traditional description based on the notion of phase shifts involving interference effects.

The time-domain approach is highly interesting from the general point of view for its direct physical interpretations enriching the image of analysed phenomena. It has proved to be a powerfull tool of fundamental phenomenological analysis of wave-field effects. It is also valuable from the point of view of technical applications in acoustic systems engineering for its calculation efficiency making possible quick and precise computations of acoustic signals in the field. Linking together multiple advantages of both the general and applied aspects, the time-domain approach and its impulse method are worth to be developed and promoted as the proper education tool in teaching diffraction-related problems.

REFERENCES

1. Lasota,H., Salamon,R., and Delannoy,B., "Acoustic diffraction analysis by the impulse response method: A line impulse response approach", J.Acoust.Soc.Am., 76, 280-290 (1984)
2. Chinchurreta,F. and Salamon,R.,: "Study of multilayer piezoelectric transducer by the difference equations method", Proc.12th Int.Cong.Acoust., L 3-5 (1985)
3. Lasota,H., "Diffraction of acoustic plane waves: A time-domain analysis", J.Acoust.Soc.Am., 78, 1086-1092 (1985)
4. Lasota,H., "Etude du champ acoustique des sources planes dans le domaine temporel", These d'etat 85-01, Univ. de Valenciennes, (1985)
5. Harris,G.R., "Review of transient field theory for a baffled planar piston", J.Acoust.Soc.Am., 70, 10-20 (1981)
6. Jessel,M., "Acoustique theorique" (Masson, Paris, 1973)
7. Bracewell,R.N., "The Fourier Transform and Its Applications" (McGraw-Hill, New York, 1978)
8. Skudrzyk,E., "The Foundations of Acoustics" (Springer-Verlag, New York, 1971)

HYDROACOUSTICAL LABORATORY AT THE TECHNICAL UNIVERSITY IN GDAŃSK

D. Ruser (GDR)
Technical University of Gdańsk
Underwater Acoustics Department

Text of the paper not available in time for printing.

Students Classes in Conditions of Minimum Laboratory Equipment

Adam Wasilewski
Faculty of Civil Engineering
Lublin University of Technology

INTRODUCTION

Acoustical laboratory classes have been included into the teaching program of the Faculty of Civil Engineering at Lublin Technical University in 1982. They are only a part of the subject called laboratory of the fundamentals of architecture and physics of building. Ten hours per semester are devoted to these classes. Lectures on acoustics (15 h per semester) were also introduced then as a complement of the same subject.

At the beginning, the laboratory equipment consisted of a level recorder and a sound level meter with octave filter. A tapping machine which is necessery to measure the impact of noise was obtained a little later. In the initial stage of organization there was no special laboratory room. In this difficult situation we started with classes, assuming that in spite of it they will give didactic results as a practical and experimental illustration of lectures.

Since that time, the laboratory has obtained its own place and a better equipment which enables us to carry out research in the field of acoustical engineering. The article primarily is concerned with the initial stage of creating the laboratory and presents experiments which were made in those conditions.

DIDACTIC PURPOSES

The main purpose of laboratory classes is a practical illustration of information contained in lectures on acoustics and presentation of typical measuring methods which are applied in the acoustics of building.

Another very important goal is facilitating the students of Civil Engineering to get more familiar with electrical measuring methods of unelectrical parameters. Acoustic measurements are a very good example. It is an important problem, in the authors opinion, because the students contact with electronic devices too rarely. We know that nowadays the more and more measurements of mechanical, physical as well as biological values are based on a analysis of electric signals from the transducer which changes unelectrical magnitude to electrical one.

The discussed laboratory classes offer one of few possibilities of contact with electronic measurements for students of Civil Engineering.

LABORATORY TEACHING PROGRAM.
PRINCIPLES OF REALIZATION.

The acoustical part of the laboratory of physics of building consists of five exercises.
1. Noise measurements
2. Measuring the reverberation time
3. Measuring airborn sound insulation
4. Measuring impact noise insulation
5. Vibration measurements.

Noise Measurements

The measurements, octave band frequency analysis and comparing results with noise criteria, are carried out in this exercise. A vibration table for mixing concrete is used as a noise source (Fig. 1). The exercise illustrates the principles of spectral analysis using constant percentage bandwith filters and A, B, C - weighting networks. The influence of time constants SLOW, FAST is presented as well as measuring of time - varying sound pressure level. The exercise is connected with first two lectures which contain elementary information about acoustical metrology. On the basis of measurements students prepare a report which contains frequency characteristics of noise, the acoustic map of the lab hall, calculated equivalent level, noise index N and conclusions.

Measuring the Reverberation Time (Fig. 2)

At the beginning, measurements of the reverberation time were carried out in the lecture auditoria and the conference room of the Faculty. A tape recorder with recorded white noise signal and typical electroacoustic set (loudspeakers and power amplifier) was used as a sound source. The reverberation time was measured then after putting up many empty bottles on the floor of the room, to show the influence of Helmholtz's resonators upon the absorption of sound [1]. The students calculate reverberation time and compare theoretical results with experimental ones. They determine the best reverberation time characteristics taking into consideration type of place. The exercise is connected with the lecture on the room acoustics.

Measuring Airborn Sound Insulation

There were carried out measurements of aconstic property of the wall between two adjacent lab halls (Fig. 3). The same sound source as in the case of measuring the reverberation time was used. The measurements were made with a sound level meter and octave filter. Total absorption value of the receiving room was determined in the measuring way.

The students prepare the measurements report which contains frequency characteristics of insulation, insulation index E_L and a comparison with standard velues. At the beginning, we used an octave filter insted of 1/3 octave one. It was not in agreement with the Polish Norm.

Measuring Impact Noise Insulation

Measurements were carried out in the way shown in Fig. 3. A standard tapping machine is the sound source and measurement of acoustic pressure level are performed basing on the same meter as in previons exercises. Total acoustic absorption of the receiving room is determined earlier basing on measured reverberation time.

Vibration Measurements

The exercise is primarily connected with the assessment of the effect of vibrations on building constructions and consists of measurements of resonant frequencies of a steel chimney model and measurements and assessment of vibrations of the building. The exercise is not concerned with the assessment of vibration effect on peaple. It will not be discused here at length since it is not directly connected with the acoustic program of the laboratory.

The above listed exercises are, as a rule, supplemented with outdoors exercises during which students can get acquainted with realization of the acoustic project. These latter exercises have recently been taking place in the studios of the radio broadcasting station in Lublin where at present sound absorbing systems are being replaced by new ones.

CONCLUSIONS

The article is concerned with the initial stage of introducing acoustic classes to the study curriculum. Practical experience suggests that even with a minimum equipment of the acoustic laboratory, it is possible to introduce classes which supplement theoretical information with practical knowledge necessery for an engineer. It can be observed that after including laboratory classes in the curriculum the following didactic effects have been obtained.

- Students show greater knowledge about acoustics and especially better understanding of physical phenomena occurring in it. This is observable at designing classes.
- Problems specially dificult for students of civil engineering such as relations between time domain and frequency domaine have become clearer after a practical demonstration of spectral analysis principles.
- After certain exercises students acquire a greater skill in using measuring equipment even when measuring system is somewhat complex which is possible with the present equipment of

the laboratory. Students themselves choose necessary apparatus and calibrate the measuring system.

Equipment of the laboratory has a tremendous influence on didactic affects. On the other hand, however, it should not be too complex since a greatly sophisticated measuring system can blur the physical sense of the examined phenomenon. It can be exemplified here with a "classical" measurement of reverberation time basing on level recorder and measurement with a digital meter.

REFERENCES

[1] Czarnecki, S., Archive of Acoustics $\underline{1}$, 5 (1966)(in polish)
[2] Ginn, K. B., Architectural Acoustics, Bruel-Kjaer, 1978
[3] Sadowski, J., Architectural Acoustics, PWN 1976 (in polish)
[4] Wasilewski, A. J., Acoustic Laboratory Classes, (manuscript)

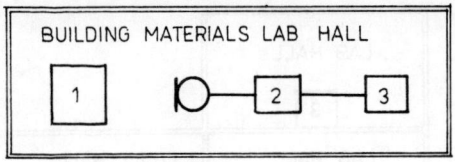

1. VIBRATION TABLE AS A NOISE SOURCE
2. SOUND LEVEL METER
3. OCTAVE FILTER

FIG. 1 NOISE MEASUREMENTS - SCHEMATIC DIAGRAM

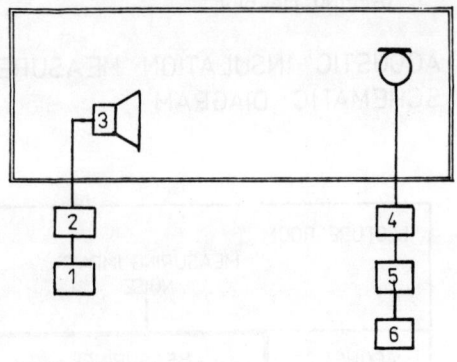

1. TAPE RECORDER WITH RECORDED WHITE NOISE SIGNAL OR NOISE GENERATOR
2. POWER AMPLIFILTER
3. LOUDSPEAKER
4. SOUND LEVEL METER
5. OCTAVE FILTER
6. LEVEL RECORDER

FIG. 2 MEASURING REVERBERATION - SCHEMATIC DIAGRAM.

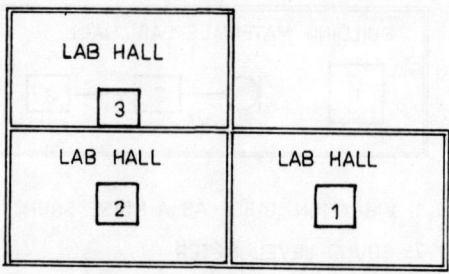

1. THE SAME SOUND SOURCE AS IN THE EXERCISE NO 2
2. SOUND LEVEL METER WITH OCTAVE (1/3 OCTAVE) FILTER
3. TAPPING MACHINE

FIG. 3 ACOUSTIC INSULATION MEASUREMENTS - SCHEMATIC DIAGRAM

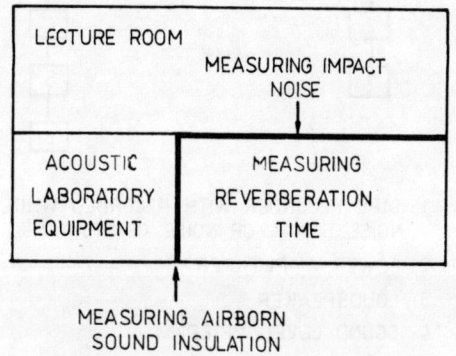

FIG. 4 ACOUSTIC LABORATORY NOWADAYS

5

POSTER PRESENTATIONS

PERSONAL COMPUTER APPLICATION TO THE DEMONSTRATION OF MECHANICAL IMPEDANCE ANALYSIS OF VISCOELASTIC SPECIMENS

W.BANDERA[x] Z.TRUMPAKAJ[+] T.ZALESKI[+]

[x] Laboratory of Applied Acoustics and Spectroscopy
[+] Institute of Experimental Physics University of Gdańsk, Wita Stwosza 57, 80-952 Gdańsk, Poland.

ABSTRACT

Two programs for the numerical analysis of mechanical impedance of a viscoelastic rod, adapted to personal computer ZX Spectrum, have been prepared for IV th year students of Applied Physics section. The simple steering of programs allows the student to simulate the experiment by himself. The calculation results are presented in the form of plots. Exemplary printouts of numerical calculations of mechanical impedance of the rod-like specimen are given on diagrams.

1. INTRODUCTION

The problem of forming an appropriate acoustic climate in rooms is directly connected with the usage of sound- and vibroisolating materials. It is known that viscoelastic properties of a material, well described by the complex elasticity modulus [1], are decisive in acoustic energy transfer through the medium at a determined density of the material. Measurements of the complex elasticity modulus are significant while determining the viscoelastic properties of such solids as polymers or porous solids of fibrous structure (mineral wool, glass wool). Up to the present day, in order to find the complex elasticity modulus, the resonance method [2] and the method based on transfer function or mechanical impedance measurement of a rod-like specimen excited to longitudinal vibrations has been used, while the end of the specimen is free or loaded by an optional mass [3,4]. The present work is a proposal of a problem exercise "Longitudinal vibrations of a viscoelastic rod", which can be carried out by the IV th year students of Applied Physics section within the Specialist Lab. In the first stage of the exercise the student makes a numerical analysis of the mechanical impedance of a rod of expected viscoelastic parameters using a simple personal computer.

In the second stage the student carries out the experiment which is the practical application of the theoretical knowledge acquired in the first stage of the exercise. The whole exercise is intended for about 30 hours of independent student's work.

2. MECHANICAL IMPEDANCE EQUATION OF A ROD-LIKE SPECIMEN

Fig 1 presents a theoretical model of a rod excited at the point x=0 into longitudinal harmonic vibrations. The end of the rod x=l is stiffly attached to an optional mass M. For deriving the mechanical impedance equation it is assumed that the cross section of the specimen is constant, the lateral dimension is much smaller than the longitudinal wavelength, lateral motion is not prevented at the specimen ends and the relation between stress and strain is linear. Besides, it is assumed that material of the rod is homogenous and isotropic. To find the expression for the mechanical impedance of rod considered, we make use of electro-mechanical analogies. The analogue of the rod presented in fig 1 is the waveguide loaded by inductivity.

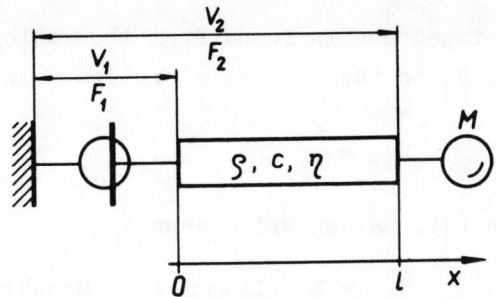

Figure 1. Theoretical model of specimen excited into longitudinal vibration and loaded by the mass M.

The connection between the force and velocity at the ends of the rod can be thus expressed by the following matrix equation:

$$\begin{bmatrix} F_1 \\ V_1 \end{bmatrix} = \begin{bmatrix} \cosh \gamma l & Z_0 \sinh \gamma l \\ Z_0^{-1} \sinh \gamma l & \cosh \gamma l \end{bmatrix} \begin{bmatrix} F_2 \\ V_2 \end{bmatrix} \quad (1)$$

where: F_1, F_2 and V_1, V_2 - complex forces and vibration velocities at input and output of the rod, l - length of the rod, γ is the propagation constant, $\gamma = \alpha + i\beta$, where α and β are the attenuation constant and phase constant, respectively, Z_0 - wave impedance, $Z_0 = \frac{i\omega m}{\gamma l}$, m - mass of the rod, $\omega = 2\pi f$, f - frequency.

The ratio of the complex force F_1 to the complex velocity V_1 determines the input mechanical impedance

$$Z_{in} = \frac{F_1}{V_1} \quad (2)$$

which, on the basis of (1) can be written as:

$$Z_{in} = \frac{Z_0 \sinh(\gamma l) + i\omega M \cosh(\gamma l)}{\cosh(\gamma l) + i\omega M Z_0^{-1} \sinh(\gamma l)} \quad (3)$$

The transfer impedance is found from the ratio of the complex force F_1 to the complex vibration velocity V_2

$$Z_{tr} = \frac{F_1}{V_2} \qquad (4)$$

and basing on (1), we can write down :

$$Z_{tr} = Z_o \sinh \gamma l + i\omega M \cosh \gamma l \qquad (5)$$

Formulas (3) and (5) allow to find, for a given frequency, the real and imaginary part of the input or the transfer mechanical impedance of the rod-like specimen, if we only know its viscoelastic properties, i. e. sound velocity c and loss factor η [3]. The calculations can be carried out at optional boundary conditions accepting different values of mass M.

3. DESCRIPTION OF PROGRAMS FOR NUMERICAL COMPUTATIONS.

In order to facilitate the numerical analysis of mechanical impedance of a viscoelastic rod for optional boundary conditions (free end of the rod, loaded by an optional mass M or stiffened), two programs, adapted to ZX Spectrum computer, have been prepared for students. The first of them "Impedance 1" has been written structurally in the Pascal language by means of a compiler made by Hisoft. This program contains the indispensable information, figures and formulas calculations are based. The program enables the calculations of mechanical impedance in the frequency range from 1 Hz to 10 kHz for a rubber-like rod of a fixed length l=0.1 m and sound velocity c=100 m/s. The computations can be made for optional η , M and f from the changeability ranges given in the program. The calculation results are presented each time in the form of plots. The program takes about 25 KB of RAM memory. The simple steering of the program (only two keys are active : "space" - option change, "enter" - choice of the indicated option) allows the student to simulate the experiment by himself.

The second program "Impedance 2" has been written in the Basic language and is adapted to calculations in the form compiled by the "TOBOS-FP" compiler of Basic, made by INEL, Wrocław, Poland. This is a floating point compilator with a four-bite arithmetic (such as Pascal) but faster than that [5]. In the fundamental loop of the program in which frequency f changes with the constant step df, the trigonometrical and exponential functions appearing in formulas (3) and (5) are calculated from the following recurrence formulas :

$$\cos(f + 2df) = 2\cos(df) \cos(f + df) - \cos(f)$$
$$\sin(f + 2df) = 2\cos(df) \sin(f + df) - \sin(f)$$
$$\exp(f + df) = \exp(f) \exp(df)$$

which increase the speed of calculations.

The original formulas (3) and (5) are so modified that indeterminancy for f=0 will not occur.

The program "Impedance 2" contains indispensable instructions, basic formulas and complementary figures. The calculation results are presented in the form of plots. The modulus and phase plots in parameter function βl as well as mechanical impedance in a complex number plane appear on one screen at a time. Program "Impedance 2" can be easily modified and allows to carry out calculations for optionally accepted parameters of the specimen. The program takes 10 KB + 12 KB TOBOS of RAM memory.

Exemplary printouts of numerical calculations of mechanical impedance of the rod-like specimen are given on diagrams.

5. CONCLUSIONS.

The above described computation programs enable the numerical analysis of mechanical impedance of a rod-like

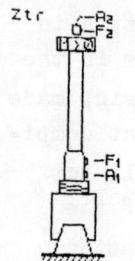

EXPERIMENTAL SCHEME
ACCELEROMETER
LOADING MASS
SPECIMEN with
 l-length
 m-mass
 η-loss factor
 c-sound velocity
IMPEDANCE HEAD
VIBRATOR

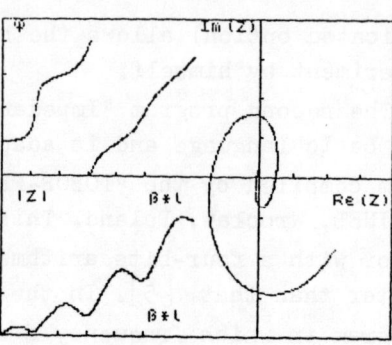

TRANSFER IMPEDANCE
M/m=2   m=.01 kg   l=0.1 m
η=0.2 c_0=100 m/s   c=101 m/s
fmin=0 Hz   fmax=1698 Hz
Scale unit:
π/2   for (β*l) and φ
100Nm/s for |Z|, Im(Z) and Re(Z)

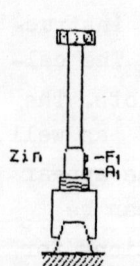

EXPERIMENTAL SCHEME
ACCELEROMETER
LOADING MASS
SPECIMEN with
 l-length
 m-mass
 η-loss factor
 c-sound velocity
IMPEDANCE HEAD
VIBRATOR

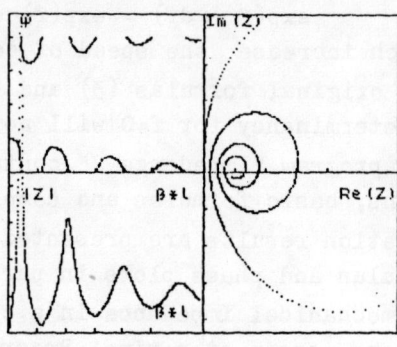

INPUT IMPEDANCE
M/m=2   m=.01 kg   l=0.1 m
η=0.2 c_0=100 m/s   c=101 m/s
fmin=0 Hz   fmax=1698 Hz
Scale unit:
π/2   for (β*l) and φ
10 Nm/s for |Z|, Im(Z) and Re(Z)

specimen with optional viscoelastic properties at different boundary conditions. The arrangement of both programs, the contained in them instructions and factual information meet the requirements of educational syllabus. Both programs "Impedance 1" and "Impedance 2" complement each other and enable the student to understand precisely the problem of longitudinal vibrations of a rod and they serve as indispensible extending of knowledge to carry out experimental work in this field.

REFERENCES

1. Zwikker,C. and Kosten,C.W., "Sound Absorbing Materials", London, (1949).
2. Rosin,G.S., "Measurement of Dynamic Properties of Acoustic Materials", Moscow, (1972) (in Russian).
3. Pritz,T., Journal of Sound and Vibration, 72 , 3 , 317-341, (1980).
4. Bandera,W., Proceedings of 3rd European Conference on NDT, 4 , 259-268, (1984).
5. Skaba,W., and Borkowski,J., Computer, 3 , 28 , (1987) (in Polish).

EDUCATIONAL RECORDING STUDIO

G. CYPUKOW & J. DYŻEWSKI
Sound Engineering Department
Institute of Telecommunications
Technical University of Gdańsk
80 - 952 Gdańsk, Poland

ABSTRACT

An educational recording studio with its parameters and equipment is presented. Course program of laboratory work at the professional studio is presented as well.

1. Sound Engineering Laboratory at the Electronics Faculty of the Technical University of Gdansk educates in the profession of Sound Engineering. The specialists of this field of technique are in demand at the market of music industry.

The sound engineering program is an intensive course at the fourth year of study, based on theoretical lectures and laboratory work at the professional recording studio.

2. The recording studio is situated in the Electronics Department building and it consists of two rooms:

a/ the recording room - which is located in a large lecture-hall (see fig. 1),
b/ the control room - small but well equipped (see fig. 2).

Fig. 1. The recording room area. View from the auditorium.

Of a volume of about 1200 m3, the recording room has a 0,7 sec. reverberation time. Because of a special sound - monitoring system all combinations from mono - to stereo - and to any kind of surround sound presentations can be performed (see fig. 3). Two pianos, the Steinway grand piano and the Bechstein piano are situated in the recording room as well.

The control room is oblong in a shape. It has panelled walls and a special, low frequencies <u>absorbing ceiling.</u>

Fig. 2. The control room. View from the door.

The centerpiece of the control room is a Fonia SMM -104s console with 16 line/mic inputs and two equalisable echo returns.

Tape machines: a 3M series 79 8 - track, Studer type J37 4 - track and two Telefunken M 15 stereo tape recorders and six-output reverberation tape machine are situated in the room. The automatic Fonia SME 121 console with built - in limiters, special cross - system, two RFT type TF - 824 equalisers complete the equipment. Two Tonsil ZG 60 monitors are driven by 60 Watts amplifiers. The block diagram of the control room and the equipment are shown on fig. 4.

The disposable microphones include AKG D 1000C, D 224C, D 202E1, D 60E1, D 190, D 24, Shure SD 565, Neumann SM 69 FET, W 87, K 84 & K 83, Sennheiser type 416 and all Polish studio - microphones.

The laboratory program gives a suitable approach to improve the sound engineering abilities of the future specialists. Every recording, mixing and sound reinforcement situation is different. That is the reason why the presented program gives such a broad range of experience including everything from the speech recording to classical orchestras and rock bands. The subjects of laboratory sessions are as follows:

1. One microphone - speech recording.
2. The mixing of various programme - sources.

3. Multi – microphones drama recording.
4. Interview recorded by a portable tape – recorder and its preparation in the studio.
5. Piano music recording.
6. Multi – track drama recording.
7. Multi – track classical orchestra recording.
8. Multitrack – rock – band recording.

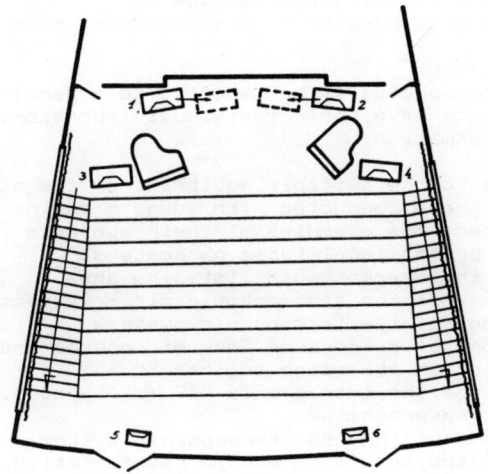

Fig. 3. The block diagram of the monitoring system situated in the control room. The loudspeakers no 1 & no 2 can be moved in space.

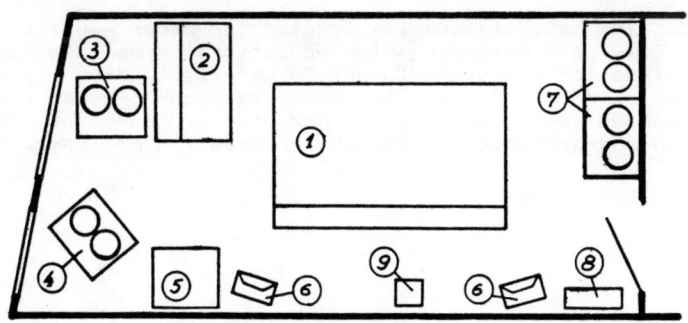

Fig. 4. The block diagram of the equipment situated in the control room.

1 - SMM 104s console
2 - SME 121 console
3 - Studer J37 four - track
4 - 3 M type 79 eight - track
5 - cross point system
6 - ZG - 60c monitors
7 - Telefunken M 15 stereo
8 - EMT reverb unit type 240
9 - TV monitor to the recording room

The students' group usually has about 3 to 6 people.
All students have their individual laboratory program with theoretical information.

4. Thanks to the suitable equipment of the studio more than ten masters degrees connected with sound - engineering problems have been awarded. The examples of their subjects are as follows:
1. Analysis of professional tapes parameters.
2. Analysis of the stereophonic listening area.
3. Investigations of the stereophonic mic - systems.
4. Investigations of the tetraphonic systems.
5. Investigations of a Pressure Zone Microphone model.
6. Doppler effect in the sound engineering.
7. Measurements of the tetraphonic "3F + T" system.
8. Experimental reverberator.
9. Investigations of the OSS stereophonic system.
10. Analysis of the recording studio reverberation properties.
11. Investigations of the strobophonic system.
12. Investigations of phase distortions in the stereophonic track.
13. Project of the dummy - head for stereophonic recordings.
14. Investigations of magnetic heads.
    etc.

5. The studio complex is used for other applications as well. Some of recordings for educational proposals and sound productions for Polish Radio and TV were made. Favorite classical solists and orchestras as well as pop groups took their performances and concerts in the studio. Some of them were recorded and presented in Polish Broadcasting programs.

RESEARCH AT THE INSTITUTE OF ACOUSTICS ADAM MICKIEWICZ UNIVERSITY IN POZNAŃ

E.Hojan   U.Jorasz

The Institute of Acoustics at the Adam Mickiewicz University in Poznań was established on the basis of the Chair of Acoustics, acting for almost 30 years and is now the only scientific institute of this profile at university level in Poland.

Two things were of substantial significance in the history of the development of the Chair of Acoustics: $1^{\underline{o}}$ its connection with the faculty of physics and $2^{\underline{o}}$ the amount of research work done on psychoacoustics. The scope of subjects covered by the scientific works of the Chair expanded early enough through adoption of problems from the border line of other scientific disciplines, which is strictly connected with the interdisciplinary character of acoustics. As result of this, a development of studies on many other scientific subjects followed, such as spectral analysis of signals, electroacoustics, molecular acoustics as well as studies from the border line of technical sciences as: architecture and town-planning and also from the border line of humanities e.i. phonetics and musicology, and also medical sciences dealing with physiology of hearing system.

As far as didactics are concerned, the Institute takes part in the fulfillment of the teaching programme for students of physics starting from the 3rd year of their studies. Since 1984 the specialization on acoustics has been created at the Adam Mickiewicz University in Poznań. The diagram below presents the programme as realised in that specialization.

P r o g r a m

	III rd year		IV th year		V th year	
	5th Semester	6th Semester	7th Semester	8th Semester	9th Semester	10th Semester
	Spectral analysis of signals	Psycho-acoustics	Molecular acoustics	Master degree seminar	
	Fundamentals of electro-acoustics	Room acoustics	Monographi-cal lecture	Monographi-cal lecture	
	Waves and vibration	Numerical methods			
	Environmental acoustics	Seminar	Seminar		
		Specialistic workshop			
		Preparation of M. Sc. Th.			

Of great importance as far as the training of scientific staff is concerned are the post-graduate studies run for years by the Institute, especially on problems of protection of environment and electroacoustics.

MICROCOMPUTER APPLICATION TO ORGAN SOUND ANALYSIS

A. KACZMAREK
Sound Engineering Department
Institute of Telecomunication
Technical University of Gdansk
80 - 952 Gdansk, Poland

INTRODUCTION

An appreciation of the quality of the organ sound by objective methods is an everlasting problem. In a former sound analysis analog methods were used. Great number of available computers caused that analog methods became obsolete methods. The presented paper contains the outline of the digital analysis of sounds applying the microcomputer Spectrum+. Principles of the analysis of the steady and the transient state of signals are described and examples illustrating the use of these methods are presented.

INSTRUMENTATION

The application of the computer is based on a suitable program, and an analog circuit is responsible only for an analog-digital conversion of the signal, its registration and moreover for sending digital samples to the computer, where they are stored and calculated by a program. Different kinds of analyses are displayed on the computer screen in the suitable program menu; the choice of the options caused an execution of such an analysis. Due to the circuit joined to the computer such a function is realized as like: amplification of the signal, synchronization corresponding to the chosen moment of the sound to be analyzed, sending samples to the computer memory. Fig. 1 presents the block diagram of the computer stand for the realization of these functions.

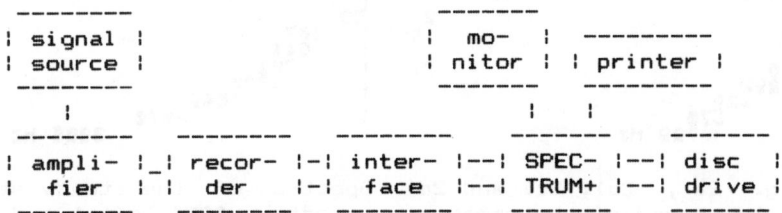

Fig. 1. Computer stand to examine acoustic signals.

The signal amplified to the maximal amplitude 5V is sent to the Bruel&Kjaer's Digital Event Recorder Type 7502. It is stored there in 4096 cells and then following samples are fetched by the program to the computer memory through the interface.

The recorder may sample the signal with the frequency in a range 0,1 - 100 kHz. Although the sampling frequency is lower, and recorded time of the signal is longer, but the analyzed frequency band is more narrow. The recorder contains the Trigger After Recording circuit allowing to record signal with delay from 0 to 9,9 time with a step 0,1 of the whole memory time; moreover it is possible to analyze longer times when several following records are sent to the computer memory.

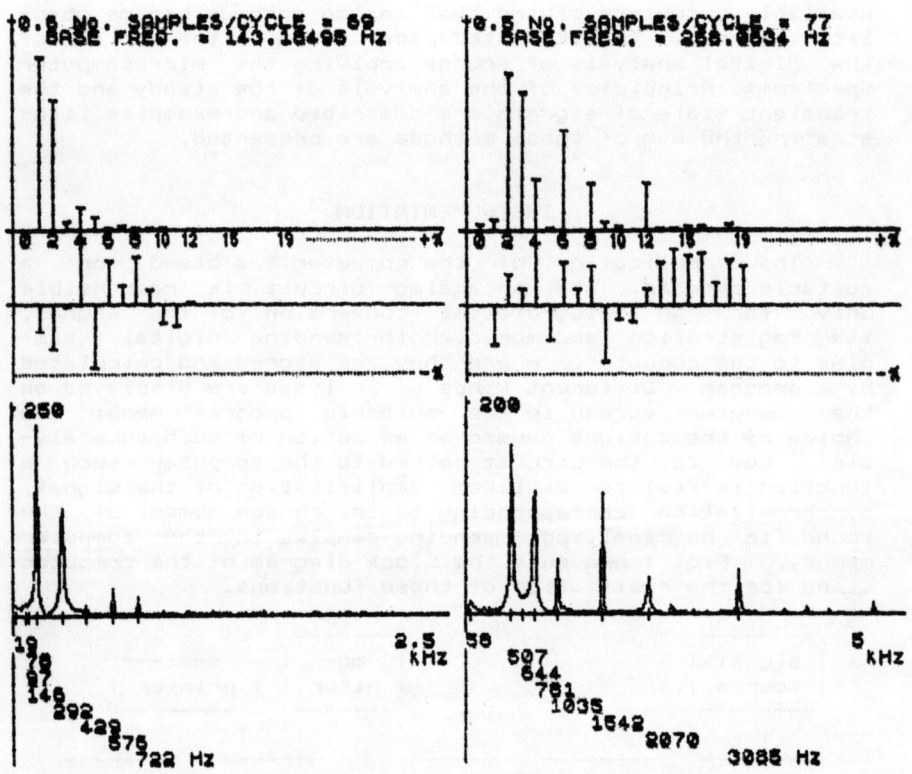

Fig. 2a), 2b), 2c) and 2d). Spectra of the steady state of the sound of the organ pipe - octava 4' (left) and the sound of the organ mixture - acuta 4-6 ch. (right) from St. Nicholas Church in Gdansk. Method 1 - above, method 2 - below.

Additional units in a describing system are as follows: the computer screen used to present the program while it works, a printer used to write all the results, a disc drive storing different data: results, diagrams, programs and samples of the sound signal, which can be analyzed later on.

PROGRAMS AND EXAMPLES OF THE FREQUENCY ANALYSIS FOR THE STEADY AND TRANSIENTS STATE

To analyze steady state it is possible to use one of the two methods:
1) based on the integral Fourier's formulas,
2) based on the FFT algorithm. [1]

Using the first program one have to choose only one period of the recorded signal, and then the program calculates particular harmonics. Subsequent harmonics obtained during calculation are subtracted from the primary signal and the result is presented on the screen. To reach maximal accuracy of this operation, the program makes an oversampling, calculating simultaneously the exact frequency base. The last information may be applied to tune musical instruments.

The second program demands no intervention before calculations, but the number of samples ought to be equal to the integer power of 2. The calculated number of points is greater than the number of harmonics and the obtained diagram looks like a diagram of a continuous spectrum. To analyze the signal with non-harmonics a longer time of the analysis and more samples are needed. For such a case a more exact identification of maxima and better appraisal of a degree of non-harmony are wanted.

Fig. 2a), 2b), 2c) and 2d) present examples of analyses of a steady state of two organ sounds obtained with the use of the described methods.

The program for transients analysis is based on results of calculations of the steady-state analysis for only one period of the signal. Number of samples contained in one period is important. Having written the whole signal in the computer memory the program analyzes subsequent parts equal to one period. The obtained results for chosen harmonics are presented as an envelope of the harmonics (evolutive spectrum). That allows to make conclusion about transient time of the sound and about relative contents of harmonics in any moment.

Fig. 3a), 3b), 3c) and 3d) illustrate an example of the analysis of a starting transient of the pipe sound, for which the steady-state spectrum has been presented on fig. 2a) and 2c).

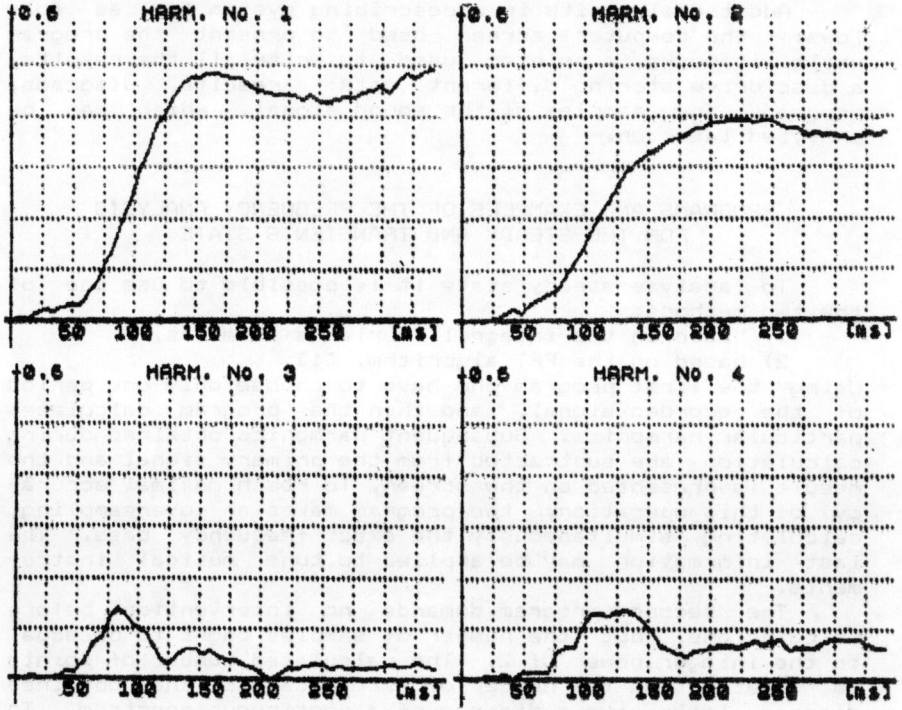

Fig. 3a), 3b), 3c) and 3d). Evolutive spectrum of the organ pipe sound - octava 4', presented by the four harmonics.

CONCLUDING REMARKS

The chief advantage of the presented method is the shortness of time needed for an analysis, the simplicity of the operating service and the great precision of calculations. The implemented programs were verified during a model investigation of an electronic organ.

REFERENCE:

[1] A.V. Oppenheim, R.W. Schafer: Cyfrowe przetwarzanie sygnalow. WKL, Warszawa 1979.

PROPOSALS OF CURRICULUM FOR HEALTH AND ENVIRONMENT PROTECTION AGAINST NOISE FOR ELEMENTARY SCHOOL

E. Kotlicka

Institute of Physics
Technical University of Warsaw
ul.Chodkiewicza 8, 02525 Warsaw
Poland
Polish Noise Abatement League

In the curriculum for elementary school, begining with the first classes, there are included themes containing information about environment protection. In the first three classes the subject is included in the environmental lessons, and in the other classes in the lessons of biology and hygiene.

The concern about the protection of the environment has grown rapidly as it has become generally recognised that the steady rise in pollution of all kinds cannot be allowed to continue indefinitely. Town development and industralization of the world not only make life more comfortable but, unfortunately, cause pollution and degradation of the environment. One of such harmful agents connected with industrial development in agglomerations is the NOISE.

As the acoustic noise there is usually understood /according to the definition [1], [2] / the unwanted acoustical phenomenon, such as an undesirable, unpleasant, annoying, harmful sound. Thus, the evaluation, if the given sound effect will be considered as a noise depends on the personal feeling. For that reason, it is important to teach, even the children, to distinguish the noise from other sounds.

In the poster from Fig. 1 there is shown, in the picture form, the possibility of presenting to pupils in elementary

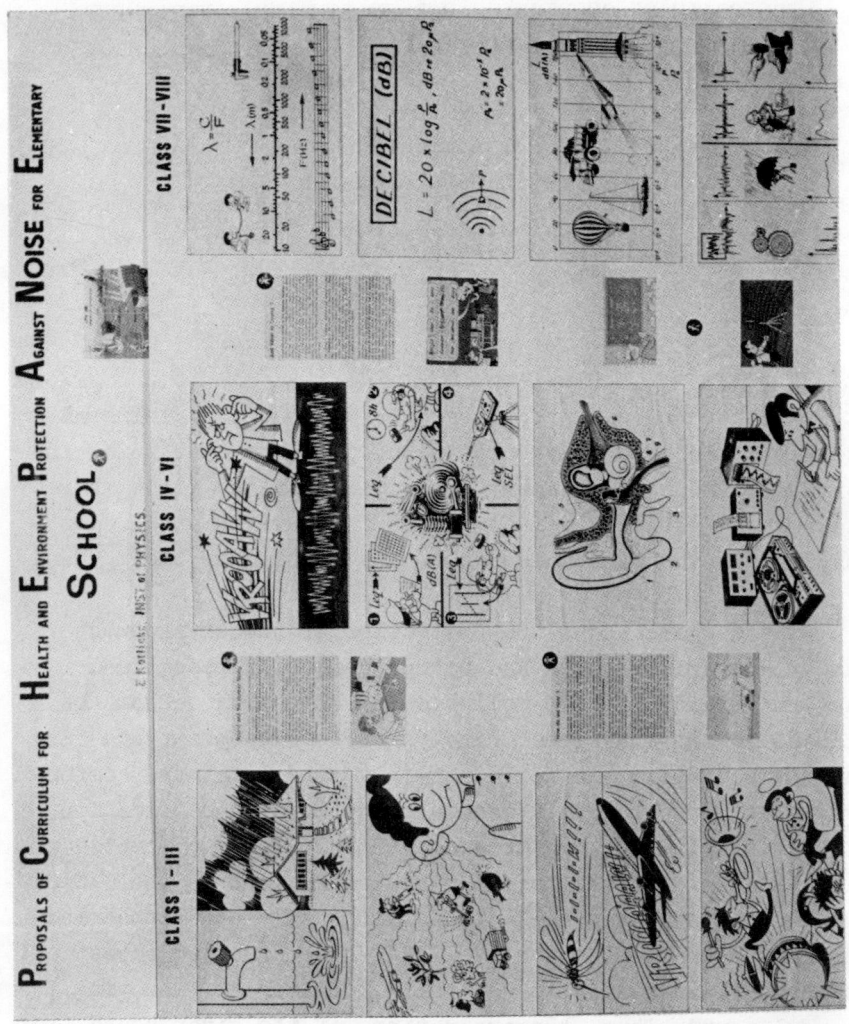

Fig. 1. The poster. The illustrations from literature published by Brüel & Kjaer.

school many date about sound and noise. The pictures in the first column show how to attract the interest of children from the first to the third class in the acoustics and noise problems. On the music lessons, there is an opportunity to deal with sounds in a wider aspect, there is the possibility to say something about other sources of sound, than the music instruments, which we can meet in every day life in the nature and produced by technical devices. The illustrations of these sources are in the first and second picture, where we have the storm, dropping tap, speaking people, flying plane, singing bird, power hammer, grass mower and car. All these sounds reach us and exert an influence on us.

The human reaction on different types of sounds can be demonstrated by comparing the sounds of buzzing mosquito and rumble of a takeing-off plane. In this case the levels of the sounds differ considerably but we feel them both as an annoying sound. The level of annoyance depends not only on the quality of the sound, but also, on the kind of information being contained in it. On the other hand, the reception of the sound depends also on our attitude to the sound which is illustrated by the last picture as they are listening to big-beat music.

If the teacher has available this kind of set of illustrations, he can, in one hour in each class, deal with the following problems:
- what we hear,
- from what sources the sounds are coming,
- what our reactions to the sounds are.

In summerizing the lessons there should be reminded to the children the rules of good manners in order to create a quiet, acceptable acoustic environment.

Continuing the programme of education in the classes from the fourth to the sixth on the lessons of biology and hygiene, subjects discused previously only qualitatively could be presented in the quantitative aspect. Thus problems of sounds in human life can be described by following examples:

- the sound and information connected with it /telephone ringing, knocking on the door, thunder, heart murmur, squak of the tires, hourly bugle-call heard from the tower of St Mary's Church in Cracow/,
- human perception of the sound /the construction of human ear, the hearing mechanizm/,
- human reactions to the sound /agents inducing loss of hearing, noisiness and annoyance/.

In conclusion of this series of lectures it should be underlined that noise can damage the human ear, and its influence is harmful for all the human body through the nervous system.

In the last two classes, seventh and eighth, the pupils get to know about such concepts as pressure, wave, frequency on lessons of physics and logarithm on lessons of mathematics and then it is possible to systematize former information about sound and noise. There can be introduced the definition of the sound as any pressure vibration /in air, water or other medium/ that the human ear can detect. There can be described a logaritmic scale of sound level measurement and useing it can be explained by the construction of the human ear and the way of perception. On this base it is possible to consider the level of most common sounds starting from the threshold of hearing to the threshold of pain /Fig. 2/. Attention should be paid to different sensitivity of the air for sounds of various frequency. In Fig. 2 there are presented contours of normal equal loudness for pure tones. Discussing these curves it should be noted that the noise of the level of 65 dB - 70 dB is typical for normal day-to-day activities and that in international standards the noise of the level of 85 dB and with prevalence of high tones may cause loss of hearing.

In lessons conclusions it should be emphasized that we are exposed to noise all the time and everywhere. Recently all over the world the acoustic climat has deteriorated

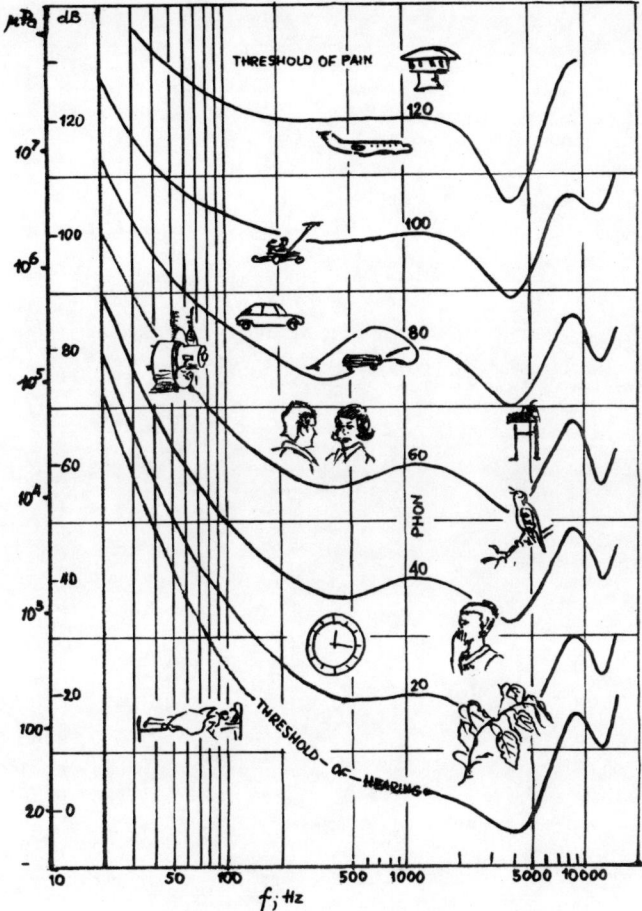

Fig. 2. Typical sound pressure levels of common noise sources and normal equal loudness conturs for pure tones.

considerably. In this connection, it is necessary to remind the children of cultural behaviour so as not to worsen the situation.

1. Hassall, J.R. and Zaveri, K., "Acoustic Noise Measurements", Brüel & Kjaer, /1979/.
2. Project of Polish Standard, PN/T-01009.

TECHNICAL COURSES IN THE DOMAIN OF AN ENVIRONMENT AND WORK-PLACE PROTECTION AGAINST NOISE AND VIBRATION

J. MOTYLEWSKI

Institute of Fundamental Technological Research
Polish Academy of Sciences
ul. Świętokrzyska 21, 00049 Warsaw
Poland
Polish Noise Abatement League

ABSTRACT

In 1971 there was set up the Polish Noise Abatement League. The main educational activity of the League covers organization of courses, conferences, exhibitions and issuing of publications dealing with protection of environment and working places against noise and vibrations. Technical courses organized in Rydzyna for persons of various specializations are of particular significance for the training activity.

The development of technical civilization brings about an increase of noise and vibration causing many unfavorable consequences in the working place, at the place of residence and relaxation.

The noise and vibrations affect harmfully man's health, considerably reducing his labour efficiency and effects of work.

In 1971 there was set up in Poland a social association named The Polish Noise Abatement League, which is a member of AICB - Association Internationale Contre le Bruit. One of the basic goals of that organization is to intensify educational activity of the society in the field of protection of the natural environment and working place against the harmful effect of noise and vibration.

The main educational activity of the League covers organization of courses, conferences, exhibitions and issuing of publications dealing with protection of environment and working places against noise and vibrations. Weekly cyclic technical courses organized at the Center of Technical Advance in Rydzyna for persons of various specializations who want to acquire more thorough knowledge in this field are of particular significance for the training activity.

There has also been started a cooperation with the information mass media such as the press, radio and TV. All the problems are presented in poster form in Fig.1. The above named weekly courses are usually organized twice a year, in spring and in autumn. According to the subject the course covers 35 - 40-hours and contains lectures, exercises, collective and individual consultations and control tests.

So far the subject of the courses covered the following:
- method and apparatus for noise and vibrations measurements,
- protection of natural environment against noise and vibrations,
- protection of working places against noise and vibrations,
- protection against noise and vibration in industrial and housing building.

For example the 40-hours course concerning the method and instrumentalization of noise measurements coverd the following problems:
- bases of acoustics,
- an approach to noise protection,
- precision sound measurements,
- measuring microphones,
- octave and 1/3-octave analysis,

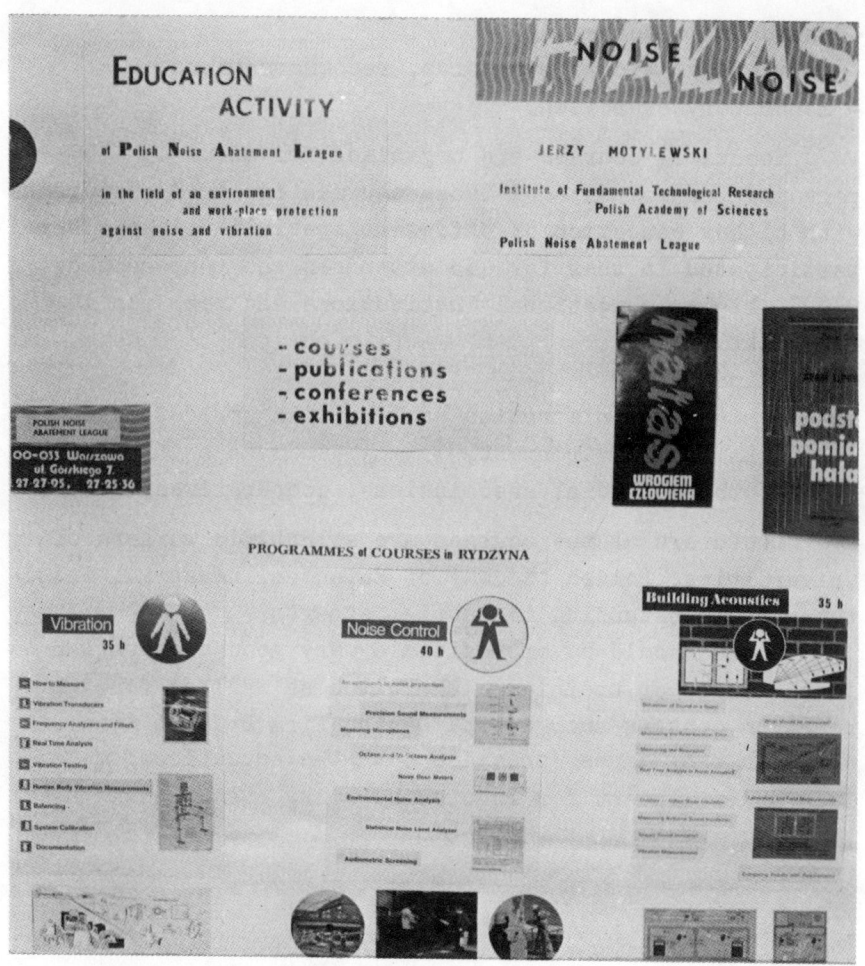

Fig.1. The poster. Education activity of the Polish Noise Abatement League.

- noise dose meters,
- environmental noise analysis,
- statistical noise level analyzer,

- audiometric screening,
- legal regulations, standards, recommendations,
- laboratory exercises.

Technical courses are organized for about 45 - 55 persons. Participants of those courses are in 85 % persons with higher education of different specializations. There participated in them for 455 attendees who represented:

9 % - higher educational institutions and research institutes,
42 % - industry,
45 % - state administration dealing with protection of natural environment,
4 % - others /social associations, cooperatives, etc./.

Lecturers at the courses are scientific workers of Universities, Polish Academy of Sciences, Industrial Institutes and outstanding industrial experts.

There should be emphasized a very good cooperation in the field of training with Polish as well as foreign research centers such as the Danish firm Brüel & Kjaer, which, in many cases, provides both the educational materials and equipment for laboratory work.

COMPUTER MODELLING OF THE BOWED STRING OSCILLATIONS

Z. PERUCKI & B. KOSTEK
Sound Engineering Department
Institute of Telecommunications
Technical University of Gdańsk
80 - 952 Gdańsk, Poland

ABSTRACT

An example of a computer modelling of the bowed string oscillations is presented. An excitation model of a vibrating string and conditions under which it works are discussed. A program on a basis of Lienard's graphical construction is applied to describe a shape of modelled string vibrations. Concluding remarks, resulting from the FFT analysis of obtained curves, are given.

INTRODUCTION

The analysis of string vibrations is a frequent subject of scientific discussions. One of the important problem is the action of the bow on the on the string is highly complicated. It should be mentioned that the bow simultaneously generates transverse, longitudinal and torsional vibrations of the string as well as of its fixed points. String vibrations are highly nonlinear due to a nonlinear relation between transverse, longitudinal and torsional elastic forces and their corresponding displacements. Moreover, the vibrating string corresponds to a distributed-constant system where the heterogeneity of the string material affects the vibration propagation. The complexity and difficulty of the bowed string vibration maintenance analysis requires the problem to be reexamined by computer-modelling with some simplifications, allowing to obtain a large quantity of results in a clear and demonstrative manner.

MODEL OF STRING EXCITATIONS

A bow-string system may be analysed as a mechanical oscillator, due to the fact that the string vibrations maintained by permanent bow motion are typical self-sustained oscillations. Basing on the electro-mechanical analogies, procedural formulae may be written (see Tab.1).

Tab.1. Function values described in analogy for the mechanical and the electrical parallel oscillator.

electrical oscillator	mechanical oscillator
I(u)-current generator	F(v)-force generator
u-voltage on resonant circuit	v -absolute string velocity
C-capacitance of oscillators	M_s-mass of the string
L-inductance of oscillators	C_s-string compliance
ϕ-magnetic flux in the coil	x -string displacement
i-variable quantity proportional to the coil current	f -variable proportional to the force affecting Cs

The corresponding equations for electrical /1/ and mechanical /2/ oscillators are presented below [1]:

$$\frac{du}{d\tau} = -\sqrt{\frac{L}{C}}\, I(u) - i, \qquad \frac{di}{d\tau} = u \qquad /1/$$

where: $\tau = t / \sqrt{LC}$; $i = \phi / \sqrt{LC}$ /2/

$$\frac{dv}{d\tau} = -\sqrt{\frac{C_s}{M_s}}\, F(v) - f, \qquad \frac{df}{d\tau} = v \qquad /3/$$

where: $\tau = t / \sqrt{C_s M_s}$; $f = x / \sqrt{C_s M_s}$ /4/

The nonlinear function F(v) of equation /3/ represents the friction force between the bow and the string as a function of the absolute string velocity [1]. Equation /5/ acts as a stimulating function, similar to other typical oscillator systems.

$$F(v) = \frac{T_o * \text{sign}(v + V_o)}{1 + k * |v + V_o|} \qquad /5/$$

where: V_o - string velocity [m/s]
         T_o - friction force [N]
         k - bow colophanying factor [.]

The damping of the free-vibrating string being small in relation to the damping brought by the bow friction force is provided to be negligible in this model. Moreover, it seemed reasonable to consider only transverse vibrations as energetically prevailing. The next assumption is that the lumped-constant mechanical system is assumed as a string model disregarding the heterogeneity of the string material.

GRAPHICS OF THE COMPUTER PROGRAM

Lienard's graphical construction [2] allows to find limit cycles of the equation /3/ by means of steady-state values of v-f. After the demonstration of the limit cycle on a phase

plane, v-f curves of both values as a function of time are
presented on the screen. Due to the automatic changes of the
scale of variables, all the diagrams may be applied on a full
screen. The three dimensional diagram presents all the
results obtained during the calculations of limit cycles in
terms of one of the parameter values-for example V_o is being
changed (see fig.1).

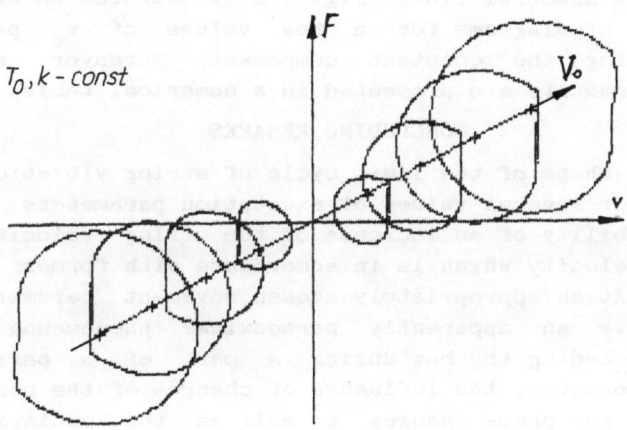

Fig.1. Shapes of limit cycle of string vibrations in
coordinates: reduced force f, absolute string velocity v,
calculated for eight values of bow velocity V_o.

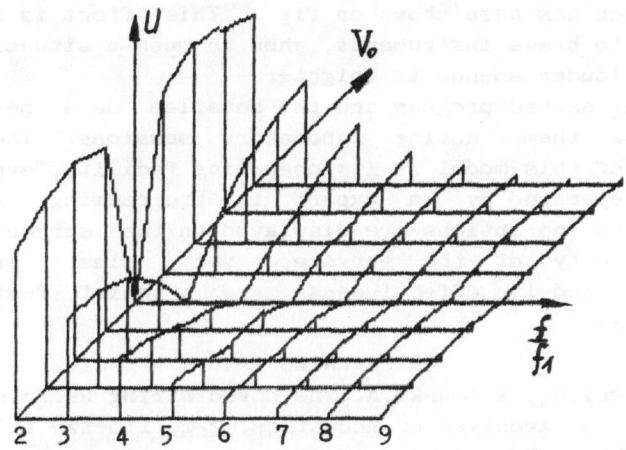

Fig.2. Amplitude spectrum of the string vibration.

To make the program more efficient results of FFT analysis of all the vector pairs are stored on a floppy disk. It makes possible to present any calculated amplitude spectra on three-dimensional diagram in terms of one of the parameters being changed. As the presented diagram is not enough distinct (higher components have lower levels), it is possible to display it in a smaller scale omitting some of the lowest spectral lines. Figure 2 illustrates an example of this kind of diagrams for a few values of v_o parameter, disregarding the constant component. Moreover, all the detailed results are presented in a numerical table.

CONCLUDING REMARKS

The shape of the limit cycle of string vibrations calculated for several values of excitation parameters confirms the possibility of an increase of the string velocity above the bow velocity which is in accordance with former investigations. At an appropriately chosen movement parameters one can observe an apparently paradoxical phenomenon of the string preceding the bow during a part of a period (see fig.1). Moreover, the influence of changes of the bow orientation on the phase changes, as well as the antisymmetrical distortion of the period of the vibrating curve. It is obvious that as the bow velocity $|V_o|$ is lowered, the amplitude of the vibration decreases and the cycle is more distorted. The author remark is that the lower the bow velocity is, the weaker the sound is, but its timbre is brighter and richer which has been shown on fig.2. This effect is reversed comparing to brass instruments, when in such a situation the timbre of louder sounds is brighter.

The presented program and the detailed data permit to demonstrate them during laboratory sessions. The main advantage of this model is its operating facility, even when it is not operated by an expert in programming. All the instructions and options are displayed on the screen. Thus, the simplicity of the service, very clear graphical application and its effectiveness, are a chief feature of this program.

REFERENCES

[1] Budzyński,G., Kulowski,A.,The bowed string as..oscillator
Archives of Acoustics, 2/2, 115-120 (1977)

[2] Kulowski,A.,Znajdowanie okresowego rozwiązania ...,
Zeszyty Naukowe PG - Matematyka, 11 (1978)

APPLICATION OF THE LOCAL KRAMERS-KRONIG DISPERSION RELATIONSHIP TO THE CALCULATION OF THE ULTRASONIC WAVES ATTENUATION IN HIGHLY VISCOUS LIQUIDS

E. SOCZKIEWICZ

Institute of Physics of the Silesian Technical University
44-100 Gliwice, Poland

ABSTRACT

The local Kramers-Kronig relationship between dispersion and attenuation of ultrasonic waves has been applied to the calculation of ultrasound attenuation in glycerol. The Isakovitch-Chaban's theory of ultrasonic waves propagation in highly viscous liquids has been used and dispersion of ultrasonic waves has been estimated in case of $\omega\tau \gg 1$ and $\omega\tau \ll 1$. In the case of $\omega\tau \gg 1$ a good agreement has been proved between values of α obtained from the Kramers-Kronig local dispersion relationship and those given by Chaban's theory, while in the case $\omega\tau \ll 1$ the agreement is only for certain range of $\omega\tau$.

1. INTRODUCTION

Acoustical properties of highly viscous liquids are similar to the properties of microinhomogeneous media e.g. emulsions [1]. Isakovitch-Chaban's theory [1] as well as the theory of Eldin et al. [2,3] consider the highly viscous media as media with locally ordered molecular arrangements. Acoustic waves disturb the equilibrium state between disordered and locally ordered molecular arrangements. The disturbed equilibrium with respect to number of surplus holes is restored by diffusion of holes and the phase shift of this process with respect to the incident wave results in anomalous absorption and dispersion of acoustic waves. The following formulae result from Chaban's theory, respectively for the velocity c and attenuation coefficient α of ultrasonic waves in the case of $\omega\tau \gg 1$:

$$c = c_\infty \left[1 - \frac{3}{8} \frac{c_\infty^2 - c_o^2}{c_o^2} (\omega\tau)^{-1/2}\right], \qquad (1)$$

$$\frac{\alpha}{\omega} = \frac{3}{8} \frac{c_\infty^2 - c_o^2}{c_o^2 c_\infty} (\omega\tau)^{-1/2}, \qquad (2)$$

and for $\omega\tau \ll 1$:

$$c = c_o \left[1 + \frac{1}{3} \frac{c_\infty^2 - c_o^2}{c_\infty^2} (\omega\tau)^{3/2}\right], \qquad (3)$$

$$\frac{\alpha}{\omega} = \frac{2}{5} \frac{c_\infty^2 - c_o^2}{c_o c_\infty^2} \omega\tau, \qquad (4)$$

where ω denotes the cyclic frequency, c_o, c_∞ -limiting values of c for $\omega = 0$ and $\omega = \infty$ respectively, $\tau = \frac{a^2}{2D}$, a denotes the radius of an inclusion with ordered molecular arrangements and D is the coefficient of diffusion.

2. THE LOCAL KRAMERS-KRONIG RELATIONSHIP

Between real $K_1(\omega)$ and imaginary part $K_2(\omega)$ of the complex adiabatic coefficient of compressibility one can write the Kramers-Kronig relationship in the form [4]:

$$K_2(\omega) = -\frac{2}{\pi} P \int_0^\infty \frac{\omega \left(K_1(\omega') - K_1(\infty)\right)}{\omega'^2 - \omega^2} d\omega', \qquad (5)$$

where P denotes the principal value of the integral. $K_1(\omega)$ and $K_2(\omega)$ are related with the wave velocity and attenuation:

$$c = \left(\varrho K_1(\omega)\right)^{-1/2}, \qquad (6)$$

$$\alpha = \frac{1}{2} \varrho c \omega K_2(\omega). \qquad (7)$$

The main drawback of the Kramers-Kronig relationship (5) is the need of knowledge of $K_1(\omega)$ and $K_2(\omega)$ in the whole range of ω from 0 to ∞. O'Donnell, Jaynes and Miller [4] have modified the formula (5) to the local form, valid for limited range of ω. Introducing a new variable $x = \ln\frac{\omega'}{\omega}$ and denoting $K_1(\omega') = D(x)$, $K_1(\infty) = D(\infty)$, one can write the eq. (5) in the form [4]:

$$K_2(\omega) = -\frac{1}{\pi} \int_{-\infty}^{\infty} \frac{dD(x)}{dx} \ln\left(\text{ctgh}\frac{|x|}{2}\right) dx. \tag{8}$$

The function $\ln(\text{ctgh}\frac{|x|}{2})$ is different from zero only in a close neighbourhood of the point $x = 0$, where it has a sharp maximum, so developing the functions $F(x) = \frac{dD(x)}{dx}$ and $\ln(\text{ctgh}\frac{|x|}{2})$ in a series, one obtains the equation:

$$K_2(\omega) = -\frac{4}{\pi}\left[\sum_{n=0}^{\infty} F^{(2n)}(0)\left(\sum_{m=0}^{\infty}(2m+1)^{-2n-2}\right)\right]; \tag{9}$$

Using the formula [5]:

$$\sum_{k=1}^{\infty}(2k-1)^{-2n} = \frac{(2^{2n}-1)\pi^{2n}}{2\cdot(2n)!}|B_{2n}|, \tag{10}$$

where B_{2n} denotes the so called Bernoulli's numbers, i.e. coefficients in expanding the function: $\frac{t}{e^t-1} = \sum_{n=0}^{\infty} B_n \frac{t^n}{n!}$, we obtained from (9) and (10):

$$K_2(\omega) = -\frac{4}{\pi}\left[\frac{\pi^2}{8}F(0) + \frac{\pi^4}{96}\frac{d^2F(x)}{dx^2}\bigg|_{x=0} + \frac{\pi^6}{960}\frac{d^4F(x)}{dx^4}\bigg|_{x=0} + \frac{51\pi^8}{483840}\frac{d^6F(x)}{dx^6}\bigg|_{x=0} + \ldots\right]. \tag{11}$$

O'Donnell has approximated the sum (11) by the first term, obtaining in this case:

$$K_2(\omega) = -\frac{\pi}{2}\omega\frac{dK_1(\omega)}{d\omega}. \tag{12}$$

The O'Donnell approximation is permissible if both the wave velocity and the attenuation coefficient are slowly varying functions of frequency and if the change in wave velocity dispersion is small over a limited frequency range. From the equations (6), (7) and (12) there results the following relation between α and $\frac{dc}{d\omega}$:

$$\alpha = \frac{\pi\omega^2}{2c^2}\frac{dc}{d\omega}. \tag{13}$$

3. CALCULATIONS OF THE ATTENUATION COEFFICIENT OF ULTRASONIC WAVES IN GLYCEROL AND CONCLUSIONS

We used the eq.(13) to calculations of the ultrasonic waves attenuation coefficients in highly viscous liquids. Computations have been made for glycerol and the Chaban's theory has been used. Limiting values c_o and c_∞ of ultrasonic waves velocity have been taken from [6] . In the case $\omega\tau \gg 1$ we used equations (1) and (13) while in the case $\omega\tau \ll 1$ the equations (3) and (13) . The obtained results have been compared with those ones resulting from Chaban's formulae (2) and (4) respectively. A good agreement between values of $\frac{\alpha}{\omega}$ calculated by means of O'Donnell formula (13) and those resulting from Chaban's formula (2) has been found in the case $\omega\tau \gg 1$. Differences between $\frac{\alpha}{\omega}$ values calculated from eq.(13) and Chaban's formula (2) don't exceed 20% of $\frac{\alpha}{\omega}$ in the ω range 10-2000. However in the case $\omega\tau \ll 1$ the formulae (13) and (4) give closed results only in some rather narrow range of ω .

REFERENCES

[1] Isakovitch M.A.,Chaban I.A.,Žurn.Eksp.i Teoret.Fiziki 50,1343-1363 (1966).
[2] Eldin G.,Laheurte J.P.,Quentrec B.,Journal de Physique 41,315-318 (1980).
[3] Zorębski E.,Ernst S.,Soczkiewicz E.,Acustica 62,151-155 (1986).
[4] O'Donnell M.,Jaynes E.T.,Miller J.G.,Journal Acoust. Soc.Am. 69,696-701 (1981).
[5] Gradsztein I.S.,Ryżik I.M.,Tablicy Integrałow,Summ,Riadow i Proizwedenij,Nauka,Moskwa 1971.
[6] Mason W.P.,Fiziczeskaja Akustika vol.2A,Mir,Moskwa 1968

ACOUSTICAL MODELLING LABORATORY

A. WITKOWSKI & M. CWIECZKOWSKI
Sound Engineering Department
Institute of Telecommunication
Technical University of Gdańsk
80 - 952 Gdańsk, Poland

ABSTRACT

In this paper the technical equipment of the room modelling laboratory, the research methods and the scope of experiments carried out has been presented. The use of the laboratory for the teaching purposes has been also discussed.

INTRODUCTION

One of the basic problems ever present in acoustics is searching for more and more perfect methods which would predict acoustic properties of newly built interiors as early as in their designing stage. In practice it proved quite effective to substitute a reduction model bearing close similarity to its original - for the real object. The perpetual technological progress, both in the sphere of measurement methods and that of measuring equipment has led to the increase of interest in this procedure - shown by acousticians and research centres [2]. The Sound Engineering Department of the Technical University of Gdańsk has also decided to build an acoustical modelling laboratory with an eye to its use in the research of the interiors carried out by the department. Apart from its application in research, the laboratory is also used as a didactic means - in teaching elements of room acoustics to the students of the University.

INSTRUMENTATION FOR ACOUSTIC MODELLING

The technical equipment used at the modelling research stand in any laboratory is generally conditioned by the adopted modelling method and the economic standing of the researcher. The laboratory described in the present paper was based on application of two modelling methods - an optical method and an acoustic method.

In the optical method the researchers used scattered light as a light source whereas their receiver of luminous energy was a photoelectric transducer. The originality of the elaborated method lies in the use of the higher number of photoelectric transducers and in securing simultaneous readings of the energy reaching particular observations points. The condition of the simultaneity was met by adaptation of the multichannel monitor for that purpose; the monitor itself was a part of the FSP-80 spectrum analyser produced by the RFT company. Graphical presentation on the screen of the energy values appearing at particular observation points enables both the immediate appraisal of the spatial distribution of energy in the field and a fast comparison of the studied variants of the model. Selection of the appropriate directional response pattern of the photoelectric transducers also enables to estimate the directional scattering of the field.

The acoustic method was based on generally known principles of representation of the acoustic field in the model as given by Spandöck [1].

The similarity between the model and its original is defined by its scale n:

$$n = \frac{\lambda_o}{\lambda_m} = \frac{f_m \cdot c_o}{f_o \cdot c_m} = \frac{L_o}{L_m}$$

where: λ_m, f_m, c_m, L_m stand for the model values of wavelength, frequency, acoustic velocity and linear measure appropriately, whereas λ_o, f_o, c_o, L_o stand for those values in the original.

As it appears from the above, the measuring apparatus used in the model studies should be adopted to the operation within a range of frequencies appropriate to the scale of the model. The equipment installed inside the model, for instance sound sources and microphones should additionally have appropriately reduced dimensions.

The block diagram of the measuring equipment used in the acoustic method is presented in Fig. 1. The absorption properties of the materials used in the acoustical adaptation of the model interior are measured in a glass model of a reverberation chamber (volume 0.25 m3).

At present the laboratory makes use of models of the following three objects:

- St. Mary's Basilica in Gdańsk; cubic capacity 97000 metres, scale 1:30;
- Auditorium of the Musical Theatre, Szczecin; cubic capacity 4000 m, scale 1:10;
- Auditorium of the Variete Theatre, Gdańsk, cubic capacity 2000 metres, scale 1:8.

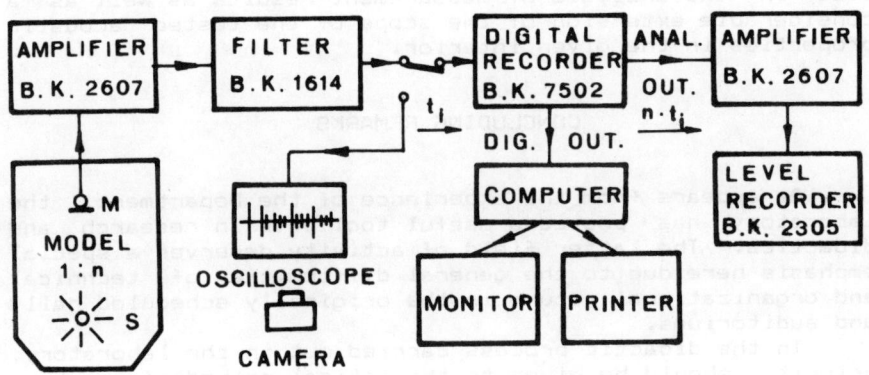

Fig. 1. The block diagram of the setup for the acoustical measurements in models.
S - the spark sound source or the loudspeaker,
M - the condenser microphone Brüel & Kjaer, type 4135 (1/4 in.).

SCOPE OF EXPERIMENTS

The following activities are performed in the laboratory:
- research of cognitive character; solution of the problems which cannot be solved theoretically, eg. search for the optimum shape of the interiors, appraisal of scattering power and absorbency of their inner structure;
- verification of suggested acoustic designs for newly built or adapted interiors;
- lectures and classes for the students of the Electronics Faculty and Faculty of Architecture in the field of room acoustics;
-preparation of M. Sc. dissertations.

The measuring equipment used in the laboratory enables the researchers to appraise the following acoustic properties of the examined interiors:
- spatial distribution of acoustic field energy at steady state,
- directional characteristics of acoustic field energy at steady state, at particular observation points,
- sound rise and decay transients.

Owing to the computer-aided system, which is just being implemented, a considerable acceleration may be expected in the analysis of measurement results as well as a considerable extension of the scope of the tested acoustic properties in the given interior.

CONCLUDING REMARKS

As appears from the experience of the Department, the laboratory has proved a useful tool both in research and didactics. The latter field of activity deserves a special emphasis here due to the general difficulties of technical and organizational nature in the originally scheduled halls and auditoriums.

In the didactic process carried out in the laboratory, priority should be given to the optical method of research due to its demonstrating value. The acoustic method, as it is more advanced technologically, may find application mainly in the realization of research proper and in preparing M. Sc. dissertations.

REFERENCES

[1]. Spandöck, F. "Die Vorausbestimmung der Akustik eines Raumes mit Hilfe von Modellversuchen", Rapp. 5e cong. Intern. d'Acoustique, Liège, II, pp. 313-343 (1965).
[2]. Witkowski, A. "The modelling in room acoustics" (in Polish), Archiwum Akustyki, vol. 17, 3-4, pp. 221-240 (1982).

PERSONAL COMPUTER APPLICATION TO THE CALCULATION OF KUTTRUFF'S "TEMPORAL DIFFUSION" OF A ROOM IN THE PREPARATION OF STUDENTS' MASTER THESES.

S.ZDRAL, K.ŚRODECKI, Z.GARNUSZ

Laboratory of Applied Acoustics
University of Gdańsk
80-952 Gdańsk
Poland

ABSTRACT

Results of the application of a personal computer to an analysis of autocorrelation function registered in a room by Kuttruff's method are presented in the paper.
The worked out calculation computer program allows to find the value of "temporal diffusion"-the parameter introduced by Kuttruff to estimate the acoustic quality of a room and to determine a possible appearance of such phenomena as colouration, echo or flutter echo in the investigated room. The program was worked out while realizing one of the master theses carried out at the Laboratory of Applied Acoustics of Gdańsk University.

1. INTRODUCTION

The application of Kuttruff's method of autocorrelation function measurement [1] to an investigation of the acoustic field in a closed room allows to simplify the calculations of all those field parameters which have been so far found on the basis of echograms.

Kuttruff's method allows to determine the acoustic quality of a room by finding the degree of sound diffusion in the room as well as by finding amplitude and time structure of reflections - particularly early reflections reaching the listener.

Basing on the autocorrelation function course of a room we can estimate the number, time distribution and energy of each particular strong reflection or reflection seqence perceived as echo, flutter echo or sound colouration [2, 3, 4].

While realizing one of the master theses, whose aim was finding the above mentioned parameters of the acoustic field by applying Kuttruff's method of autocorrelation function measurement of a room, a simple personal computer was used to analyze the registered in the room autocorrelation functions.

2. THE CALCULATION PROGRAM FOR AN ANALYSIS OF AUTOCORRELATION FUNCTION OF A ROOM.

A ZX Spectrum microcomputer coupled with a 12-bit analog-to-digital converter was used. In order to introduce the 12-bit words representing the sampled values of the autocorrelation function, a special program of aquisition and unpacking of the data was worked out. Basing on thus introduced data, there was found, by means of the main program, the precise value of "temporal diffusion" Δ -the parameter introduced by Kuttruff for the quantitative estimation of the registered correlograms [1]. The calculation program also allows to compare the calculation results with Bilsen's weighting function [2] (approximated by six sections for calculation needs) and on the basis of that we can estimate a possible appearance of the perceptible flutter echo or sound colouration in the tested room. The program allows to plot the course of autocorrelation function. The essence of the calculation program consists in the formation of an arranged table of data $A(I)$ in the computer's memory, where I is the successive number of a sample of the investigated autocorrelation function course.

By a simple comparison of succesive values, the maximal value $MAX(O)$ and ascribed to it time $T(I)$ is selected out of the set formed in this way. On account of the fact that the appearance of lateral maxima in the autocorrelation function is interesting, we should first determine the width of the correlation interval of its course. It is, actually, for the digital form, a range of the table $A(I)$ searching. The width of this range is equal to the difference of the number of the first sample after MAX(0), for which the following ones have bigger value, and the number of the sample for MAX(0).

The searching of the table A(I) is started from the moment of the appearance of MAX(0) after leaving the first range. From thus found local maxima MAX(I) and corresponding to them time moments T(I) there is formed table B(I) arranged according to decreasing values.

In the next step the elements of table B(I) are compared respectively according to the fixed amplitude and time criteria, and on the basis of the comparison result, there is calculated the temporal diffusion coefficient DELTA from one of the below given dependances:

$$DELTA = MAX(0) / MAX(1) , \qquad (1)$$

where: MAX(0) - main maximum of autocorrelation function, MAX(1) - lateral maximum of autocorrelation function, succesive as to the magnitude after the main maximum, or

$$DELTA = MAX(0) / RMS , \qquad (2)$$

where : RMS - root mean square value from the lateral part of autocorrelation function in the case when the distinct lateral maximum does not appear.

The next step is the comparison of the calculated value DELTA with Bilsen's weighting function and the printout of the calculation results.

3. AN EXAMPLE OF CALCULATION PROGRAM FUNCTIONING.

Fig. 1 presents an exemmplary printout of the results of autocorrelation function analysis registered in a room by Kuttruff's method.

The printout contains :
- in the headline the description of the measurement point location and parameters of the filter used for the measurement.
- listed in columns some of the first characteristic points of the analyzed course (of the maxima) arranged according

to decreasing values and corresponding time moments.
- results of data processing - value DELTA with a comment.
- plot of autocorrelation function with a description.

4. CONCLUSIONS

The presented calculation program enables a precise autocorrelation function analysis of a room, registered by Kuttruff's method.

The program proved to be very useful in the educational process to illustrate phenomena appearing in a room, perceived as echo, flutter echo or sound colouration.

REFERENCES

1. Kuttruff,H., Acustica Vol.16 (1965/66).
2. Bilsen,F.A., Acustica Vol.19 (1967/68).
3. Srodecki,K. and Nienaltowski,B., Proc.of The XXVII Open Seminar on Acoustics, 4 , 30-33 (1980)-in Polish.
4. Srodecki,K., Proc.of The XXX Open Seminar on Acoustics, 1, 119-122 (1983) - in Polish.

AUDYTORIUM 3

Measurement Point 6

Center frequency CF=500Hz
Band-width B=115Hz

Maximum	Time (ms)	Probe
MAX0=1217	TMAX0=0	IMAX0=336
M(1)=252	T(1)=26.25	546
M(2)=244	T(2)=18.25	482
M(3)=184	T(3)=107.5	1196
M(4)=176	T(4)=84.625	1013
M(5)=164	T(5)=50.75	742
M(6)=132	T(6)=40.25	658
M(7)=128	T(7)=101.125	1145
M(8)=126	T(8)=194.5	1892
M(9)=115	T(9)=26.25	546
M(10)=104	T(10)=64.5	852

DELTA=4.8293651

Point is situated above Bilsen's curve

Lenght of the analyzed signal t=470ms

Autocorrelation function plot

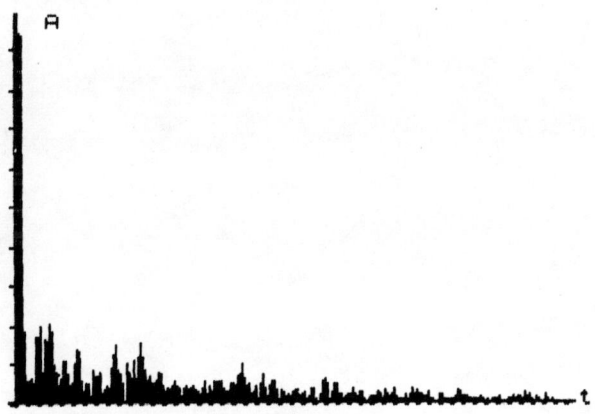

t axis: 1 div.=10ms
A axis: values normalized to 1

Fig. 1

6

ROUND-TABLE DISCUSSIONS AND CONCLUSIONS

ROUND-TABLE DISCUSSIONS

Two round-table discussions were foreseen in the Conference programme. The first one was devoted to an exchange of views and opinions concerning educational methods, and the other one intended to summarize debates and to formulate general conclusions. Those discussions were scheduled on the first and the third conference day, the second one being entirely reserved for visiting laboratories and the Old City. The time foreseen for discussion, as it already appeared on the first day, was insufficient.

However, due to the cordial atmosphere of the debates, many spontaneous discussions developed between the lectures, as well as outside the lecture-hall , e.g. during coffee breaks, and lunch, and on several other occasions. Thus, the final discussion was to some extent prepared by the participants in advance.

The atmosphere of the conference is adequately depicted by a video-report made during all the sessions, debates and discussions, as well as during the sightseeing tours etc. Several fragments of live-recorded discussions have been included into the edited VHS-cassette, enabling everybody to hear and see the face of every speaker, to look at his audience and observe its reactions. More detailed information about that video-report will be available soon after the edition of this book, either from the Publishing House, or directly from the Editors.

As the number of debaters was considerable and their pronouncements rather broad and intense, so the reporting of a full written text of discussion would take too much space here. Therefore, the Editors have prepared a summary, basing on the recorded pronouncements of all the debaters. The task was easier than expected, due to a general concurrence of the expressed opinions.

The first part of the summary contains opinions concerning educational aspects of discussions, irrespectively of the date of pronouncements. The second part summarizes general conclusions concerning the Conference as a whole, so as they were outspoken during debates. Thus, the sequence of debaters is modified, however, all of them are mentioned, their names being quoted in brackets.

The discussion started by two pronouncements of its moderators: Professor Daniel Sette and Professor Ignacy Malecki.

I

There is a common feeling of the community of acousticians, which may be considered as the main reason of this Conference, namely that the subjects of acoustics are less and less taught at the universities. This impedes a proper organization of teaching in highly specialized topics of applied acoustics. All scientists-acousticians are now dissatisfied with the feeble position that acoustics actually plays in the educational process all over the world. On the other hand, there are growing demands from industry and communities for well educated acousticians, prepared for professional activity in various fields of acoustics. This discussion should point out the aims and means of action to restore the proper position of acoustics in modern processes of university education [Sette].

Now, forecasting is indispensable for showing the image of the society in the year 2000, and the future position of acoustics. The position of education in acoustics depends on the position of acoustical science. So the role of research and development in acoustics is inextricably tied with all the problems of teaching, which are to be discussed here. New specialities in the field of acoustics should be taken into consideration, as well as the influence of other fields of science on acoustics applications, as e.g. of informatics, of medicine, of environmental problems etc. At any rate , the necessity of integration among all acousticians is of an increasing importance [Malecki].

On the other hand, considering the problem from the point of view of electronics engineering, tightly bound with acoustics in the field of telecommunications, echo location etc, the role of the digital signal processing can not be overestimated in the education of acoustical engineers. It should be remembered that among all the signals applied in practice, acoustical signals are the most difficult to be analyzed and processed [Jagodziński].

Future envisaged teaching programs in acoustics should contain as an important subject the automatic speech recognition and the man-machine voice communication [Tarnoczy]

The education in acoustics should be maintained as general as possible, moreover, students should not be allowed to specialize too soon [Rossing]

The important property of acoustics which should be preserved in education is its flexibility. It should be remembered that the actual students will be professionally ahead of their teachers in the near future. It should be considered in conceiving future programs [Bjorno].

Education in acoustics should not be orientated towards actual demands of employers but rather towards future needs, which are still unknown. Thus, the importance of a certain degree of generality in teaching programs [Løchstøer].

Acoustics will be more and more required by industry, technology etc., thus more hours for university education in acoustics will be necessary. This is a difficult problem for the university staff. National and international acoustical societies should help in solving that problem, in promoting the role of acoustical education by widening the exchange of information about experience and achievements of particular countries and of individual universities [Kolmer].

The importance of the interest in studying acoustics with young students, which should be encouraged at the start of their education ought not to be overlooked. The interdisciplinary character of acoustics and its flexibility is to be shown at an early stage of education. Otherwise, many capable students will be lost as potential acousticians [Walkling].

The preparation of acousticians for solving specialized engineering problems by application of acoustical methods is a very important requirement of education in acoustics. The problem of oil detection in water may serve as a typical example. In the absence of appropriately prepared acousticians, problems of that kind are being solved by other, less effective engineering methods [Jones].

Among the sources of interest which may attract students to study acoustics, several applications should be considered e.g. the use of ultrasound in medicine, the exploration of sea resources, acoustical testing of structures and other applications of ultrasonics. When graduated they obtain well paid professional positions. It influences the organization of education, especially in India [Bindal].

When speaking generally about teaching programs in acoustics we must be aware of their unavoidable variety. Their levels, their teaching profiles can not be unified because of specific needs existing in every country, or every part of the world. Thus, flexibility in acoustics education should be maintained. Moreover there are distinct differences in the teaching programs in socialist countries relative to free market countries, caused by otherwise organized industry and thus, other requirements from young acousticians [Malecki].

Teaching programs obviously depend on directions of research undertaken at a given university or educational centre. The larger are the discrepancies in research the bigger differences in teaching programs. Therefore a strong coupling between such centres ought to exist and an exchange of information should be well functioning. Moreover, a problem of equipment needed for the efficient teaching emerges [Śliwiński].

Even in the absence of modern equipment we must prepare students for their future job. Thus, the fundamental education in physics, mathematics, signal processing, system theory, and informatics should accompany the education in acoustics, to enable the graduated acousticians to undertake any job actually needed in the above mentioned field of applications [Schommartz].

An important problem is to keep students' interest in their future job. To balance the theoretical subjects which prevail during the first two years of studies, a chance to students to take part in research work should be given. Their interest will be growing and their situation will improve [Bjørnø].

A certain lack of information and lack of knowledge as to what acoustics is, may cause young people to ommit acoustics as their line of professional life. Although we are not responsible for informing, we may act to improve this situation Workshops on acoustics, mentioned during an earlier discussion may function also as means of information exchange [Sette].

There is a possibility of improving the information about prospects in acoustics, influencing the ICA as well as national acoustical societies towards an appropriate action. Besides, a very useful proposition was made during this discussion, namely, to organize a kind of an international or regional workshop on acoustical education. It is a completely new idea and such a workshop might be a perfect occasion to exchange experience in teaching and at the same time an opportunity for dissemination of information about acoustics, its role and importance. This conclusion should be presented to ICA [Kolmer].

Some kind of introduction of young people into acoustics is necessary before they enter the university. French experience proves that such functions might be fulfilled by means of interactive museums, like e.g. in Paris, or by scientific centres led by industry. Demonstrating the acoustic phenomena and explaining their meaning, as well as physical laws governing them, may easily arouse interest for acoustics [François].

An obstacle hindering the development in acoustics is economical one. How much is spent up to there, for an improvement of acoustical methods and for acoustical specialization, and how much is still needed for supplying people working in acoustics [Jones]

As to the workshop question there is a necessity to limit the topics so that a group of interested participants might enter sufficiently deep into problems. On the other hand it must be a workshop devoted entirely to teaching [Sette].

Many well organized workshops or summer schools in the field of acoustics in various countries have been successful. The workshop on teaching should not be less successfull [Malecki].

Other means should be considered earlier than a workshop organization, e.g. well prepared textbooks, matched to the level of students, appropriate papers, visual aids, recordings etc. A workshop without good textbooks may turn out to be a failure [Bjørnø].

A series of papers under the title 'Impact' have been edited by UNESCO. Other series entitled 'The New Trends of Science' is actually being edited. Similarly we can initiate a series of monographs on acoustics supported by ICA and sponsored by UNESCO. This proposal should be sent to the ICSU chief-editor. [Malecki].

Another question to be answered is how to teach modern mathematics connected to acoustics, as e.g. harmonic equation, Fast Fourier Transform etc. Here, new methods of teaching are needed, with preference to the time domain approach [Lasota].

It should be noticed that many scientists-acousticians who teach acoustics now, were not educated in acoustics. They are however performing very well, having learned acoustics alone. On the other hand, attention must be paid not to replace teaching by applying only measuring instructions and laboratory manuals, even as well prepared as e.g. by Bruel & Kjaer. So it is to be taken into consideration when discussing "how to teach acoustics" [Kyncl].

A long experience proves that combined, two-person, research-teams, are the most effective when composed of an electronics engineer and of a physicist. The work either of an electronics engineer alone, or of a physicist alone is less efficient. Being together they influence and stimulate each other [Filipczyński].

Electronics-engineers are necessary in teams working with acoustic problems. The most important elements of the electroacoustic channel are of electronic nature, as e.g. transducuers etc. Connections between acoustics and electronics are very strong; they must be considered in teaching acoustics [Jagodzinski].

II

General opinions were expressed mostly during final session of the Conference, moderated by the Editors of these Proceedings.

First of all, a consensus of opinion has been stated about the necessity to preserve bonds linking acoustics and physics. The commonly accepted requirement to maintain teaching programs as general as possible corroborates that opinion.

On the other hand, the need for strong connections with different fields of applied acoustics and of mixed applications was repeatedly stressed. Only applications may bring to university units teaching acoustics the necessary funds for modern equipment as well as attract young able students, willing to be engaged in acoustical research. In this context, bonds connecting acoustics, and especially sound engineering, with broadcasting, television and film techniques, as well as with music recording industry, were many times emphasized.

Thus, the two demands opposed to each other have been acknowledged: a) teaching programs possibly general, containing all fundamental problems of physical acoustics; b) teaching programs highly specialized, concentrated on selected problems of engineering, medicine, music, technology, etc. Every conception of a modern teaching program in acoustics must consider both demands and find a compromisingly balanced scheme of education. Here again the notion of 'flexibility' of acoustics finds its justification.

Parallel to the opinions expressed during discussions, similar arguments are to be found in the presented papers: invited lectures, contributed papers, poster presentations. Both tendencies are almost equally represented by particular authors, depending on their personal position and role played in the process of education.

Besides discussion on teaching programs several opinions concerning the important question of the advanced equipment, necessary for modern teaching, were expressed. The development of acoustics is not possible without an advanced instrumentation. Moreover, the teaching of acoustics requires also appropriate tools and adequate methods. This was presented in several papers and posters devoted to applications of small computers to teaching purposes. The elegant and convincing contribution has been presented by prof. Heckl, who showed how efficiently may films and computers be used to teach some fundamental laws and phenomena of acoustics. The educational value of computer simulations was clearly proved and exemplified by Prof. Strong. The educational role of well conceived demonstrations was shown especially in the lectures of Prof. Walkling and of Professors Houtsma and Rossing. Many other demonstrated examples provided valuable hints or suggestions for teaching acoustics.

Those presentations showed the need for an enlarged exchange of experience in teaching tools and methods. It evoked again the discussion on a workshop devoted to teaching acoustics. It stressed the need for an augmented exchange of information.

The debaters concurrently have authorized the Editors to

formulate the subsequent letters:
1. Letter to International Commission on Acoustics asking for sponsorships: for an intented workshop on teaching acoustics, and for a series of monographs on acoustics.
2. Letter to "Proceedings of IEEE" or to other scientific journals, informing broad circles of scientists and academic teachers about the results of this Conference.

CONCLUDING REMARKS

It may be worth noticing and emphasizing again, the value of direct contacts between acousticians of different countries and part of the world, as well as the value of open discussion. Such experience should encourage prospective organizers of similar conferences undertaken on world or regional scale.

Summarizing the direct effects of the Conference:

1. The 44 papers have been presented during debates (see the Conference Programme enclosed in Appendix 1, p. 441), namely:
> 25 invited lectures
> 7 contributed papers
> 12 poster presentations

2. The 94 participants attended the debates (see the List of Participants, p. 453)

3. The 43 papers have been submitted to the present Proceedings, namely:
> 26 invited lectures
> 7 contributed papers
> 10 abbreviated poster presentations

4. Discussion has been recorded and a concise abstract has been prepared for the Proceedings.

5. A video-report has been made during the debates using the VHS-cassette system.

6. The present Proceedings are prepared to print by the Editors with the help of several members of the Organizing Committee. The assistance of K.Cisowski, I.Cypukow, J.Dyżewski, A.Kaczmarek, B.Kostek, A.Kulowski, J.Kuźnicki, B.Linde, Z.Perucki and A.Witkowski is gratefully acknowledged.

7. According to the agreement with Publishing House all contributors to these Proceedings receive their copy free of charge. These copies will be mailed to the contributors

addresses by the Editors. Additional copies will be distributed by the Publishing House and sold by appropriate book-stores.

8. These Proceedings' aim is, among others, to help organizing further international activities devoted to the improvement of education and acceleration of development in acoustics.

The Editors

7

APPENDIX 1

**Programme Executed
During the Conference**

JASTRZEBIA GORA

A C T U A L P R O G R A M M E
19th May Tuesday

9.30 — Opening of the Conference

Morning Session 1
Chairman prof. F. Kolmer

10.15-10.45 — I. Malecki (Poland)
Place of acoustics in the technical universities.

10.45-11.15 — R. Lehmann (France)
Education in acoustics in high schools.
(presents P. François)

11.15-11.30 — Coffee break

Morning Session 2
Chairman prof. H.W. Jones

11.30-12.00 — A. Zarembowitch (France)
Concept of old physics still active in the modern development of acoustics.

12.00-12.30 — D. Sette (Italy)
Physical acoustics in the Summer School in Varenna.

12.30-13.00 — P. François (France)
Pedagogical tools and knowledges in acoustics.

13.00-13.30 — V.N. Bindal (India)
Status of teaching acoustics in India.

13.30-15.00 — Lunch break

ACTUAL PROGRAMME

Afternoon Session 1

Chairman - prof. A. Zarembowitch

15.00-15.30 - T. Rossing (USA), A. Houtsma (Holland)
New series of auditory demonstrations on psychoacoustics.

15.30-15.45 - A. Rakowski (Poland)
Interest in music and progress in acoustics.

15.45-16.00 - G. Budzyński (Poland), G. Papanikolau (Greece), M. Sankiewicz (Poland)
How to teach sound-engineering.

16.00-16.15 - H. Harajda (Poland)
Selected problems of teaching acoustics in luthery studies on secondary and higher level.

16.15-16.45 - T. Tarnoczy (Hungary)
Acoustical education in Hungary.

16.45-17.05 - Tea break

ACTUAL PROGRAMME

Afternoon Session 2

Chairman - prof. V.N. Bindal

17.05-17.20 - V. Chalupowa, K. Sobotkova (Czechoslovakia)
Acoustics education at the electrotechnical faculty at the Technical University of Prague.

17.20-17.50 - W. Løchstøer (Norway)
Acoustics and physics.

17.50-18.20 - L. Filipczyński (Poland)
Ultrasonics and medicine.

18.20-18.50 - R. Millner (GDR)
Acoustical foundation in the training of medical students and postgraduates.
(presents H.J. Hein)

19.00-20.00 - Supper break

20.00 - Round Table Discussion on educational programmes of acoustics.

Moderators - prof. D. Sette, prof. I. Malecki

MODIFIED PROGRAMME
19th May Tuesday

Afternoon Session 2

15.00-16.00 — Technical break (due to lack of electricity)

Afternoon Session 1

16.00-17.00 — Poster session

Afternoon Session 2

Chairman — prof. A. Zarembowitch

17.15-17.45 — T. Rossing (USA), A. Houtsma (Holland)
New series of auditory demonstrations on psychoacoustics.

17.45-18.00 — A. Rakowski (Poland)
Interest in music and progress in acoustics.

18.00-18.30 — T. Tarnoczy (Hungary)
Acoustical education in Hungary.

18.30-18.45 — V. Chalupowa, K. Sobotkova (Czechoslovakia)
Acoustics education at the electrotechnical faculty at the Technical University of Prague.

18.45-19.15 — L. Filipczynski (Poland)
Ultrasonics and medicine.

19.00-20.00 — Supper break

20.00 — Round Table Discussion on educational programmes of acoustics.

Moderators — prof. D. Sette, prof. I. Malecki

ACTUAL PROGRAMME
20th May Wednesday

All Conference participants are kindly requested to declare their will to take part in a full-day excursion by bus to Gdańsk.

- 8.00 – Departure (all groups together)
  Visits to acoustic laboratories (in parallel groups).
  Three options (declare one, please):
  A. Labs at the Institute of Experimental Physics at the University of Gdańsk,
  B. Labs at the Technical University of Gdańsk,
  C. Labs at the Ship Technology Centre, Gdańsk.
- 13.30 – Lunch at the University canteen (all groups together)
- 14.45 – Departure to the Old City (to bus parking place)
- 15.00 – Meeting with the City Authorities at the Town-Hall
- 16.15 – Presentation of the Great Organ at St. Mary's Basilica
- 16.45 – Sightseeing walk around the Old City
- 17.45 – Coffee at the 'Palowa' House
- 18.30 – Departure from the parking place to Jastrzębia Góra
- 20.00 – Common supper at 'Celuloza' House

A C T U A L   P R O G R A M M E
21st May Thursday

Meals: 8.00 - Breakfast, 14.00 - Lunch, 19.00 - Supper

MORNING SESSION 1
Chairman - prof. L. Bjørnø

9.00 - 9.30 - W. Løchstøer (Norway)
Acoustics and physics.

9.30 -10.00 - R.A. Walkling (USA)
Acoustics for everyone; a general course in acoustics.

10.00-10.15 - Z. Engel (Poland)
Education in vibroacoustics.

10.15-10.30 - A. Kulowski (Poland)
The ray method as a teaching tool.

10.30-11.00 - Coffee break

MORNING SESSION 2
Chairman - prof. W. Løchstøer

11.00-11.15 - H. Lasota (Poland)
Diffraction as time-space phenomenon - educational aspects.

11.15-11.30 - D. Ayers (USA)
Fourier methods in physics; a course in waves at the senior level.

11.30-12.00 - W.J. Strong (USA)
Computers in modern acoustics.

12.00-12.15 - Break

MORNING SESSION 3
Chairman - prof. W.J. Strong

12.15-12.45 - M. Heckl (West Berlin)
The use of films in teaching acoustics.

12.45-13.15 - H.W. Jones (Canada)
Waves, acoustics and the student or practical studies in acoustics.

13.15-13.45 - R. Millner, H.J. Hein (GDR)
Acoustical foundation in the training of medical students and postgraduates.

13.45-15.00 - Lunch break

AFTERNOON SESSION 1
Chairman - prof. R.A. Walkling

15.00-15.15 - W. Majewski, J. Zalewski (Poland)
Academic programs in acoustical sciences at the Technical University of Wroclaw.

15.15-15.45 - G. Schommartz (GDR)
TE-LAB a successful philosophy of unified lab's at the Department of Technical Electronics.

15.45-16.00 - A. Wasilewski (Poland)
Student classes in condition of minimum laboratory equipment.

16.00-16.30 - K. Breitschwerdt (FRG)
Phonon propagation in disordered media.

16.30-17.00 - Tea break

AFTERNOON SESSION 2
Chairman - prof. M. Heckl

17.00-17.30 - L. Bjørnø (Denmark)
Advanced education in industrial acoustics in Denmark.

17.30-17.45 - A. Śliwiński (Poland)
Acoustical education and industrial application.

17.45-18.00 - G. Budzyński (Poland), G. Papanikolau (Greece), M. Sankiewicz (Poland)
How to teach sound engineering.

18.00-18.15 - Z. Jagodziński (Poland)
Specific problems in education and development in underwater acoustics.

18.15-18.30 - D. Ruser (GDR)
Hydroacoustical laboratory at the Technical University in Gdansk.

18.30-18.40 - E. Kotlicka (Poland)
Proposals of curriculum for health and environment protection against noise for elementary school.

POSTER SESSION

18.30-19.00 - Discussion (continued from Tuesday)

19.00-20.00 - Supper break

CLOSING SESSION

20.00-21.00 - Final Discussion and Closing of the Conference

8

APPENDIX 2

Letter to ICA

The President
of the International
Commission on Acoustics
Professor Henry Myncke,
Laboratorium voor Akoestiek
en Warmtegeleiding,
University of Leuven,
Celestijnenlaan 200 D,
B - 3030, Heverlee,
BELGIUM

Dear Sir,

Authorized by the participants of the ICA Conference on Prospects of Modern Acoustics - Education and Development held in Gdansk - Jastrzebia Gora in Poland on May 19 - 21, we kindly inform You about the outcoming conclusions of the round table and final discussion of the Conference.

The debaters expressed concurrently the opinion, that the Conference had been a useful and important step in preserving the actual state and the quality of acoustical education all over the world.

Continuing this initiative, the debaters proposed two prospective activities which will expand the achived results:

1. To organize periodically the international or regional workshops devoted to educational problems in acoustics. Such workshops may concern either selected problems of teaching as e.g. the use of computers in teaching aids etc., or more general problems such as exchange of experience in organization of teaching processes, teaching programs, syllabuses etc.

2. To introduce into the ICSU-Monograph Series entitled "The New Trends of Science" a new series devoted to "Prospects and Development of Modern Acoustics" as a means for wide popularization of acoustics as a science of interdisciplinary aspects and applications.

The support and promoting action of ICA is decisive for both activities and thus, highly expected. We suggest the above proposals should be introduced into the program of ICA activities for the new term of the commision.

With best regards from all the participants of the Conference.

Your sincerely,

Antoni Śliwiński

Gustaw K.E. Budzyński

9

INDEX

INDEX

The Conference on
PROSPECTS IN MODERN ACOUSTICS-EDUCATION AND DEVELOPMENT
Gdansk-Jastrzębia Góra, 19-21 May, 1987

LIST OF PARTICIPANTS   PAGE

R.D. Ayers   Brigham Young University   23
                    c/o Department of Physics
                    and Astronomy
                    Provo, Utah 84602, USA

K. Badziąg   Gdansk University
                    Institute of Experimental
                    Physics
                    Wita Stwosza 57
                    80-952 Gdansk, Poland

W. Bandera   Gdansk University   383
                    Laboratory of Applied
                    Acoustics and Spectroscopy
                    Wita Stwosza 57
                    80-952 Gdansk, Poland

V.N. Bindal   National Physical Laboratory   35
                    Hillside Road
                    New Delhi - 110 012, India

L. Bjørnø   Technical University of Denmark   57
                    Industrial Acoustics
                    Laboratory, Bldg. 352
                    DK-2800 Lyngby, Denmark

M. Borysewicz   Gdansk University
                    Institute of Experimental
                    Physics
                    Wita Stwosza 57
                    80-952 Gdansk, Poland

K.G. Breitschwerdt Universität Heidelberg   323
                    Institute of Physics
                    Grabengasse 1
                    6900 Heidelberg, FRG

G. Budzynski   Technical University of Gdansk   69
                    Sound Engineering Department
                    80-952 Gdansk, Poland

V. Chalupova	CVUT - FEL Department of Physics Suchbatarova 2 166 27 Praha 6, CSSR	345
K. Cisowski	Technical University of Gdańsk Sound Engineering Department 80-952 Gdańsk, Poland	
I. Cypukow	Technical University of Gdańsk Sound Engineering Department 80-952 Gdańsk, Poland	391
M. Cwieczkowski	Technical University of Gdańsk Sound Engineering Department 80-952 Gdańsk, Poland	421
A. Dobrucki	Technical University of Wroclaw Institute of Telecommunication and Acoustics B. Prusa 53/55 50-370 Wrocław, Poland	
Z. Dukiewicz	Gdańsk University Laboratory of Applied Acoustics and Spectroscopy Wita Stwosza 57 80-952 Gdańsk, Poland	
J. Dyżewski	Technical University of Gdańsk Sound Engineering Department 80-952 Gdańsk, Poland	391
Z. Engel	Academy of Mining and Metallurgy Institute of Mechanics and Vibroacoustics A. Mickiewicza 30 30-059 Kraków, Poland	79
S. Ernst	Silesian University Institute of Chemistry Szkolna 9 40-006 Katowice, Poland	
Z. Engels	Gdańsk University Institute of Experimental Physics Wita Stwosza 57 80-952 Gdańsk, Poland	

L. Filipczyński	Institute of Fundamental Technical Research of Polish Academy of Sciences Świętokrzyska 21 00-049 Warszawa	15,19
P. François	Department " Acoustique et Vibrations" L'Avenue General de Gaulle, 92141 Clamart, France	93
J. Gätke	Wilhelm-Pieck-Universität Rostock Sektion Technische Elektronik DDR - 2500 Rostock Albert-Einstein-Strasse 2, DDR	
J. Gudel	Musical Academy of Gdańsk Institute of Theory of Musics Puławskiego 18/20 81-762 Sopot, Poland	
H. Harajda	High Pedagogical School Physics Department pl. Słowiański 6 65-069 Zielona Góra, Poland	349
B. Hałaciński	Technical University of Warsaw Institute of Physics Chodkiewicza 8 02-525 Warszawa, Poland	
M. Heckl	Technische Universität Berlin Fachbereich 21 Umwelttechnik Str. des 17 Juni 135 1 Berlin 12, West Berlin	101
H-J. Hein	Martin-Luther-Universität Institute für Angewandte Biophysik 4014 Halle/S, DDR	193
E. Hojan	Poznań University Institute of Acoustics Matejki 48/49 60-769 Poznań, Poland	395
A. Houtsma	Institute for Perception Research Insulindenlaan 2 Eindhoven, Holland	117

Z. Jagodziński	Technical University of Gdańsk Underwater Acoustics Department 80-952 Gdańsk, Poland	15,19, 129
H.W. Jones	Technical University of Nova Scotia Head of the Department of Engineering Physics P.O. Box 1000 Halifax, Nova Scotia Canada B3J 2X4	137
U. Jorasz	Poznań University Institute of Acoustics Matejki 48/49 60-769 Poznań, Poland	395
A. Kaczmarek	Technical University of Gdańsk Sound Engineering Department 80-952 Gdańsk, Poland	399
Z. Kleszczewski	Silesian Technical University Institute of Physics Krzywoustego 2 44-100 Gliwice, Poland	
F. Kolmer	VUZORT Plzenska 66, L51 24 Prague 5 Czechoslovakia	15
M. Kosmol	Gdańsk University Institute of Experimental Physics Wita Stwosza 57 80-952 Gdańsk, Poland	
A. Kołodziejski	Naval High School 81-919 Gdynia, Poland	
B. Kostek	Technical University of Gdańsk Sound Engineering Department 80-952 Gdańsk, Poland	413
E. Kotarbińska	Technical University of Warsaw Institute of Radioelectronics Nowowiejska 15/19 00-665 Warszawa, Poland	
A. Kulowski	Technical University of Gdańsk Sound Engineering Department 80-952 Gdańsk, Poland	355

E. Kotlicka	Technical University of Warsaw Institute of Physics Chodkiewicza 8 02-525 Warszawa, Poland	403
E. Kozaczka	Naval High School 81-919 Gdynia, Poland	
T. Kwiek-Walasiak	Poznań University Department of Ergonomics Kantaka 2 61-812 Poznań, Poland	
Z. Kyncl	CVUT Suchbatarova 2, 16627 Prague 6, Czechoslovakia	
H. Lasota	Technical University of Gdańsk Underwater Acoustics Department 80-952 Gdańsk, Poland	363
A. Latuszek	Technical University of Warszawa Institute of Physics Chodkiewicza 8 02-525 Warszawa, Poland	
L. Lipiński	Gdańsk University Laboratory of Applied Acoustics and Spectroscopy Wita Stwosza 57 80-952 Gdańsk, Poland	
B. Linde	Gdańsk University Institute of Physics Wita Stwosza 57 80-952 Gdańsk, Poland	
W. Løchstøer	University of Oslo Box 1048 - Blindern 0316 Oslo 3, Norway	15, 169
A. Majewski	Technical University of Wrocław Institute of Telecommunication and Acoustics B. Prusa 53/55 50-317 Wrocław, Poland	185

I. Malecki	Institute of Fundamental Technical Research of Polish Academy of Sciences Świętokrzyska 21 00-049 Warszawa, Poland	15,173
A. Miśkiewicz	Musical Academy of Warsaw Department of Musical Acoustics Okólnik 2 00-326 Warszawa, Poland	
J. Motylewski	Institute of Fundamental Technical Research of Polish Academy of Sciences Świętokrzyska 21 00-049 Warszawa, Poland	409
H. Møller	University of Aalborg Institute of Electronics Systems Strandvejen 19 DK-9000 Aalborg, Denmark	
B. Nienałtowski	Gdańsk University Laboratory of Applied Acoustics and Spectroscopy Wita Stwosza 57 80-952 Gdańsk, Poland	
A. Opilski	Silesian Technical University Institute of Physics B. Krzywoustego 2 44-100 Gliwice, Poland	
B. Osiecki	Musical Academy of Warsaw Department of Musical Acoustics Okólnik 2 00-326 Warszawa, Poland	
Z. Perucki	Technical University of Gdańsk Sound Engineering Department 80-952 Gdańsk, Poland	413
A. Rakowski	Musical Academy of Warsaw Department of Musical Acoustics Okólnik 2 00-326 Warszawa, Poland	203

J. Ranachowski	Institute of Fundamental Technical Research of Polish Academy of Sciences Świętokrzyska 21 00-049 Warszawa	
J. Renowski	Technical University of Wrocław Institute of Telecommunication and Acoustics B. Prusa 53/55 50-317 Wrocław, Poland	
T. Rossing	Northern Illinois University Physics Department De Kalb, IL 60115 USA	117
D. Ruser	Technical University of Gdańsk Underwater Acoustics Department 80-952 Gdańsk, Poland	371
M. Sankiewicz	Technical University of Gdańsk Sound Engineering Department 80-952 Gdańsk, Poland	69
G. Schommartz	Wilhelm-Pieck-Universität Rostock Sektion Technische Elektronik Albert-Einstein-Strasse 2 DDR - 2500 Rostock	225
D. Sette	Universita "La Sapienza" Dipertimento di Energetica Via A. Scarpa 14 00161 Roma, Italy	15,213
J. Smurzyński	Musical Academy of Warsaw Department of Musical Acoustics Okólnik 2 00-326 Warszawa, Poland	
K. Sobotkova	CVUT FEL Department of Physics Suchbatarova 2 16627 Praha 6, Czechoslovakia	345
E. Soczkiewicz	Silesian Technical University Institute of Physics B. Krzywoustego 2 44-100 Gliwice, Poland	417

E. Stadnik	High Pedagogical School Physics Department pl. Słowiański 6 65-069 Zielona Góra, Poland	
W. Straszewicz	Technical University of Warsaw Institute of Radioelectronics Nowowiejska 15/19 00-665 Warszawa, Poland	15
W.J. Strong	Bringham Young University Department of Physics and Astronomy 296 Eyring Science Center Provo, Utah 84602, USA	227
J. Sułocki	Gdańsk University Laboratory of Applied Acoustics and Spectroscopy Wita Stwosza 57 80-952 Gdańsk, Poland	
C. Szmal	Technical University of Wrocław Institute of Telecommunication and Acoustics B. Prusa 53/55 50-317 Wrocław, Poland	
A. Śliwiński	Gdańsk University Institute of Experimental Physics Wita Stwosza 57 80-952 Gdańsk, Poland	237
K. Środecki	Gdańsk University Laboratory of Applied Acoustics and Spectoscopy Wita Stwosza 57 80-952 Gdańsk, Poland	421
V. Tandara	Offentlich Besteller und Vereidigter Sachverständiger für Verbrennungsmaschinen und Lärmminderung bei Kraftfahrzeugen Knobelsdorffstrasse 11 D-1000 Berlin 19, West Berlin	
T. Tarnoczy	Hungarian Academy of Sciences Acoustic Research Laboratory P.O. Box 132 H-1502 Budapest 112, Hungary	265

O. Tichy	Film and TV Faculty AMV Smetanovo nabr 2 11000 Praha 1, CSSR	
Z. Trumpakaj	Gdańsk University Institute of Experimental Physics Wita Stwosza 57 80-952 Gdańsk, Poland	383
M. Urbańczyk	Silesian Technical University Institute of Physics B. Krzywoustego 2 44-100 Gliwice, Poland	
R.A. Walkling	University of Southern Maine 96 Falmouth Street Portland, Maine 04103 USA	273
A. Wasilewski	Technical University of Lublin Faculty of Building and Sanitary Engineering Nadbystrzycka 40 20-618 Lublin	373
J. Werle	University of Warsaw Institute of Theoretical Physics Hoża 69 00-691 Warszawa, Poland	15,17
A. Witkowski	Technical University of Gdańsk Sound Engineering Department 80-952 Gdańsk, Poland	421
I. Wojciechowska	Gdańsk University Institute of Experimental Physics Wita Stwosza 57 80-952 Gdańsk, Poland	
R. Wyrzykowski	High Pedagogical School Institute of Physics T. Rejtana 16a 35-959 Rzeszów, Poland	299
O. Zadrażil	CVUT Electrotechnical Faculty Suchbatarova 2 16627 Prague 6, Czechoslovakia	

T. Zaleski	Gdańsk University Institute of Experimental Physics Wita Stwosza 57 80-952 Gdańsk, Poland	185
E. Zalewska-Paciorek	Academy of Mining and Metallurgy Institute of Mechanics and Vibroacoustics A. Mickiewicza 30 30-059 Kraków, Poland	
J. Zalewski	Technical University of Wrocław Institute of Telecommunication and Acoustics B. Prusa 53/55 50-317 Wrocław, Poland	383
A. Zarembowitch	Laboratoire "Dynamique cristalline et ultrasons du D.R.P.", L.A. No 71 Associate au CNRS Universite Paris VI 4 Place Jussieu 75230 Paris, France	311
S. Zdral	Gdańsk University Laboratory of Applied Acoustics and Spectroscopy Wita Stwosza 57 80-952 Gdańsk, Poland	425
J. Żera	Musical Academy of Warsaw Department of Musical Acoustics Okólnik 2 00-326 Warszawa, Poland